PRA
SKIN

'A thorough overview and detailed critique of contemporary "race science," and a poignant description and assessment of scientific racism... an easy-to-read overview of the latest academic research in genetics, evolutionary anthropology, archaeology, and palae-ontology that explains disparities in human intelligence... but not in terms of the construct "race." As a seasoned science journalist, Evans explains complex concepts and theories with clarity and precision, and he brings moments of levity to this challenging subject matter.'

Choice

'A world in thrall to far-right politics and ethnic nationalism demands vigilance. We must guard science against abuse and reinforce the essential unity of the human species. I am grateful that in Evans we have someone conscientious, brave and willing to do that.'

Nature

'With the probability that existing political and social tensions will be exacerbated by climate change, rising migration and conflicts over resources, these kinds of books are sorely needed.'

Literary Review

'Excellent! In *Skin Deep,* Gavin Evans lucidly and comprehensively demolishes the rationale and evidence of the so-called "race science" employed by some of our most privileged and respected scholars.'

Stephen Oppenheimer, author of *Out of Eden:
The Peopling of the World* and *The Origins of the British*

'Well-researched and richly rewarding, *Skin Deep* methodically dismantles the quackery and junk science that seeks to justify economic inequality by appeals to racial difference. Given the insidious rise of pseudoscientific racism, *Skin Deep* is timely and urgent, its patient scholarship a fine antidote to the quackocracy's shrill rhetoric. In this spirit it will appeal to readers who enjoyed Ben Goldacre's *Bad Science* and *I Think You'll Find It's A Bit More Complicated Than That.*'
Robert Newman, comedian and author

'This book contains a wealth of information, old and new, about race and its underlying genetics. Everyone interested in the distinction between knowledge and prejudice in this sensitive and challenging subject should read it.'
Jonathan Bard, Emeritus Professor, University of Edinburgh, and author of *Principles of Evolution*

ABOUT THE AUTHOR

Gavin Evans was born in London and grew up in Cape Town, where he became intensely involved in the anti-apartheid struggle. He studied economic history and law before completing a PhD in political studies, writing extensively on race and racism. He lectures in the Culture and Media department at Birkbeck College, London, broadcasts regularly for the BBC World Service and is a keen marathon runner.

SKIN DEEP

DISPELLING THE SCIENCE OF RACE

GAVIN EVANS

A Oneworld Book

Published by Oneworld Publications, 2019
This paperback edition published in 2020

ISBN 978-1-78607-811-7
eISBN 978-1-78607-623-6

Typeset by Tetragon, London
Printed and bound in Great Britain by Clays Ltd, Elcograf S.p.A.

Oneworld Publications
10 Bloomsbury Street
London WC1B 3SR
England

Stay up to date with the latest books,
special offers, and exclusive content from
Oneworld with our newsletter

Sign up on our website
oneworld-publications.com

MIX
Paper from
responsible sources
FSC® C018072

To my mum, Joan, and my late dad, Bruce, who set such fine examples in opposing racism.

CONTENTS

FOREWORD

A few tales from my early years growing up in apartheid South Africa:

1981: I was hitchhiking to Zimbabwe and was picked up by a middle-aged, English-speaking couple who proceeded to discuss the mental capacity of African people. The woman explained: 'The problem with the black mind is that it can only learn things one way and when things don't work that way it gets confused. You can teach him that the wheel goes clockwise when it goes forward but if one day the wheel goes anti-clockwise, he won't understand. It's a known fact and that's why he won't ever invent anything and he'll always rely on us.'

1975: A group of my white South African classmates was mimicking African accents, prompting a discussion of why they spoke differently. 'It's because of their brains and their mouths – it's impossible for them to learn to talk properly,' one thirteen-year-old volunteered.

1972: Mr B, the deputy head at my Cape Town state primary school, was teaching us about the hardy heroism of the godly Voortrekkers, who conquered the treacherous Zulus at the Battle of Blood River on their Great Trek north. He drew our attention to the implications of the humanity of his hardy ancestors. 'The reason we have so much trouble with our natives today,' he said, 'is because unlike the Americans and Australians we didn't wipe ours out. They criticise us but think about it: we wouldn't have so many problems if we had.'

1970: Mr O, my teacher, watched H, a Jewish boy, examine his change to see if he had enough for a tuck-shop bun. 'Counting your Shekels again, hey H,' Mr O said knowingly, nodding his head while stroking the tip of his nose. The bell had already rung so he couldn't beat H this time. Usually he managed to find an excuse; at least once a week in woodwork class he'd send H off to find a two-by-four plank, which he would then break on the poor boy's bum. Every year he seemed to pick a fresh Jewish boy as his favoured victim.

Each of these vignettes suggests a racism resting on a belief that there are profound deficiencies among population groups – the unintelligent, linguistically challenged blacks, the treacherous Zulus, the miserly, money-grabbing Jews. This, if you think about it, is what all forms of racism, whether 'scientific' or otherwise, have in common: the notion of innate distinctions between 'us' and 'them', invariably to 'our' advantage.

I was born in London but grew up mainly in a state that elevated faith in profound difference into a doctrine of racial dominance. As with so many forms of racism, it was spawned by conquest and colonialism, with first Dutch and then British colonists dispossessing the majority of its land, consolidated by segregationist white self-rule from 1910. Nazi-influenced racial science flourished in the 1930s and 1940s, informed by a faith in IQ scores, 'ethno-psychology' and mythology about the 'native brain'.[1] It began to be applied after the apartheid government came to power in 1948, through policies that led to the forced removal of 3.5 million black people from their homes to make more room for white people. Each race group had its own in-built characteristics and was ascribed its own destiny. Whites comprised less than 15 per cent of the population but owned 87 per cent of the land, more than 90 per cent of the wealth, and all the power.

Growing up as a member of the 'European' population meant all this land, and all the privileges accompanying it, was taken for granted. The contact most whites had with black people was with their servants or workers and most seemed comfortable with this system, which helped to mould their views of the world. It was the only life they knew; the appropriate order of things. White children would socialise solely with other white children. The only black people they knew worked for their parents and they found it hard to consider the inferior position of these employees as anything other than natural.

Under apartheid, black Africans were divided into tribal categories, allocated ten little parcels of land to express their 'national' identity, and had only contingent rights to live in white South Africa (contingent on having a 'pass' permitting them to work there). 'Coloureds' (mixed race people) and 'Asiatics' (people with ancestry in the Indian subcontinent) had more rights in 'white' South Africa than Africans but considerably fewer rights than 'Europeans'. These divisions and subdivisions, and the living conditions they created, reinforced perceptions of the 'other'. White South Africans therefore tended to view black South Africans as less intelligent, lacking powers of invention or innovation, oversexed, dishonest and, essentially, child-like; 'Coloureds' as lazy and alcoholic; and Asiatics as sly and scheming. 'Europeans' were not legally subdivided but the backwash of the historical conflict among them lingered; hardly surprising since the Brits were responsible for history's first concentration camps. Twenty thousand Boer children and eight thousand Boer women died in these camps between 1900 and 1902, along with twenty thousand black servants, who are rarely mentioned.

When I was at school and we played rugby against an Afrikaans school, we might absorb insults such as *rooinek* (red neck), *pommie*, *sout piel* (salty penis) and even *poes gesig* (cunt face), and we were viewed as softer, weaker and less patriotic. We, in turn, imitated their accents in a tone of derision, told jokes about thick Afrikaners, referred to them as 'Dutchmen', 'rocks', 'rockspiders' and 'crunchies' and generally viewed them as lacking in brain power and sophistication. Even among English speakers there were subcategories of prejudice; people of Portuguese descent were sometimes called 'sea kaffirs' and, of course, the usual range of anti-Semitic prejudices were never far from the surface. This was the world I grew up in: a place that defined its existence in race categories.

However, my family history offered me a glimpse of a different view. My father, a Jewish South African, was raised an atheist but converted to Christianity in his twenties. He went on to become an evangelical Anglican clergyman. His relatively liberal mother despised the National Party government, and his own experience of anti-Semitism during the war reinforced his hatred of prejudice. After returning to South Africa in the early 1960s, his first church was in an area where his 'coloured' congregants were being forcibly removed from their homes to make way for white people. His colleagues included black clergymen who were regular visitors to our house but had to live under the country's notorious pass laws.

The result was that I grew up with more exposure to non-racial ideas and to black people than was common for white children at that time. From my late teens I became active in anti-apartheid activities, both legal and illegal, which meant working closely with black activists, some of whom became comrades, friends and eventually family members. It also meant spending time in black townships and 'homelands'; access that was further facilitated by my academic research and journalism.

The more I witnessed and experienced the brutal inhumanity of this system, the more I came to despise the deep-rooted racism at its heart. What had once been a mainly cerebral contempt became ever more visceral and personal. I would bristle whenever racism raised its head, so often in casual, unconscious forms, such as apparently well-meaning 'madams' referring to their domestic workers as the 'girl' and 'garden boy' and addressing them in slow voices of command.

I regarded these as peculiarly South African failings, way behind attitudes in the enlightened world. By the time I travelled to Texas in 1978, for a year as an exchange student, I'd read widely on the American Civil War and the segregation of the Deep South but I assumed this stuff was in the distant past, a view affirmed by the apparently easy racial mixing I was delighted to observe at my Texan high school and university. But along the way I discovered I didn't have to scratch too deep to find the spirit of Jim Crow. I remember one discussion I had with a Texan rancher who told me he had 'nothing against niggers and Mescans' and illustrated this point by reminding me that one of the young ropers who used his rodeo arena was black, and that his oldest son was dating a 'Mescan girl'. What he really hated was 'uppity niggers', which brought him to the subject of Martin Luther King. 'You know son,' he drawled.

'If at that time, nine years back, they'd asked me to contribute to a fund to shoot that uppity nigger son-of-a-bitch, I would've paid up, no problem at all.' He thought about this for a moment and nodded his head. 'Still would, still would.'

I also harboured enlightenment illusions about the British before returning to live in London twenty-seven years ago. I knew all about the brutality of colonialism, about 'No dogs, no blacks, no Irish' and about the National Front but I assumed this was the past or limited to a diminishing rump. And for the most part that proved to be true. Now and then, however, I'd find myself enraged by public displays of racism, sparking street confrontations with the culprits. And I soon discovered such attitudes could extend to more salubrious surroundings.

My first foray into investigative journalism after returning to London in the early 1990s focused on the Conservative Monday Club, a fringe group on the edges of the Tory Party, which met in the House of Lords. One of its ennobled leaders took me into his confidence, expressing his sadness at the decline of apartheid rule. He proceeded to tell me how cold European weather created the conditions for the evolution of the European brain, leaving Africans behind, which was why 'we' had the Industrial Revolution and 'they' had tribalism. What exasperated me so much about this kind of thinking was that it so precisely echoed the stuff we'd exposed and fought against in apartheid South Africa. I thought it had gone away, shown up for its illogicality, and yet for this tweedy lord it took nothing more than my reassuringly pale skin and South African accent to bring it to the surface.

Early in the new millennium I noticed new variants of this old cancer on both sides of the Atlantic. At first it was just fringe players such as Richard Lynn, the University of Ulster evolutionary psychologist, who wrote that white and Asian people were inherently more intelligent than black people, with Bushmen and Pygmies at the bottom of the intellectual pile; all based on his take on IQ scores. Reading Lynn alerted me to a network of far-right academics and publishers who were relentlessly pushing out their papers in tame house journals and looking for entrée into the mainstream media.

I began to pry deeper and realised I needed to examine the underlying premises of these beliefs. My postgraduate academic background was in economic history and law, and my PhD was in politics. Other than taking a module in neurology when studying psychology as an undergraduate, my reading on biology hadn't progressed much beyond high school level. But these new expressions of scientific racism – when combined with my parallel intellectual interest in genes and gender – nudged me to read on. I delved into more serious scientific books and academic papers on genetics and neuroscience, periodically probing the brains of friends who were biologists, neurologists and psychiatrists when I needed help. Along the way I returned to the source – Charles Darwin and Alfred Russel Wallace – and branched out to devour books and academic papers on IQ theory, twin and adoption studies, and on archaeology, anthropology and palaeontology.

While I was undertaking this process of auto-didactic research, a new wave emerged, this time from closer to the establishment than the likes of Lynn (though often drawing on his writing). Their thin-edge issue was an ostensibly innocent claim about superior Ashkenazi Jewish intelligence, first made in papers by Lynn and a trio of anthropologists from Utah. It was based on an ahistorical snapshot of Ashkenazi IQ scores and a misreading of Ashkenazi disease profiles and genetic history. Yet the doyen of evolutionary psychology, Steven Pinker, piled in with a positive endorsement. He was joined by others such as the super-blogger Andrew Sullivan, the political scientist Charles Murray and the journalist Nicholas Wade. The notion that different race groups had different innate mental attributes, including intelligence, was edging its way into the mainstream. Wade, a former *New York Times* science correspondent, wrote a particularly noxious book claiming that African tribalism, English enterprise, Chinese conformity and Jewish business sense had a genetic basis.[2]

By then, I'd started writing about this stuff; newspaper articles and blogs showing why views such as these were unscientific and dangerous. Eventually, I wrote a book, *Black Brain, White Brain*, published in South Africa in 2014, which dissected the key arguments of race science and showed why they were mistaken and based on key errors regarding the archaeological record, IQ theory and biology. Hardly surprisingly, the book and the media exposure that came in its wake drew a fair amount of heat from people who hated seeing their cherished prejudices challenged. For example, the YouTube version of a minor television interview I did in South Africa drew 100,000 unique hits and some very racist comments.[3] It seemed clear that the international far right were finding each other on the Web, and the issue of race and intelligence was the one that really got them going.

In the penultimate chapter of that book, I wrote that publicity for the claims of race science came in waves, and that we could expect more of the same in the future. But I hoped I was wrong or that at least this latest wave, launched by the Ashkenazi fallacy, had been seen off for a while. Sadly, the opposite happened. The election of Donald Trump gave a huge boost to the American alt-right (alternative right) with its race-obsessed agenda. Through YouTube, Reddit, 4chan and other social media platforms, and through its blogs, podcasts and websites, the alt-right has relentlessly pursued the cause of race science. Those such as Wade and Murray, whose work had been eviscerated through peer review, were given new life. Sullivan returned to the fray, joined by others from outside the alt-right faithful, such as the writer and podcaster Sam Harris and the YouTube pontificator Jordan Peterson; all backing the claims of race science advocates and all attacking their critics as politically correct and intolerant. The sad truth is that the revival of race science has been far more expansive, determined and vigorous than I anticipated.

However, there have been several far more lasting developments in this field, which have largely stayed under the radar, over the last few years; in particular, a stream of exciting new discoveries by archaeologists and geneticists. These have helped to shift many assumptions about our shared origins, pushing our roots back 120,000 years and changing our understanding of how we populated the world. They have also upset the few remaining shibboleths of those clinging to traditional ideas about race groups.

This book picks up on what is new and interesting in a range of complementary areas of research – evolutionary theory, genetics, biological anthropology, archaeology, IQ studies and twin and adoption studies – to present a fresh picture of what we know about humans and intelligence. It also takes readers down the dark warrens where the alt-right breed, showing how the ideas of race science blend with more traditional and visceral forms of racism to produce a truly dangerous brew.

This evidence should close the door on the thoroughly unscientific idea that different population groups have significantly different, biologically innate, mental and emotional attributes. But that won't happen. There are too many vested interests, too many reputations resting on pseudoscience and far too much anger and hate circulating on the Web for a mere book to tip the balance. The racist world view is alive and well and kicking in Trump's nativist America, in Hungary and Italy and Brazil, in Brexit Britain and beyond, spurred on by the same faux-scientific racial prejudice that inspired the fathers of apartheid. This book is for those who instinctively reject racism but who have not known how to fight back when confronted with its claims to be authentically scientific.

1

WHAT IS SCIENTIFIC RACISM?

In 1977 I was president of a society, the 'Grey Union', at my South African state secondary school. One of my jobs was to organise an Education Week for the school's seven hundred pupils, involving talks at morning assembly and lunch time. For one of these, I invited a young sociologist from the local University of Port Elizabeth to address the school assembly, the idea being that he would tell us about the latest sociological thinking on our country. After his speech he fielded a question about whether black people were naturally as intelligent as white people. 'No,' he said without hesitation. 'IQ tests here and in America show they have IQs about fifteen points below ours. You can draw your own conclusions.'

I'd already formed a sceptical impression of what IQ tests really measured, and through my parents' church connections I'd met enough black people who were so much cleverer than me to know that this could not be true. It would be very difficult for anyone to spend time with the wit and wisdom of, say, Desmond Tutu and draw the conclusion suggested by this young lecturer. My reaction was to double down on my dismissal of IQ as a means of measuring intelligence and to doubt that it reflected anything useful.

In retrospect, the sociologist's claim shouldn't have been too surprising. His university was led at that time by a member of a secret, race-obsessed, nationalist society, the *Afrikaner Broederbond*, and its lecturers tended to toe the line. Naively, I'd expected something different, and wondered how this young man had reached this obviously ridiculous conclusion. I didn't realise that he was drawing on the discredited research of the American psychologist

Arthur Jensen, the man most responsible for reviving racist psychology after the post-war lull. Jensen was celebrated in apartheid educational circles and I suspect that the many critiques of his methodology and conclusions that were already in academic circulation by 1977 were not on the University of Port Elizabeth's curriculum.

DEFINING RACISM AND RACE SCIENCE

I'll discuss Jensen's ideas in Chapter 12 but I mention him here because it was my first unambiguous exposure to what we now call race science or scientific racism. Before defining this, it is worth saying a bit more about racism more generally. It's a newish term that was coined in the 1930s, took off in the 1970s and had its first definition in the Oxford English Dictionary, in 1989, as a synonym for the older term, 'racialism'. The OED's current definition is that it is a belief that the 'members of each race possess characteristics, abilities or qualities specific to that race, especially so as to distinguish it as inferior or superior to another race or races'. Webster's takes a different angle: 'A belief that race is the primary determinant of human traits and capacities and that racial differences produce an inherent superiority of a particular race'. Wikipedia opts for the simpler 'belief in the superiority of one race over another'.

All these definitions put the emphasis on belief, which is appropriate. It follows that a racist is someone who holds these beliefs; who embraces the idea that different 'races' tend to have different collective characters, personalities or potentials. I stress this essence because there is a view, often heard in the United States but less elsewhere, that black people can't be racist because they don't possess power (which parallels a view that women can't be sexist because under the patriarchy only men have power). I believe this is wrong, and will discuss it further in Chapter 4, but for now I'll repeat that racism is all about beliefs. Power is something different, although obviously racist beliefs held by those with power are likely to be more dangerous than such beliefs among the powerless. Just one preliminary example: anti-Semitism is a form of racism. A power-less person, regardless of race, who believes, say, that a secret cabal of Jewish bankers, politicians and industrialists controls the world and, as a result, looks askance at Jews, is a racist. The consequences of that racist belief would be very different, of course, if that anti-Semite had power (as in Nazi Germany).

Scientific racism is a variant of these beliefs. One might say it is the attempt to attach the categories of science to racist beliefs, to give them ballast, but one needn't be that cynical. It is more likely to be the genuine belief that this is where science leads. As we shall see in Chapter 4, the eighteenth- and nineteenth-century scientists who believed the mental capacity of different races could be found by measuring their skulls or weighing their brains were perfectly sincere in their wholly mistaken views. Today we might say the same thing about those who believe, say, that different average IQ scores among different population groups tell us something profound about their innate intellectual potential. Regardless of motivation, such thinking fits squarely in dictionaries' definitions of racism. This, incidentally, would be true even if such ideas were correct, although I will show in later chapters why they are profoundly mistaken; in essence, unscientific.

This raises an obvious but tricky question: is someone a racist if they hold the idea that different population groups have different innate, average intelligence? A few of those advancing such views are indeed happy to own up to racism. One of those is Richard Lynn, the University of Ulster evolutionary psychologist, who has no hesitation about calling himself a 'racialist', a 'racist' and a 'scientific racist'.[1] But most of those who advocate race science, including several who enthusiastically quote Lynn, deny they're racists, preferring to view themselves as intrepid truth-tellers who follow science wherever it may lead. To say they are racists would put the likes of Steven Pinker, Andrew Sullivan, Jordan Peterson, Sam Harris and Nicholas Wade, along with older hands such as Charles Murray, in a particularly odious circle of hell. I would prefer not to go that far, because I do not know what goes on in their every secret place. It is perfectly possible that all or some of these men treat black, white, Asian and Hispanic people the same, perhaps even that they have close friends who are not white, and that their belief that some population groups are, on average, less intelligent than others has no influence on the way they treat individual people from any of these groups. What I will say, however, is that some of the beliefs they advance are indeed racist, and that the adjective 'scientific' does nothing to mitigate this verdict.

What would be some examples of contemporary scientific racism? I've already mentioned, in the foreword, Nicholas Wade's views on innately tribalist Africans, enterprising Brits, bright but conformist Chinese and capitalistic Jews. I could add a few other prominent claims made by various university-based academics over the past decade or so: Europeans and Asians evolved to be more intelligent than Africans because of their exposure to ice age conditions 45,000 years ago; a gene variant that makes sub-Saharan Africans less intelligent than everyone else; the smartest people on earth today are Ashkenazi Jews, followed by East Asians and white Europeans and Americans; the dumbest are Bushmen and Congo Pygmies, followed by Australian Aboriginals and Ethiopians; poor people are poor because they're stupid, which is why there are so many underclass black people; the prime cause of poor health all over the world is low IQ, which is why Africa suffers; infectious diseases have affected the genomes of Africans, making them less intelligent; sub-Saharan Africans haven't evolved to possess a work ethic. We will tackle all these ideas head-on.

THE TWENTY-FIRST-CENTURY REVIVAL OF RACE SCIENCE

The complementary ideas of a link between race and intelligence and between race and character have long pedigrees, probably even longer than slavery and colonialism. But because such thinking is seldom aired in polite circles nowadays, it is tempting to think that, Steve Bannon aside, it is confined to the anonymous midnight trolls who furiously patrol racist websites, or the backwoodsmen of Confederate America, and that it is no longer something significant to bother about. But there are good reasons to bother because after a post-Holocaust lull, scientific racism has returned in a full-fledged, brazen form and its current alt-right wave is still building its momentum.

It is hard to pick a precise starting point but 2007 is as good as any. That was when James Watson, one of America's greatest living scientists, a Nobel laureate and co-discoverer of the structure of DNA, attracted headlines for trumpeting his belief that black people were inherently less intelligent than white. Having previously advocated eugenic solutions to weed out less intelligent people, he started speaking out on race in 2000, when he announced there was a link between darker skin and higher libido. Seven years later he went significantly further, saying that the idea that black and white people shared 'equal powers of reason' was a delusion and that 'people who have to deal with black employees find this is not true.'[2] Subsequently, following an outcry, he apologised but made it clear that he hadn't changed his mind, noting that the desire of society 'to assume that equal powers of reason are a universal heritage of humanity' was 'not science and that it was not racist to question this'.[3] Watson has since been quoted as suggesting that Jews are smarter than everyone else, and that Indian Brahmins had been naturally selected for both intelligence and servility and East Asians for conformity.[4] In 2019, in a PBS television documentary on his life, he said his views were unchanged, explaining that 'there is a difference on the average between blacks and whites on IQ tests. I would say the difference is genetic.'[5]

Since 2007, ideas such as Watson's have begun to proliferate on the Web, often finding their way into the mainstream media. They were given a huge boost by the rise of the American alt-right in the wake of Donald Trump's election. Watson aside, their purveyors are usually people outside the 'hard' sciences; a mixture of evolutionary psychologists, journalists, social theorists and media personalities, who believe that different population groups have different innate mental and emotional assets and who feel that these ideas are being suppressed by the political correctness of a self-serving liberal elite.

THREE PILLARS OF RACE SCIENCE

This group's first contention is that our brains, like our bodies, have continued to evolve in response to different environmental conditions, leading some ethnic groups to develop superior intelligence and different character traits. Some confidently predict that significant genetic markers related to brain power will be found to differ substantially between races. The reason: the extreme challenges created by ice ages in Europe were not faced by those living in warmer climates, and these cold challenges prompted further evolution of the brain after groups of humans left Africa 50,000 years ago. Richard Lynn, for example, wrote that in ice age Europe 'less intelligent individuals and tribes would have died out, leaving as survivors the more intelligent.'[6]

Second, they claim their perspective is borne out by archaeology. Evidence cited for this conceptual leap includes the flowering of cave art and other creative innovations in parts of Europe, some of it dated to be more than 40,000 years old. Some, such as Nicholas Wade, have argued that the diverging evolution of character traits and of intelligence has continued even over the last thousand years.

Third, they claim proof of hardwired racial differences in intelligence comes from IQ tests, which they believe can measure innate 'general' intelligence. We are told that the reliability of these tests as an accurate measure of intelligence is proved by studies of twins, which show that IQ is highly heritable. There is indeed variation in IQ scores when assessed on a population basis. For example, Asian Americans have higher average scores than white Americans, who have higher average scores than African Americans. Some USA-based writers – who include media luminaries such as the evolutionary psychologist and popular science author Steven Pinker – suggest that this racial variation in IQ proves the point that the brains of different groups have evolved differently. They frequently add that those who ignore this evidence are obdurately turning a blind eye to scientific fact.

THE FALLACIES OF RACE SCIENCE

One area of confusion relates to the fact that in certain ways human bodies have continued to evolve over the millennia. This can be illustrated by looking at diseases that are more common among some ethnic groups than others, such as sickle cell anaemia among those with sub-Saharan ancestry, Tay–Sachs disease among Ashkenazi Jews, and so on. Some populations have also evolved certain physical capacities, such as survival at altitude and, more widely, the ability to digest lactose. Other examples include skin colour, eye colour, hair type, the presence or absence of an eyelid fold, average height, bone density and body type. Today, scientists can identify a person's regional historic origins and population mix by examining genetic markers in their DNA. The recent capacity to sequence whole genomes has expanded this ability, for example the discovery that around 50,000 years ago early human migrants to Eurasia interbred with Neanderthals, and those in New Guinea, Australia and the Philippines with Denisovans (another, recently discovered, extinct human group).

Despite these differences, all humans are remarkably similar, in the sense that there is very little genetic variation among us. The small amount of Neanderthal or Denisovan DNA possessed by non-Africans appears to make little physical or other difference to them, although Denisovan DNA does seem to have contributed to helping Tibetans live at altitude and Neanderthal DNA to have helped Europeans live in cold climates. Because of our relatively recent common ancestry – the first humans like us emerged in Africa just 300,000 years ago – humans share a remarkably high proportion of their genes, compared to other mammals. The single subspecies of chimpanzee that lives in Central Africa, for example, has significantly more genetic variation than the entire human race. Richard Dawkins, the British ethologist, put it like this: 'We are indeed a very uniform species if you count the totality of genes, or you take a truly random sample of genes.'[7]

The problem with drawing analogies between lactose intolerance and human intelligence is that we are not comparing like with like.

Intelligence is complex, and not just because it is an abstract notion that is hard to define and comes in a variety of guises. More than ten thousand genes are implicated in the development and functioning of the brain, and neuroscientists believe that a network of perhaps thousands are implicated in intelligence. Scientists may be able to identify scores or hundreds of genes that appear to have a limited bearing on performance in IQ tests, but the quest for a single intelligence gene, or even a handful, has proved quixotic. Even if slight distinctions were one day established, it is highly unlikely that these would follow the traditional boundaries of 'race'.

The American palaeo-anthropologist Ian Tattersall, widely acknowledged as a world expert on the Cro-Magnons (early European-based cave-dwelling humans), says that long before humans left Africa for Eurasia they had reached the end of the line for significant evolution of their brains. 'We don't have the right conditions for any meaningful biological evolution of the species,' he said. 'In order to get the fixation of evolutionary novelties ... you need to have small isolated populations. Large, interbreeding populations are just not the right place for innovations to become fixed.'[8]

Contrary views, such as James Watson's, that claim significant racial or geographical genetic differences in intelligence and character, are dismissed by most biologists focusing on genetics and human evolution. Craig Venter, the American scientist who led the privately funded effort to decode the human genome, noted in response to Watson's outburst that 'skin colour as a surrogate for race is a social concept, not a scientific one.' And he added: 'There is no basis in scientific fact or in the human genetic code for the notion that skin colour will be predictive of intelligence.'[9] Later, in his autobiography, he went further, dismissing the concept of race altogether by saying it had 'no genetic or scientific basis'.[10]

There is also no evidence that icy European weather prompted further evolution of the brain. We could just as well claim that the heat of Africa, Asia and the Middle East had the same nudging effect by pointing to the early emergence of agriculture, writing and city states in the warmth of Mesopotamia, China, Egypt, Nubia,

Anatolia and India. And the focus on European cave wall paintings as proof of superior intelligence is also highly selective. Cave wall art emerged in Australia and parts of Africa at roughly the same time as in Europe, but artistic expression started far earlier. Caves in the Cape province in South Africa have evidence of symbolic geometric art dated at 77,000 years old, and of the use of carefully blended paint and sophisticated tool-hardening techniques dated at 100,000 years old. Both achievements would be impossible without recourse to language. Other evidence of modernity from this period includes beads used for adornment, fish hooks, arrow heads and animal traps, and signs of land division, long-distance trade and burial of the dead. It seems clear that the humans who left Africa at least 70,000 years ago, and eventually ended up in Australia, or who headed towards Europe 15,000 years later, must have had brains capable of symbolic, artistic, self-conscious behaviour, of scientific experiment and future-planning. In all probability, they were humans with brains very much like ours.

The idea that the proof of racial differences in intelligence can be drawn from IQ tests is also spurious. Such tests measure the capacity to cope with a certain kind of abstract logic; they are therefore useful to assess aptitude for certain jobs, university courses and so on. However they do not, and cannot, measure general intelligence. In fact, 'general' intelligence doesn't really exist.

And yet claims of a link between brain power, IQ and race roll on. The most prominent and persistent over the last decade or so have related to the idea that Ashkenazi Jews are inherently more intelligent than anyone else (rivalled only by those of East Asian origin, we're told). I will devote a whole later chapter to this idea, partly because it is currently the smiling face of race science: everyone knows Jews are smart, so what's the problem? The problem is that it is a cat's paw issue – if Ashkenazi Jews are accepted as being naturally smarter than everyone else then it is only logical to say that other groups are naturally dumber. It is perhaps for this reason that it is promoted with such vigour by advocates of race science.

Those currently pushing the Ashkenazi fallacy include evolutionary psychologists, anthropologists, journalists and YouTube stars.

But even the finest scientific minds are capable of profound errors when tackling subjects beyond their calling. James Watson's DNA-unravelling Nobel Prize-winning colleague, Francis Crick, believed that life on earth was directed by an advanced extra-terrestrial civilisation. And like Watson, he believed blacks were less intelligent than whites. And we could back-pedal all the way to the grandfather of them all, Isaac Newton, whose immense contribution to science did not hold him back from devoting years of attention to finding the Philosopher's Stone, to the Biblical Apocalypse and to occult studies. In a sense, the determinedly contrarian, octogenarian Watson was in fine company when, in addressing an area of learning way beyond his own expertise, he came out with his dubious views. Expanding on his notion that the brains of Africans and non-Africans evolved differently, making the former less intelligent, he argued that there is 'no firm reason to anticipate that the intellectual capacities of peoples geographically separated in their evolution should prove to have evolved identically'.[11] In the chapters that follow, I will show that there are, in fact, firm scientific reasons to anticipate precisely this. Or at least not to anticipate substantial differences in the evolution of their brains.

Recent research in genetics, evolutionary theory, archaeology, palaeontology, biological anthropology, psychology, sociology and IQ studies all pulls in the same direction when it comes to understanding human intelligence: it might be influenced by a number of factors including the genes inherited from parents, *in utero* experience, health and nutrition, early education, wealth or poverty and, in particular, exposure to abstract logic in your formative years. But not by 'race'.

It is now forty years since I heard an apartheid-era sociologist telling my school assembly that we should draw our own conclusions on why there were racial differences in average IQ scores. If he's still around today and following the evidence where it leads, he should come to the opposite conclusion: that there is a complete absence of evidence that race is important when it comes to innate intelligence potential. The chapters that follow explain why.

2

ARE WE SMARTER THAN OUR ANCESTORS?

1960: a Moroccan miner is picking away at a cave wall, looking for the mineral baryte, when he hits a white rock. Closer inspection reveals it's a skull. He removes it with care and presents it to his engineer, who keeps it for a while as a souvenir. But one day, when visiting the town of Rabat, the engineer decides to hand it over to experts at the university. They're intrigued, and set about examining it, but their archaeological dating techniques are primitive, especially when it came to ancient bones, and their knowledge about human origins is also lacking. Their best guess is that the skull may be 40,000 years old and possibly Neanderthal, which seems odd, because Neanderthals never set foot in Morocco.

Skip forward forty-four years, to 2004. The Jebel Irhoud Moroccan cave remains of interest to those excited by old bones. More skulls and tools have been discovered over the years, which is why the French paleoanthropologist Jean-Jacques Hublin is there with a team of scientists. Among an abundance of fossils, one find particularly excites them: five skulls from people who appeared to have died at about the same time.

Alongside the skulls, in the same layer of rock, they uncover more human bones and teeth, and scores of carefully crafted flint blades, including spear blades designed to be mounted on wooden shafts. Some of these blades show signs of being burned, perhaps thrown into cooking fires after use, or maybe used as the points of barbecue forks. This offers the scientists the means to date them, and they also cross-check with a technique of calculating age by assessing the fossilisation of dental tissue. These test results provide a shock: the tools and skulls are around 315,000 years old, which puts back the date of the first known appearance of modern humans (*Homo sapiens*) by about 120,000 years.

The skulls look pretty much like present-day humans from the front; they suggest wide faces with heavy brows and smallish chins. 'The face is that of somebody you could come across in the Metro,'[1] said Hublin, who is now director of the department of human evolution at Leipzig's prestigious Max Planck Institute for Evolutionary Anthropology. Skull measurement also suggests their brains were the same size as those of contemporary humans, but with one key difference: they had longer, lower braincases, more typical of earlier hominins than the more rounded braincases of all modern humans; seen from the side, they would appear slightly odd.

It might be that the different skull shape suggests different cognitive capacities among these very early humans but we can't be sure. Certainly, aspects of their behaviour appear to be as sophisticated as those of later humans. These gazelle-cooking cave dwellers seemed capable of making a variety of tools and weapons, including wooden-handled spears, and their flint came from another site nearly forty kilometres away, which suggests the ability to plan and to travel long distances to seek resources.

Over the next thirteen years Professor Hublin's team continued to work on the site, making further finds and analysing the bones and tools, using new techniques of genetic analysis, until they were ready to go public. In 2017, the team published a paper in the journal *Nature* that shocked the archaeological world and overturned conventional wisdom on human origins; not only of the dating of modern human origins but also of the distribution. The other huge implication of this discovery is, as the paper's authors put it, that 'these data suggest a larger scale, potentially pan-African origin.'[2] This quote relates to an alternative to the conventional wisdom that all humans had a single common African ancestor, and raises the possibility that, instead, we had many common ancestors from different parts of Africa.

Adding to this pan-Africanist picture are recent discoveries at an archaeological site called Olorgesailie, in a dry basin of what was once a lake in the Eastern Rift Valley of Kenya. The first interesting artefacts were discovered in 1919 but real excavation started only in 1943, directed by the pioneering British-Kenyan archaeologists Mary and Louis Leakey, with Italian prisoners of war providing the hard physical graft. The site became famous for its hoard of hand axes made between about 600,000 and 1.2 million years ago, along with fossils of a variety of extinct animals, including giant baboons and some bones and part of a skull of a *Homo erectus* child, dated at nearly one million years old.

The paleoanthropologist Rick Potts, who leads the Smithsonian's Human Origins Program, began working in Olorgesailie in 1985, for what he thought was a three-year stint that has ended up passing the thirty-three-year mark. In recent years his team has made a series of remarkable discoveries, including carefully crafted human tools made from a volcanic black rock, obsidian. Potts's team was able to date these tools by analysing radioactive isotopes of uranium and argon contained in the obsidian; they dated to between 305,000 and 320,000 years old. In other words, they are about the same age as the human skulls and tools found in Morocco. And yet these tools were typical of the Middle Stone Age period, previously thought to have started well over a hundred thousand years later.

The hoard included spear tips, blades, scraping tools and awls (small pointed piercing tools). But the most intriguing discovery related to the stuff they were made from. The closest source of obsidian is 100 km away as the crow flies, provided the crow flies over rugged mountains. And yet, huge amounts were imported. Potts and his team considered all the possible explanations and concluded the only answer was long-distance trade, which would mean this trading network preceded any other known network by more than 100,000 years. Further evidence of trade came from other discoveries, including lumps of manganese ore and iron oxide that had been ground into powder with awls, almost certainly to create paint, which suggests symbolic activity.

Potts has no doubt about the significance of these finds. 'What we're seeing in Olorgesailie is right at the root of *Homo sapiens*,' he said in a 2018 interview. 'It seems that this package of cognitive and social behaviours was there from the outset.'[3] The American archaeologist Alison Brooks, who together with her colleague Sally McBrearty has done so much to draw the world's attention to the wealth of prehistoric evidence of modern human behaviour in Africa, studied the artefacts at the Kenyan site: 'There was an argument that *Homo sapiens* came along and then developed all these things but now it seems that the behaviour and the morphology came along together,' she said.[4] What Brooks and Potts are suggesting is that the cognitive capacity for modern human behaviour was there right from the start; from the time the first anatomically modern humans evolved.

WHO DID WE COME FROM?

It's time to reverse gear and consider the deep origins of the hominin family. Humans, chimpanzees and bonobos had a common ancestor more than six million years ago, but it was another four million years before the first hominins – *Homo habilis* and *Homo gautengensis* – evolved 2.5 million years ago, after which several sub-species popped up in Africa, soon to migrate to Asia and Europe. In looking at our ancestral cousins, the starting point is usually said to be *Homo erectus*, so named because they walked upright, but there is a view that several other early hominins, previously identified as other species, may really have been early examples of *Homo erectus*. Five *Homo erectus* skulls and skeletons, discovered along with their stone tools in Georgia in 2005, were of individuals thought to have been killed by sabre-toothed tigers between 1.77 and 1.85 million years ago. An interesting detail is that there were significant differences among the group; had they been found separately, they might well have been identified as different species. This led palaeontologists to question whether African skulls previously discovered were really examples of different hominin branches rather than part of the variety in *Homo erectus*. However, the British anthropologist Chris Stringer has his doubts:

Africa is a huge continent with a deep record of the earliest stages of human evolution and there certainly seems to have been species-level diversity there prior to two million years ago. So I still doubt that all of the 'early *Homo*' fossils can reasonably be lumped into an evolving *Homo erectus* lineage. We need similarly complete African fossils from two to 2.5 million years ago to test that idea properly.[5]

Anyway, *Homo erectus* arrived on the scene more than two million years ago and soon migrated over the Levantine corridor and the Horn of Africa, spreading to Asia (China, India, Indonesia and Java) and through parts of Europe, living in hunter-gatherer communities, using fire, travelling by raft, making tools and weapons and caring for the old and the ill. Over time, they developed brains at least three-quarters the size of modern humans'.

The next big hitter in our past is *Homo heidelbergensis* (thought to have evolved from *Homo ergaster*, an African offshoot from *Homo erectus*), which shows further skull expansion, suggesting a larger brain relative to body size. These hominins first emerged about 700,000 years ago and were still around less than 200,000 years ago. In this time, they spread to Europe (their name comes from the discovery of one of their skeletons in a cave in Germany) and through much of East and Southern Africa. They hunted with stone-pointed wooden spears, used cutting tools, and may have been the first in the human family to bury their dead. They may have painted with red ochre pigment, and also had the genetic and auditory equipment to be at least anatomically capable of speaking and hearing words, suggesting linguistic capacity. *Homo heidelbergensis* was the direct evolutionary ancestor of the Neanderthals and of the Denisovans, and probably of us.

As we've seen, the origins of modern humans have been moved back, but even before 315,000 years ago, archaic *Homo sapiens* were involved in symbolic behaviour, including painting, using tools and creating weapons. Archaeologists in Zambia, for example, have found pigments and paint-grinding equipment at least 350,000 years old.

There is much debate about where and why the evolutionary change to greater intellect happened. On the why question, some have emphasised physiological factors such as evolving to walk upright or evolving to have opposable thumbs, while others have looked to the environment. A recent environmental theory is that sudden shifts in the climate of the East African Rift Valley created a need to adapt to dramatic environmental change. 'It seems modern humans were born from climate change,' said Mark Maslin, professor of geography at University College London, who co-authored a geographical study on climate change in this period.[6] His fellow author, Dr Susanne Shultz, from the University of Manchester, added that 1.9 million years ago a number of new species appeared, which they believe was directly related to new ecological conditions in the Rift Valley, in particular the appearance of deep freshwater lakes. 'Among these species was early *Homo erectus* with a brain 80 per cent bigger than its predecessors,' she said.[7] Another compelling theory is that the social skills demanded for living alongside other hominins in communities provided the impetus, with the brightest and most innovative in the group more likely to flourish and to produce more offspring. But this is entirely speculative; it might be that the strongest and most attractive had more children than the smartest.

All these reasons relate to natural selection but it's possible that the growth of the modern human brain was at least partly the result of a coincidence of genetic acquisitions through genetic drift (a random process, often occurring in population bottlenecks). Whatever the causes, the human brain evolved rapidly over a short space of evolutionary time. Two and a half million years ago hominin brains were not much larger than those of chimpanzees. By 315,000 years ago the brain had trebled in size and it's possible it came with the full range of intellectual and creative potential we have today.

ONE AFRICAN EVE OR MANY?

The Kenyan tools and trade networks, the Moroccan skulls and a 195,000-year-old Ethiopian skull, among others, all provide evidence that modern humans were evolving in multiple sites around Africa. Another example is the Florisbad Skull, found in the Free State province of South Africa in 1932 along with tools from the Middle Stone Age period. One of the interesting things about this skull was that it had a brain volume of 1,400 cm^3, larger than that of modern humans (this does not in itself suggest greater brain capacity; it might be that its bearer had a bigger body). However, because it had some archaic features, and was first classified as belonging to a near ancestor, it was dubbed *Homo helmei*. In the 1990s it was dated, using the technique of electron spin resonance, and found to be 260,000 years old. Following the Moroccan discovery, it has been reclassified as 'archaic *Homo sapiens*' or just plain '*Homo sapiens*'.

Until recently the debate around the geographical origins of humans seemed to have been settled. The idea so decisively seen off was one suggested by the American anthropologist Milford Wolpoff; that modern humans evolved more or less simultaneously in different parts of the world.[8] If he was right, we'd anticipate big genetic differences, possibly relating to intelligence, although Wolpoff refuted this, referring to the gene flow resulting from interbreeding between people. 'Humans today are widely variable but without enough population differences to be considered subspecies or races,' he said.[9]

However, all recent genetic evidence favours the idea that humans have their evolutionary roots in Africa. The fossil record shows modern humans going back more than 300,000 years in Africa, but the record of skulls and bones from Australia, China, India and Europe is far more recent, in most cases after *Homo erectus* is thought to have died out. The genetic record also shows that modern humans are closely related to each other and to what the British paleoanthropologist Stringer called 'all the fossil people', in reference to the ancient skulls from Europe, Africa, China and Australia that he has studied. 'If modern humans had evolved all over the world,' he wrote, 'you would expect them to have deep roots back to their ancestors in that place but [the skulls] weren't showing that. They were showing a compact group, as though they had originated from the same place.'[10]

This conclusion was bolstered in 1987 by DNA evidence said to suggest that a common, female ancestor of all humans alive today (popularly known as 'mitochondrial Eve') lived around 200,000 years ago, probably in East Africa. The dating is imprecise because it is based on an assessment of the expected rate of human genetic mutation, which is uncertain. The current best guess is 160,000 years. Either way, it was impossible to reconcile with the Wolpoff theory, which would suggest our last common ancestor lived around two million years ago.

The term 'mitochondrial Eve' is a bit misleading, because it conjures images of a single, intrepid, pioneering, human female ancestor. However, she would have lived in a community, and there were certainly *Homo sapiens* communities in other parts of Africa whose lineages did not survive. Some have questioned the assumption that mitochondrial DNA evidence alone confirms the 'out of Africa' thesis. For example, the geneticist Alan Templeton noted that mitochondrial DNA patterns could not be equated with actual human populations and suggested the data could also be used to support multi-regionalist theory.[11]

The idea that Africa was the human cradle has been reinforced by evidence that, compared to other mammals, there is little genetic diversity between different human populations (if we'd evolved wherever *Homo erectus* travelled, we'd expect far more diversity) and, even more decisively, by evidence that genetic diversity is significantly higher among African populations than anywhere else in the world. This is what you'd expect if small groups of Africans migrated relatively recently, because the gene pool of the billions of descendants of these migrants would be drawn from these small groups rather than from Africa as a whole, or from whole regions of Africa.

There are two key caveats to the out-of-Africa idea. The first is decisive: genetic evidence that humans interbred with Neanderthals, and with another hominin, the Denisovans, after migrating from Africa. Non-African people have small traces of Neanderthal and/ or Denisovan DNA in quantities that vary depending on the regional origins of their ancestors. The second is more speculative but gaining ground: the increasing evidence, particularly from the fossil record, that modern humans evolved all over Africa. This would not in itself discredit the original out-of-Africa theory of a single small population responsible for populating the world. However, if modern humans were making appearances all over Africa it seems unlikely that only one ancestral line survived.

Several experts have shifted their thinking recently. The University of Oxford archaeologist Eleanor Scerri drew together leading anthropologists, archaeologists, geneticists and climatologists to review the evidence, and the result was a World Health Organization-funded paper that has shaken their world. These twenty-three authors 'challenge the view that our species, *Homo sapiens,* evolved within a single population and/or region of Africa' and instead argue that humans 'evolved within a set of interlinked groups living across Africa, whose connectivity changed through time'.[12] They believe that the whole of Africa, not just one small corner, is the cradle of modern humanity.

One of their arguments relates to the oldest African population, the San of Southern Africa (also called 'Bushmen'). The authors note that the San have the highest levels of genetic diversity among all human populations today, which suggests an early divergence from common African ancestors somewhere between 150,000 and 300,000 years ago. This divergence lasted until the arrival from East Africa of Bantu-speaking people less than 2,000 years ago; the San shared genes with them, and later with Europeans and mixed-race people. A similar perspective is offered by the Harvard geneticist David Reich, who was part of a team that studied the genetics of the San and other populations in 2016. He suggests that the separation began around 200,000 years ago and was mostly complete by 100,000 years ago, with evidence of only limited gene sharing with other African populations after that. Reich says it is possible that 'Pygmy' groups from Central African forests have a similarly distinctive ancestry.[13] It is worth remembering that the earlier dates of divergence pre-date the current best guess for 'mitochondrial Eve', of 160,000 years ago.

The twenty-three WHO-backed scientists put forward several other paleontological, genetic and environmental arguments for the African multi-regionalist view, one being that early *Homo sapiens* fossils 'do not demonstrate a similar linear progression towards contemporary human morphology'. In other words, they don't all look alike; instead, they exhibit 'remarkable morphological diversity and geographical spread'.[14] They show that these early humans were spread all over Africa, rather than being concentrated in one place, and that their skulls looked different from place to place, even though they started using similar, more sophisticated, tools at similar times. The authors paint a picture of 'semi-isolated' populations adapting to local ecologies but ultimately connecting with each other, prompting a 'sporadic gene flow'. Incidentally, the time period they're talking about starts at least half a million years ago, but not until about 100,000 years ago did all or most humans have the same rounded skull shape we have today. In the preceding 400,000 years there was a more diverse range of human size and shape, including skull shape and size. Most of these groups evolved in isolation for long spells, each developing distinct features; more distinct than the differences between modern humans. But as the environment changed and impenetrable areas became more penetrable, people moved, had sex, and exchanged both genes and ideas. And eventually a more cohesive humanity emerged.

MODERN HUMANS AND ARCHAIC HOMININS INTERBREED

The twenty-three authors also raise the possibility that 'African archaic interbreeding'[15] might help explain the slightly different skull shapes found in different parts of Africa. Recent genetic analysis shows that it is more than a possibility. About half a million years ago, *Homo sapiens* began to separate from *Homo heidelbergensis* but the two groups continued to coexist for hundreds of thousands of years. There is now compelling genetic evidence that humans bred with an archaic ghost population probably around 35,000 years ago,[16] although this date is far from certain. The result is that around 2 per cent of the genetic ancestry of some contemporary African populations derives from this source, about the same amount as the Neanderthal genetic contribution to non-Africans,[17] and with close matches to the Neanderthal DNA sequences in non-Africans, which is hardly surprising when you consider that Neanderthals also descended from *Homo heidelbergensis*. This interbreeding may have continued, at least in isolated pockets: 11,000-year-old human remains found in West Africa show archaic, pre-*Homo sapiens* features.[18]

Modern humans in Africa also coexisted with at least one other hominin: the recently discovered, small-statured and small-brained hominin *Homo naledi*, several of whose remains were found in 2015 in a cave in the Gauteng province of South Africa, along with their tools. In 2017 these skeletons were dated at between 236,000 and 335,000 years old. So far, there is no evidence that *H. sapiens* interbred with them.

It is now certain from full genome genetic analysis that humans interbred with Neanderthals soon after leaving Africa, probably in the Middle East around 49,000–54,000 years ago, and this contributed to non-African DNA (2 per cent on average). Pockets of interbreeding continued after humans arrived in Europe 43,000 years ago; the evidence for this is drawn from a skeleton in a Romanian cave showing 6 to 9 per cent Neanderthal ancestry, suggesting a Neanderthal ancestor about six generations before.[19] However, these later pairings do not appear to have contributed to modern human DNA, possibly because lower fertility among human-Neanderthal hybrids meant most lineages died out. One contribution the Neanderthals made to modern European and East Asian DNA relates to keratin proteins, naturally selected because they provide protection in cold environments.[20]

Around 45,000 years ago humans bred with another recently discovered offshoot from *Homo heidelbergensis*, the Denisovans, of which one branch lived in parts of Europe and another in New Guinea and the Philippines. Between 2 and 6 per cent of the DNA of people with deep historical roots in some of these areas comes from the Denisovans (the higher end found among people in New Guinea). We know far less about the Denisovans than we do the Neanderthals because the archaeological record is sparse; a few teeth, some finger bones, a toe bone and a bracelet. However, we do know from whole genome DNA analysis that like humans and Neanderthals, they branched off from *Homo heidelbergensis* about 400,000 years ago and were still around 40,000 years ago. We also know that while they had much in common with Neanderthals and interbred with them, they were genetically distinct. From their teeth we believe they had a vegetable-rich diet and from the bracelet we assume they were capable of symbolic thinking. One of their contributions to modern humanity is a gene mutation, carried by modern Tibetans, that helps people living at high altitude.[21]

Modern humans also coexisted with *Homo floresiensis* (popularly known as the 'hobbit' because the first skeleton found was only 1.1m tall) who lived on the island of Flores in Indonesia between 190,000 and 50,000 years ago. They went extinct about the same time as modern humans arrived in the area, suggesting that we may have been responsible for their demise. Recent genetic analysis has shown that *Homo floresiensis* was not a diminutive version of *Homo erectus*, but rather a distinct species of hominin. It is now thought that their ancestors were among the first to leave Africa, perhaps two million years ago, and that they evolved further in Asia. Despite their small bodies and brains, they had sophisticated stone tools, probably used fire for cooking and hunted cooperatively. The latest discovery of hominins who coexisted with humans came from the 67,000-year-old remains of several individuals found in an island cave in the Philippines. *Homo luzonensis* were small, tool-using tree-climbers who seem to have reached their home by raft. It is also likely that modern humans coexisted with *Homo erectus* (who became extinct about 140–150,000 years ago), when *H erectus* was living exclusively in Asia and modern humans were beginning to venture out of Africa.

So for about 90 per cent of human existence, we have coexisted with at least seven other species of hominin and interbred with at least three. Given the rate of recent discoveries, it seems likely that we will find other unknown species.

YUVAL NOAH HARARI AND THE 'COGNITIVE REVOLUTION' MYTH

The fishermen who lived in a cliff-top cave on the Indian Ocean side of South Africa liked to paint. They used large shells to collect colourful dirt, pounded it, ground it and enriched it with iron oxide to create an ochre-coloured powder, which they carefully blended with mammal bone marrow and finally with charcoal. Next, they liquefied the mixture, and stirred it, before using spatulas and home-made crayons to decorate their tools, their artwork, their beads and probably themselves. In this sense, they behaved pretty much like cave artists in Europe or Australia or elsewhere in Africa. Except for one detail: the blended ochre work of South Africa's fishermen has been dated as at least 55,000 years older than any other cave-art workshop previously discovered.

What is clear from the excavation of the Blombos Cave – about 320 kilometres east of Cape Town – is that this small community of hunter-fisher people living about 100,000 years ago knew a great deal about seeking out and storing substances, and in combining them they must have had a basic knowledge of chemistry. They used their paint not just for utilitarian ends but with decorative purpose. They engaged in creative, self-aware behaviour that would have involved discussing ideas, planning and abstract thinking. The earliest dis-covery of residues of ochre used by cave people goes back at least 160,000 years and as we've seen, people in Kenya were extracting red paint from iron-bearing rocks more than 300,000 years ago, but the Blombos discovery, which involves blended ochre paint, is even more significant, because it reveals symbolic activity, which would neces-sarily involve the use of complex language, creative thought, planning and conception. As the lead researcher, Christopher Henshilwood, put it: 'We're pushing back the date of symbolic thinking in modern humans – far, far back.'[22]

Among the many astonishing discoveries from Blombos made by archaeologists over the past two decades are two large ochre plaques inscribed with geometric designs of interlocking triangles and hori-zontal lines, dated at 77,000 years old; twice the age of anything similar found in Europe.[23] We don't know what these engravings meant but finds at other coastal caves in the area have revealed similar symbols, suggesting they were widely used at the time, perhaps to keep records. The American paleoanthropologist Ian Tattersall, who has long studied European cave art, said this constituted 'the most remarkable early evidence of symbolic activity'.[24] Later, archaeologists discovered a broken-off fragment of rock that had been drawn on with an ochre crayon in a criss-cross pattern that looks as if it's part of a larger design. In 2018, after seven years of tests, this was dated as 73,000 years old, making it the world's oldest rock drawing. 'There's no doubt that it's a symbol that meant something to the people who made it,' Henshilwood was quoted as saying. 'It's a symbol that's been repeated over and over again.'[25]

Other discoveries include 77,000-year-old spearheads shaped using a technique called 'pressure flaking', previously thought to have first emerged in Europe 20,000 years ago, and engravings on bone dated at about 70,000 years.[26] Blombos has also produced beads from necklaces and other ornaments, showing that members of this community adorned themselves, which provides further indication of the nature of their self-conscious intelligence.[27] Archaeologists working there have, in addition, uncovered a range of other tools, weapons and implements, including bone-point hooks used to catch large catfish, thought to be more than 70,000 years old. This suggests that these people had a lifestyle similar to that of the cave dwellers of Europe tens of thousands of years later.

Although Africa is under-researched, Blombos is hardly an isolated case. Other examples come from a cave at Pinnacle Point on the south coast of South Africa, where in 2009 a team of researchers found stone tools that had been heated to more than 300°, using a technique of burying them under a fire, which made them easier to shape. Tools showing clear evidence of this form of heat treatment were found to be 72,000 years old but others, which were more ambiguous, went back 164,000 years. Until these discoveries, the consensus among archaeologists was that this form of heat treatment emerged in Europe around 20,000 years ago. The team also found scraped and ground ochre used for painting, barbed bone points for shell fishing and small sharpened blades for arrows and spears, dated at 71,000 years old; at least 11,000 years older than previous examples.[28]

Examples of symbolic behaviour from ancient times in Africa are emerging at a rapid clip as archaeologists, palaeontologists and geneticists combine. There's evidence of decorative objects and body ornaments such as pendants, bangles and beads that pre-date those of Europe by tens of thousands of years. Pierced shells for necklaces, dated at 70,000 years, were found in the Porc-Epic Cave in Ethiopia, a similar age to those in the South African caves, while beads and other ornaments have been found in Morocco that might be 130,000 years old. At least 100,000 years ago people began burying their dead, and more than 60,000 years back grave gifts began to appear, reflecting belief in an afterlife. Evidence of the symbolic organisation of space, dating back 100,000 years, was found at the Klasies River mouth in South Africa, along with evidence of controlled burning, an early agricultural technique to stimulate corn production, and grindstones to process the corn, while in Nazlet Sabaha in Egypt, people were mining chert, to make tools, 100,000 years ago.

All over Africa the archaeological record shows economic activity and technological innovation going back well over 100,000 years. Sophisticated bone tools, dated at 80,000 years, of a form previously thought to have been introduced in Europe, were found at a site in Katanda in the Congo. Small tools that are components for larger tools have been found at several sites dating back 65,000 years. This technique only became commonly used in Europe around 20,000 years ago. Recent discoveries in Kenya suggest long-distance trade goes back more than 300,000 years. There's also evidence from around 100,000 years ago of hunting techniques such as pitfall traps, clothes made from animal hide, and buttons and needles made from bone.[29]

One reason why these are such recent discoveries – most of them made since the new millennium – is that until recently there was not much happening in terms of archaeological digs in large parts of Africa. (The same applies to genetic analysis, which is strongly Eurocentric.) This is one reason why much of the evidence comes from South Africa; more archaeologists work there. As the biological anthropologists Sally McBrearty and Alison Brooks put it: 'Africa is vast, researchers are few and research history is short.'[30] They also suggest that a factor in our skewed perception of the African record is the reliance on research conducted in caves (such as Blombos), which provide ideal conditions for preservation but are hardly typical, because 'the vast majority of African occupation sites are in open-air contexts.'[31] What is clear, however, is that evidence of innovative, creative, self-aware, symbolic behaviour goes back a very long way: at least 100,000 years and, from the recent Kenyan and Moroccan evidence, perhaps more than 300,000 years.

This brings me to the subject of the Israeli historian Yuval Noah Harari, whose bestselling book *Sapiens: A Brief History of Humankind*, published in 2014, declared that 70,000 years ago humans began a 'cognitive revolution'. He starts with the daily use of fire for cooking around 300,000 years ago. This led to the evolution of a shorter intestinal track, meaning more energy to spare, and larger brains, heralding the arrival of *Homo sapiens* 200,000 years ago.[32] But although these early *H. sapiens* had big brains, their brain structure was 'probably different from ours'. Harari adds: 'They looked like us but their cognitive abilities – learning, remembering, communicating – were far more limited. Teaching such an ancient s*apiens* English, persuading him of the truth of Christian dogma or getting him to understand the theory of evolution would probably have been hopeless undertakings.'[33] But this all changed 'as a matter of pure chance'[34] because 'accidental mutations changed the inner wiring of the brains of *sapiens* enabling them to think in unprecedented ways and to communicate using an altogether new type of language.' He adds: 'We might call it the Tree of Knowledge mutation.'[35] After this, *Homo sapiens* just like us started 'doing very special things'.[36]

A similar view was advanced by Harari's intellectual hero, the geographer Jared Diamond. In his bestseller, *Guns, Germs and Steel*, he proposed that humans experienced a 'Great Leap Forward' 50,000 years ago.[37] This was prompted by the 'perfection of the voice box and hence the anatomical basis of modern language' and also 'a change in brain organisation'.[38] Before then, he tells us, humans had modern bodies but not modern brains, which would suggest that the *Homo sapiens* who made their way to Australia 70,000 years ago were less than fully human in the modern sense and even those who started their trek through Eurasia 60,000 years ago would have missed the 'Leap Forward' boat.

This pair were preceded by the anthropologist Richard Klein, who proposed that modern human behaviour was prompted by a single mutation of a brain gene about 50,000 years ago. Humans were incapable of modern behaviour before then.[39] Klein could produce no convincing evidence of this gene mutation, because there was none. But he was writing before the work of McBrearty and Brooks was published, as was Diamond, and before the evidence from the Blombos and Pinnacle Point caves emerged, so perhaps they can be partially excused. Not so Harari. It is clear he was writing in complete ignorance of any of the African discoveries that had been prominently published in several books and academic papers more than a decade before *Sapiens*.

Harari chooses fire as the catalyst for bigger brains, saying its daily use started only 300,000 years ago, and that *Homo sapiens* only started 200,000 years ago. Today, most scientists working in this field put the starting time for the evolution of *Homo sapiens* at 500,000 years ago, and recent discoveries have shown 'modern' human behaviour going back more than 300,000 years. There is clear evidence that *Homo erectus* used fire for cooking (in other words, 'daily use') 600,000 years ago, and perhaps as long as two million years ago. *Homo heidelbergensis* also used fire for cooking, and the Moroccan *Homo sapiens* cave-dwellers were cooking with fire 315,000 years ago. Even the small-brained *Homo floresiensis* cooked using fire. As the prime candidate for getting the brain ready for Harari's 'pure chance' cognitive revolution, fire is a dud. But really, there's no evidence for any genetic kick-start or 'great leap forward'.

The Harvard geneticist David Reich shows in his book on ancient DNA that not only does the archaeological case fail but so does the genetic. One candidate proposed as the spark for the cognitive revolution is the FOXP2 gene, which is important for language and speech. But this evolved 1.9 million years ago.[40] Reich adds that if there were a single mutation essential to modern human behaviour, you'd expect it to be more common in some populations than others. But this has not happened. As Reich puts it: 'This seems hard to reconcile with the fact that all people today are capable of mastering conceptual language and innovating their culture in a way that is a hallmark of modern humans.'[41]

A major premise of two bestselling books on human origins is, quite simply, wrong. But Harari and Diamond are far from the worst offenders. The view that the Eurasian experience prompted an intelligence spurt has an even more dubious past. One example comes from the American anthropologists Henry Harpending and Gregory Cochran, who speculate that genetic changes prompted the explosion in European, Middle Eastern and Asian cultures *after* their arrival in Eurasia, although they add that '[o]bviously, something important, some genetic change, occurred in Africa that allowed moderns to expand out of Africa and supplant archaic species.'[42]

This would not explain why previous hominin populations did the same. Migration is usually explained in terms of the need for more hunting and gathering space when populations are rising, not in terms of superior intelligence. As I will discuss later in this book, Harpending and Cochran are at the forefront of race science. They achieved notoriety for their contribution to a paper arguing that Ashkenazi Jews are innately more intelligent than anyone else, and Harpending has also argued sub-Saharan Africans have not evolved for hard work.[43]

A related view, held by almost all those who pursue a race science agenda, is that cold climates prompted the evolutionary advance in intelligence for those who migrated. The argument, advanced most avidly by the overtly racist psychologists such as Richard Lynn and J. Philippe Rushton, is that the cold posed challenges that required long-term planning, nudging selection for intelligence-related genes. Aside from the fact that their argument is contradicted by their own IQ data – for example, one lower-IQ group in Lynn's IQ-of-Nations schema is the Inuits (Eskimos) – there's the question of why cold rather than heat would prompt evolution for advanced cognition. Did the Neanderthals or Siberian Denisovans who lived through ice ages in Europe for 400,000 years evolve to have more intelligence than us? And when it comes to the idea that higher intelligence got us moving, what of *Homo heidelbergensis, Homo erectus and Homo floresiensis*? Some of them left Africa nearly two million years before modern humans did (the remains and tools of a hominin – probably early *Homo erectus* – were recently found in China and dated at more than two million years old[44]).

Most biological anthropologists who speculate about the environmental and social factors that might have nudged evolution for higher intelligence tend to focus more on the impact of communal living: when our hominin ancestors started living in small communities they would have needed social skills of various kinds to prosper, to attract mates and to pass on their genes, which would have applied as much in warm climates as cold. In other words, when it comes to the role natural selection plays for intelligence in people, where to look is at other people, and our hominin ancestors started living in communities millions of years before arriving in Eurasia.

The challenges of a warm climate – such as having to survive drought, floods and longer-term climate change – seem no less intelligence-demanding than the need to survive cold winters. In any event, the idea of Africa being one happy, warm, easy, undemanding kindergarten is ridiculous. Before the migration to Eurasia, people like us were living in caves by the sea, beside rivers and lakes, in the grassland savannahs, forests, semi-desert areas, snowy mountain regions, tropical forests and many more environments, and they'd been migrating up and down Africa for hundreds of thousands of years.

It is also worth remembering that Africans have more genetic variation than people in the rest of the world, because small groups migrated, so if there were to be natural selection for genes implicated in higher intelligence, this would most likely have happened in Africa. One view, supported by archaeologists such as Henshilwood, is that there was indeed a leap forward (a 'leap' spanning tens of thousands of years) and that it happened not in Europe but in Africa, starting at least 160,000 years ago.[45]

The strongest argument against the Eurasian view comes from evidence that I've already touched on. McBrearty and Brooks, who examined this African record of innovation, adaptation and inventiveness, argue that the 'human evolution' view of a great leap forward in Europe prompted by some genetic advance suggests a 'profound Eurocentric bias and a failure to appreciate the depth and breadth of the African archaeological record'.[46] They believe the rapidly increasing volume of evidence from archaeological research shows that our pre-human ancestors were well on their way to becoming fully human, in terms of cognitive potential and behaviour, 300,000 years ago.[47]

Ian Tattersall, a specialist in European cave dwellers, agrees, rejecting the idea of human evolution as 'a gradual progression from primitiveness to perfection', arguing that 'this conceptual hold-over from the past is clearly in error. We are *not* the result of constant fine-tuning over the aeons.' He suggests that the earliest anatomically modern humans were born with the full intellectual potential of later humans but this lay fallow until unleashed by cultural stimuli. He also believes there has since been no significant evolutionary advance in the human brain.[48] These scientists and archaeologists all agree that by the time *Homo sapiens* emerged with their current morphology, they had also attained their full intellectual potential.

THE SIGNIFICANCE OF ROCK ART AND OTHER INNOVATIONS

Much of the focus on the Eurocentric view has come from the flowering of artistic expression in Europe after the Cro-Magnon arrivals. Until recently, the oldest known example of rock art was that in the El Castillo cave in Spain, where one motif was found to be at least 40,000 years old, dating from soon after *Homo sapiens* arrived in that part of Europe. However, in 2018 that was overtaken by the announcement of the discovery of the 73,000-year-old ochre line drawing on the rock fragment at Blombos. Also in 2018, a painting of a banteng (South Asian wild cow) was discovered on a cave wall in Borneo. Dated as at least 40,000 years old (some say 52,000 years), this is the oldest known example of figurative rock art. Other examples of rock art from the Chauvet Cave in France have been dated at 35,000 years, and rock art really takes off from then.

But Europeans were not alone in painting their caves at this time. Rock paintings of animals at the Apollo 11 Cave complex in Namibia may be more than 27,000 years old. It seems likely that as more

African archaeological sites are excavated, even older African examples will emerge, complementing the 2018 Blombos discovery. As it stands, the discoveries over the past two decades at Blombos suggest that artistic expression in Africa predates the earliest European cave art by at least 35,000 years.

However, there's an obvious flaw in using artistic expression to discern intelligence. At any moment in history some societies have forged ahead in terms of technological and cultural sophistication, while others retain more ancient lifestyles. Writing was 'invented' in the heat of Mesopotamia (today's Iraq) more than 5,000 years ago, when the people of Britain were still hunter-gatherers. In parts of the Amazon and Africa today there are still hunter-gatherers who ignore or avoid modernity; people continue to live traditional lifestyles and reject modernity for all sorts of reasons, which tells us nothing about their innate cognitive capacity.

Two books on contemporary hunter-gatherer communities brought this home to me. In *Don't Sleep, There are Snakes: Life and Language in the Amazonian Jungle*,[49] the American linguist Daniel Everett describes more than two decades spent living among the Piraha people in a remote part of the Amazon. The chief academic interest in the book comes from what it reveals about language. The Piraha language seems to defy Noam Chomsky and Steven Pinker's view of innate, universal grammatical systems involving multi-clause sentences; the syntax of the Piraha tonal language appears to be very different.

The Piraha culture also throws into question the common view, held both by religious people and genetic determinists, that religion and belief in an afterlife are 'hardwired' human universals. The Piraha have a rich fantasy life and believe in spirits but have no ancestor beliefs, no belief in an afterlife, no creation myths and no sense of a god. They, and their language, show little interest in anything other than the present tense: their attitudes to sex, love, death and parenting are unique. They also have no chiefs, no numbers, no art, no writing and little interest in other cultures and yet they are a vibrant community, routinely described by visitors as extraordinarily happy and content. According to Everett, suicide is unknown, as is any sign of depression. Perhaps it is not so surprising that other than using shotguns for hunting and motorboats for some transport, they show no desire to adopt the more Westernised lifestyles of surrounding tribes. But they are not lacking in cognitive potential. Although Piraha adults showed no inclination to learn to count or read, their young children seemed as perfectly capable of learning these skills as children elsewhere. Archaeologists of the future who mistakenly believed that lack of 'achievement' was a sign of cognitive limitation might conclude they had stumbled on a 'backward' tribe but when they checked their results with genetic specialists they would discover that the Piraha are the close cousins of a neighbouring tribe known for their ambition, business acumen, love of wealth and learning.

The other book was *Intimate Fathers: The Nature and Context of Aka Pygmy Paternal Infant Care*[50] by the American anthropologist Barry Hewlett. As with the Piraha, this predominantly hunter-gatherer pygmy tribe showed no inclination towards artistic expression or technological innovation. And yet in all sorts of other ways they were innovative, not least in how they brought up their children. Hewlett's study exposes the solipsism of the view that current Western cultural norms and values are natural and universal. He found that Aka fathers spent five times as much time with their infants as the Western average, and held, or remained within arm's length of, their babies almost half of the time. They often hugged their infants close to their bodies for an hour or more, offering them their nipples to suck, comforting, cleaning and singing to them at night. Like the Piraha, they did not hit their children or shout at them. The parents readily interchanged roles, with the women playing a significant role in small game net hunting (as did the Piraha women) even when pregnant. Hewlett explains: '[T]here's a level of flexibility that's virtually unknown in our society. Aka fathers will slip into roles usually occupied by mothers without a second thought and without, more importantly, any loss of status – there's no stigma involved in the different jobs.'[51]

I cite these examples of contemporary hunter-gatherers not to glorify their rustic lifestyles. Existential threats, as well as high child and adult mortality rates, suggest otherwise. In both cases the tribespeople show the same range of human nature as the rest of us – deceit, jealousy, drunkenness and violence are certainly in their range, as are love, gentleness, humour and kindness. I use them for two other reasons. First, to make the point that when looking at the prehistoric world 'achievements' such as technological innovation and artistic expression might give us strong hints of cognitive potential but their absence does not in itself imply the opposite. Second, it's all too easy to make the error of assuming that our own way of doing things (such as in gender relations, sexual relations or parental relations) is the norm and therefore natural when really, it is culturally contingent. The Piraha show that even the structure and form of language is, in the long run, environmentally and culturally contingent.

When the fisherfolk of Blombos and Pinnacle Point were mixing their paints, carving their geometric plaques, threading their beads and using extreme heat to shape their blades and tools, other humans might still be behaving like *Homo heidelbergensis*. They all had the same big, complex brains, with similar innate potential, but their embrace of symbolic behaviour and of new technologies was uneven. It is likely that this potential spread through the increasingly varied and complex use of language and through travel and trade which, in itself, prompted changes in forms of communication including language, rather than through any biological advance. In other words, the human creative potential contained in those large brains emerged more quickly in some areas than in others. Or perhaps just differently.

When did people with intelligence like ours first emerge? When we combine the expanding archaeological record with our rapidly increasing knowledge of genetics it would seem likely that modern human intelligence goes very far back, perhaps all the way back to the arrival of the first anatomically modern humans. We have been fully human in every sense, including cognitive potential and all our immense creative and destructive glory, for far longer than generally assumed.

3

WHY DID HUMANS MIGRATE?

When I first became interested in human origins, I had a vague idea that some of us left Africa around 50,000 years ago and settled the world, after which we remained more or less in place until the empire-building of Mesopotamia, Egypt and Rome, then slavery and colonialism, mixed things up. I'd also assumed there was a one-way flow, and that Europeans gradually evolved to have lighter skins because of the cold.

I soon discovered each of these premises was flawed. For one thing, modern humans were making sorties out of Africa for more than 130,000 years before they settled in Europe, and they never stopped moving, both within Africa and beyond. The current geographical spread of humanity bears little resemblance to that of pre-agricultural times. There were huge population movements throughout Europe, and similar processes in Africa, with one group replacing another, wiping them out or simply absorbing them through the sexual dominance of the invading males. Better-armed herders, with rigidly patriarchal structures, replaced more egalitarian hunter-gatherer and farming cultures. I also discovered that while the evolution of lighter skin, hair and eyes had something to do with the cold, it was far more complex than I'd previously assumed, and that the genetic roots for light skin came from Africa.

HUMANS SETTLE THE WORLD

People, like animals, migrate because they run out of space. They need room to hunt and gather or to herd and farm, so whenever there's a population spike some need to move. There are other reasons, such as climatic change and conflict with rival groups, but migration has nothing to do with intelligence. Like our evolutionary predecessors, we *Homo sapiens* have migrated right from the start: out of Africa, around the world and back again.

Evidence of the scope of these early migrations is mounting. Finds of human-made stone tools suggest migrations via the Nile Valley into what is now Israel and the United Arab Emirates about 125,000 years ago. More significantly, a modern human jawbone and teeth found in a cave in Israel were between 177,000 and 194,000 years old; at least 50,000 years older than any previous out-of-Africa discovery. Adding to the recent hoard, in 2016 a fossilised finger bone, dated at 85,000 years old, was found in Saudi Arabia, while forty-seven modern human teeth, found in a cave in southern China, were dated at between 80,000 and 125,000 years old based on the stalagmites in that part of the cave.

What happened to these early out-of-Africa migrants? Perhaps it is not surprising that these small groups didn't take root for long, given the kind of challenges they faced finding food and shelter and protecting themselves from animal predators that included packs of sabre-toothed tigers. Many were wiped out by the Toba (in Sumatra, Indonesia) volcanic eruption 75,000 years ago, which caused a volcanic winter for between six and ten years and prompted a sharp human population decline; five thousand years later the total human population had been reduced to a few thousand.

We now know that others were out-hustled by Neanderthals, refuting the belief that human-Neanderthal conflicts always went one way. A group of humans reached the Levant around 80,000 years ago and ran into the Neanderthals. It appears from the archaeological evidence that while the Neanderthals survived, the humans either didn't, or they retreated. This does not necessarily mean that the Neanderthals were driving the humans off or killing them. Maria Martinon-Torres, a University College London anthropologist who was part of the team that made the discovery, said there was no evidence of physical confrontation. 'It was a matter of who was best able to exploit resources,' she said. 'I think we underestimated them. They were not grunting, ignorant cavemen. They were our equals.'[1]

HUMANS REACH AUSTRALIA – BY BOAT

Around 70 to 75,000 years ago another group left Africa and moved down the southern coastline of Asia, some settling in India (whose population still carries their genetic imprint), in the direction of South-East Asia and Oceania. Some found their way to China, and much later to Japan and Korea. At least 65,000 years ago others reached Australia, at one point crossing 250 kilometres of open water in ocean-going boats or rafts. It was previously assumed they settled about 45,000 years ago (Yuval Noah Harari uses this date[2]) but their arrival has been pushed back by discoveries in Madjedbebe of surprisingly sophisticated axes, spear tips and seed-grinding tools, as well as huge quantities of ground ochre, dated at 65,000 years old.

It is said that Australian Aboriginals have the oldest continuous culture in the world but this is not quite accurate. For one thing, the San (Bushman) people of Southern Africa have an even older lineage. For another, it is off the mark to talk of a single, continuous Aboriginal culture. There were at least 250 Aboriginal languages, each belonging to a different 'nation', when the Europeans arrived. But the most common misconception is that the native Australians were a backward people, perhaps even a different branch of humanity. In the nineteenth century, it was widely believed that the biological gap between Europeans and Tasmanians was so wide that if they bred, they would produce infertile offspring ('mulism'). The Tasmanians were wiped out, due to European diseases and the genocidal policies of the British, but their genes continue to flow down the generations because of the number of mixed-race British-Tasmanians who survived.

Racist perceptions of native Australians persisted into the second half of the twentieth century. In 1962, the then president of the American Association of Physical Anthropologists, Carleton Coon (a rather unfortunate name for a racist), asked:

If all races had a recent common origin, how does it happen that some peoples, like Tasmanians and many of the Australian aborigines, were still living in the nineteenth century in a manner of Europeans of over 100,000 years ago? [A reminder: Europeans of 100,000 years ago were Neanderthals, not humans] Either the common ancestors of Tasmanians-cum-Australians and that of the Europeans parted company in remote Pleistocene antiquity or else the Australians and Tasmanians have done some rapid cultural backsliding, which archaeological evidence disproves.

Coon was saying that the native Australians had always been backward, and remained so. In his bid to show they were mentally inferior to whites, he measured cranial capacity and the inclination of the jawbone and decided they were 'still in the act of sloughing off some of the genetic traits which distinguish *Homo erectus* from *Homo sapiens*'.[3] He conceded that some evolution had taken place 'but as one would expect in a marginal area of the southern hemisphere, its overall rate cannot have been rapid.'[4]

Coon was writing in an era when Australian Aboriginals did not yet have the same franchise as whites; that did not happen until 1967. It was only in the early 1970s that the forced adoption of mixed-race Aboriginal children – a eugenic policy launched in 1909 – ended. Even as late as 1974, this perspective had not vanished from 'respectable' Western academia. The Oxford biologist John R. Baker wrote that the skulls of 'Australids' carried the imprint of primitiveness, adding that they 'may serve as a reminder of a stage in the evolution of more advanced forms'.[5] Hardly surprisingly, Baker's book *Race* was endorsed by the neo-fascist British National Party. After Carleton Coon's death in 1981, it emerged that he had given covert backing to the US segregationist cause.[6]

Since the 1970s an impressive flow of data has emerged showing that the culture of these early immigrants to Australia was more complex and technologically innovative than previously thought. As well as evidence that they arrived by island-hopping using boats or rafts, which must have involved planning, navigation and technological innovation, their prehistorical cultural range can be surmised from cave art and artefacts discovered over the last decade. A few examples: finely drawn charcoal lines portraying a moving figure, dated at 28,000 years ago; a 35,000-year-old edge-ground axe of a kind that only emerged thousands of years later in other parts of the world; a red ochre painting portraying two emu-type birds with long necks (while dating of ochre is tricky, they've since been identified as a giant bird species that became extinct at least 40,000 years ago).

HUMANS REACH EUROPE AND SEE OFF THE NEANDERTHALS

Another wave of migration from Africa, which probably started nearly 60,000 years ago, took humans (known today as Cro-Magnons) across the Sinai Peninsula, throughout Asia and then, about 43,000 years ago, to Europe. There they coexisted with Neanderthals for around 4,000 years and, as we've seen, bred with them on a limited scale early in their migration, leaving a lasting genetic imprint. When resources were scarce, it seems they possibly killed Neanderthals in skirmishes, and ate them; there is evidence of Neanderthal bones that have been cut and scraped with Cro-Magnon blade tools, which strongly suggests cannibalism. Some writers have suggested this was an early example of genocide but there is scant evidence for this. The Neanderthals were already in sharp decline when the out-of-Africa humans arrived. It may be that they were killed off by a combination of inbreeding and new diseases for which the Neanderthals had no immunity, or perhaps it had more to do with competition for game and hunting grounds.

The Neanderthals, who lived in Europe, the Middle East and Asia for 400,000 years, weren't the subhuman oafs of popular mythology. They had larger skulls and brains than modern humans and although bigger brains aren't necessarily an indication of greater cognitive potential, aspects of their archaeological record, including the occasional use of pigments and body ornaments, show they were at least capable of imitating the human migrants; and recent finds suggest this was probably more than mere imitation. Three cave paintings in Spain (red motifs and hand stencils) have been dated at between 64,000 and 68,000 years old.[7] This dating – from more than 20,000 years before the Cro-Magnons' arrival – makes it clear the painters were Neanderthal. Genetic analysis shows they had the FOXP2 gene, which is associated with language, while recent computer modelling of their hyoid bone (in the neck) indicates they were capable of speech.[8] Given that *Homo sapiens* interbred with them, it seems likely that there was some form of communication between the two groups. They also cared for their sick and elderly, buried their dead and made stone tools similar to those of the Cro-Magnons, using techniques that demanded skills, dexterity and planning.

However, some question this contemporary picture of Neanderthals as our intellectual equals. The Cro-Magnon expert Ian Tattersall points to a lack of evidence of systematic symbolic activity among Neanderthals. '[S]ymbolism is highly unlikely to have been a routine or important factor in the Neanderthals' existences,' he notes, comparing them to the new arrivals from Africa who 'led lives that were drenched in symbol' and had 'acquired a fully modern sensibility' before they left Africa.[9]

One view is that the Neanderthals, who survived in Europe for so long, lacked the versatility and cunning to cope with the out-of-Africa immigrants, and also their dexterity. The Cro-Magnons brought from Africa the ability to sew, which would have allowed them to keep warm while hunting in the winter. They had an additional advantage in competing for scarce resources; their lighter bodies meant they were more mobile and required less food. They may also have had more sophisticated organisation and planning. Whatever the reasons, the human population replaced the already-declining Neanderthal population within about four thousand years of arriving in Europe.

The sad death of the last Neanderthal, around 39,000 years ago, was soon followed by the take-off of Cro-Magnon artistic expression: sculptured figurines, musical instruments and an impressive variety of rock paintings of animals and humans. The detail, scale and variety of these artistic forms has led some to conclude that this was the period in which the truly modern human brain kicked in. But as we saw in the last chapter, this view has also been discredited from the abundant evidence of far earlier artistic expression in Africa.

FROM HUNTING AND GATHERING TO WRITING AND SHOOTING

Agriculture and animal husbandry began more than 11,000 years ago but there are examples of crop growing in Africa, such as planned burning of the land to encourage corn growth and the use of tools to process corn, tens of thousands of years earlier. And hunter-gatherers were not all 'cave men'. Long before sustained agriculture took off, at least 13,000 years ago, some communities lived in larger village-type groups. In a few areas, they were also involved with more organised forms of religious practice, including building 'cult of the dead' monuments.

The most remarkable of these is the Göbekli Tepe in south-eastern Anatolia (Asian Turkey), a fifteen-metre structure built of stone, containing two hundred huge T-shaped pillars, some adorned with carvings of animals. The archaeologists excavating the site believe it functioned as a spiritual centre more than 13,000 years ago. Its most decorated part, which is three hundred metres in diameter, has been dated at 11,600 years old and its final layer was completed 10,000 years ago, which means its origins predate the Great Pyramid's by 7,000 years. They also estimate that carrying the pillars (which weigh up to twenty tonnes) from nearby quarries would have involved five hundred people, who would have to have been organised and fed. And yet there's no sign of sustained settlement. In other words, the people who built it were hunter-gatherers, or lived a predominantly hunter-gatherer lifestyle. It would seem that, at least in Anatolia, monumental building projects preceded agriculture and urban living by thousands of years. As the head archaeologist Klaus Schmidt put it: 'First came the temple, then the city.'[10]

Among the first to move from hunting and gathering to agriculture were indeed the Anatolians, along with those living in the Fertile Crescent of western Asia and Egypt, and also India. They, in turn, were followed by Mesopotamia, parts of China, Mexico, Syria, the

Jordan Valley, New Guinea, Andean South America and the Sahel region of Africa, which includes Nubia. There are a number of theories about why this change happened but most experts agree there is more than one explanation, including climate change (hot, dry seasons encouraging annual plants that die off and leave seeds and bulbs), the specific weather and soil conditions in these regions, the seeds available there and the presence of animals that were relatively easy to domesticate.

But some, or all, of these factors were present in parts of the world for 95 per cent of human history without a sustained shift to agriculture. In other words, none are enough to explain why people like us lived for 300,000 years as hunter-gatherers and then, within a few thousand years, adopted agricultural lifestyles. Explanations must therefore include factors in these communities. Some may have been idiosyncratic but when looking for a general cause, population pressure is an obvious candidate. Hunter-gatherer lifestyles are hard to sustain once the population reaches a level where there are not enough edible plants or animals to feed the growing community. The search for new food sources prompted migration all over the world but eventually the most desirable areas would have faced population growth if counter-pressures (disease, wild animals, climate change, fighting) weren't enough to keep it down. Overpopulation might have provided the impetus to cultivate plants and herd animals to supplement hunting and gathering, until it became the main source of sustenance. Farming and herding in turn prompted the development of larger, more settled and more stratified communities and the necessity for more trading.

And so, the first urban societies emerged about 7,000 years ago in several areas of the world. This in turn nudged some communities to begin recording transactions. Writing for the record emerged, using clay tokens or beads on a string for counting, followed by numbers and proto-writing (picture writing systems) which eventually progressed to the writing of language in Mesopotamia 5,200 years ago, leading to cuneiform, a representation of the Sumerian language, a few hundred years later. This was followed by other forms of writing in several parts of the region, including hieroglyphics in Egypt (influenced by cuneiform), and independently in India/Pakistan, China and Turkmenistan, and among the Olmecs in Mexico and the Mayans in Guatemala. The Nubians were the first black Africans to use writing, about 2,500 years ago; initially the hieroglyphics of their Egyptian neighbours and later their own alphabet for their Meroitic language.

ONE POPULATION REPLACES ANOTHER

Peruse sub-Saharan African history and prehistory and you'll see a story of one population replacing or absorbing another. The 'first people' of Southern Africa, the San (Bushmen), had the region all to themselves for tens of thousands of years. Then, a little more than 2,000 years ago, they were joined by their close genetic cousins, the Botswana-based Khoekhoe, who'd made the switch to a pastoral, cattle-herding, hut-living lifestyle. These two together are now known as the Khoesan. Around 1,800 years ago the first Iron Age Bantu-speaking farmers arrived, having gradually moved south from what is now Nigeria and the Cameroon in search of more verdant farmlands (they also moved in the direction of East Africa). In South Africa, there is evidence of cooperation and interbreeding, and also of conflict over land as the farmers' numbers increased, which inevitably led to the Khoesans' demise as they were eliminated from large tracts of the country.

When the Europeans arrived in the seventeenth century they wiped out the Khoekhoe people, mainly by passing on European diseases. The remaining San were hunted down and confined to the Kalahari Desert and remote parts of Botswana, Namibia, Zimbabwe and Angola. The Europeans then trekked north, using their guns to subjugate the Iron Age, Bantu-speaking farmers. By then, the admixture of these groups and of Malayan slaves had created the mixed-race 'Coloured' population, which today numbers more than five million. A 2010 genetic study of 'Cape Coloureds' (the largest 'Coloured' population) showed that their ancestry was Khoesan (32–43%), Bantu (20–36%), European (21–28%) and South/South-East Asian

(9–11%). Examination of the maternal mitochondrial line showed that the majority of the maternal line was Khoesan, while the analysis of the Y chromosome showed the male line was mostly European or European-African.[11]

A similar process happened earlier in Eurasia, with agricultural migrants devastating, often replacing, the hunter-gatherer communities. The result, as the Harvard geneticist David Reich put it, was that 'the people who live in a particular place today almost never exclusively descend from the people who lived in the same place far in the past.'[12] Reich shows that modern Europeans did not descend primarily from hunter-gatherers from their countries or regions but from different populations of agriculturalists, starting with one lot who migrated from Anatolia and another lot (who themselves were a mixture of Iranian farmers and Eastern European hunter-gatherers) that later moved from the steppe north of the Black and Caspian seas. The latter group replaced existing populations of hunter-gatherers and farmers, possibly by spreading air-borne pneumonic plague, to which steppe people had greater immunity. Throughout Western Europe, farmers squeezed hunter-gatherers to the fringes but the farmers were themselves largely replaced by these newcomers. From a different angle, just over 5,000 years ago the ancestors of 'native' northern Europeans had not yet made their appearance.[13]

The Iberian Bell Beaker strand of the steppe people (named because of their distinctive form of pottery) arrived in Britain 4,500 years ago and largely replaced the Stonehenge-building farmers who had arrived via Anatolia about 6,000 years ago (and had themselves edged out the existing hunter-gatherer population). Over time there were further waves of immigrants, some more peaceful than others. The Celts began arriving about 4,000 years ago, followed by the Angles and Saxons (from 1,600 years ago), then the Vikings (1,200 years ago) and Normans (from 1066 CE). Yet all these people shared a common genetic past, much of it drawn from the herders who migrated from the Russian steppe 5,000 years ago. In fact, many of those who followed over the next 900-plus years – including European Jews, French Huguenots, Dutch, Italians and those from the Indian subcontinent – shared substantially similar genetic heritages.

Some discoveries from full genome analysis of ancient DNA rub against nationalist sentiments. One example relates to the Nazi belief that the origins of the Germanic people and languages lay exclusively in the Corded Ware culture, which had deep roots in the German domain. Now, however, DNA analysis shows that the genetic origins of this culture lay in the Russian steppe, with its ancestry drawn from an Iranian population and Eastern European hunter-gatherers,[14] which would not have pleased Hitler. Likewise, some Hindu nationalists in India believed there had been no significant contribution to Indian culture from beyond South Asia. DNA analysis shows this is a fallacy; about half the ancestry of contemporary Indians is drawn from various migration waves from Anatolia, Iran and the Eurasian steppe over the past 9,000 years.[15]

DNA analysis and archaeological research can also prick religious bubbles, including those of the Abrahamic religions. Israeli archaeologists have long chipped away at the notion of their scriptures as history. One book, *The Bible Unearthed*,[16] written by the archaeologists Israel Finkelstein and Neil Asher Silberman, had a devastating impact on fundamentalist interpretations, showing there was no evidence that the Jewish people were enslaved in Egypt, of the subsequent exodus or the existence of Moses and Joshua. The Bible tells of God's command to the Israelites: 'You shall not leave alive anything that breathes' and shows how Joshua's men did just this in what, in today's language, amounts to large-scale ethnic cleansing. The picture presented is of the total slaughter of Canaanite men, women and children and their replacement by the Israelites. Finkelstein and Silberman point to the lack of evidence of any such conquest and the compelling evidence that the emerging Israelite population had never left the area in the first place. This has been reinforced by recent genetic research comparing full genome analysis of five 3,700-year-old Canaanites with that of contemporary Lebanese people. It reveals that the Canaanites not only survived, but flourished, and that their DNA is today found among their Lebanese descendants.[17]

As the son of an Ashkenazi Jewish father, I was particularly interested in the genetic origins of European Jewry. I will go into more detail on this question but in short it relates to a key argument made by those claiming Ashkenazim are innately of superior intelligence: that they were genetically isolated. DNA analysis has shown the opposite. A major genetic study of mitochondrial DNA (the maternal line) found that European women, not women from the Levant (Middle East), were the main female founders of the Ashkenazi population.[18] 'These results point to a significant role for the conversion of women in the formation of Ashkenazi communities, often started by male traders,' the authors note.[19] They conclude that 81 per cent of Ashkenazi female ancestry was European, with just 8 per cent from the Levant.[20] However, Y chromosome genetic studies suggest that the Middle Eastern contribution to the male line was far more substantial.[21]

This pattern of the male line showing a stronger connection to the migrating population is repeated over and over. Among Ashkenazim it involved male traders with roots in the Middle East having children with European women, who then converted. In many other cases, conquest was involved, with women among the spoils. One example comes from the horse-riding, Bronze Age Yamnaya herders, who spread from the Russian steppe, displacing both the hunter-gatherers of Central Asia and the farmers of northern Europe, and introducing a far more patriarchal, militaristic and wealth-divided society.[22] The skeletons beneath the burial mounds they left behind show battle scars, and they went to their graves with battle axes and daggers. According to the archaeologist Marija Gimbutas, their arrival, starting around 5,000 years ago, marked the decline of the 'Old Europe' in which women had played a more central role. Instead, we see a far more male-centred society in which power was concentrated in small coteries of elite men, who spread their genes vigorously.[23]

Similarly, when Genghis Khan established his thirteenth- and fourteenth-century Mongol empire, which spread from China to the Caspian Sea, he and his small number of military leaders left a significant impact on the genomes of billions of people currently living in the lands he conquered. A Y-chromosome study by the geneticist Christopher Tyler-Smith and his team suggests that around

8 per cent of males living there in the region show genetic markers of one male from Mongol times; possibly Genghis Khan himself.[24]

This pattern is repeated wherever conquest, colonisation and slavery prevailed. DNA analysis of contemporary African-American populations suggests the genetic contribution of European-American men is four times that of European-American women,[25] and it would have been higher in the slave era because of the routine rape of slave women by slave owners and their sons, or the use of slave women as concubines. One example comes from the slave-owning Thomas Jefferson's 'mistress', Sally Hemings. She had three white grandparents but because her mother was a slave, so was she. Hemings, who was thirty years younger than America's third president, might have been as young as fourteen when she started her relationship with him; she went on to have six of his children.[26]

THE EURASIAN BACKFLOW

It is often thought that the flow of African-related migration went one way, outwards, but genetic analysis shows there was significant traffic in both directions. This is most obvious in North Africa, where lighter skins and European-like facial features are commonplace. Studies of North African genomes confirm this, suggesting substantial backward migration from the Middle and Near East and from Eurasia starting 30,000 years ago, with a major gene flow 12,000 years ago, and several since.[27]

Less widely known is that East Africans also carry significant proportions of Eurasian DNA. Farmers living in what is now Jordan and Israel spread into present-day Ethiopia and other parts of East Africa. Analysis of the full genome of a 4,500-year-old man found in the Mota Cave in Ethiopia helped scientists to understand the genetic changes in the population in that area. They discovered a large-scale genetic flow from the Middle East to East Africa around 3,000 years ago, leaving a far bigger Eurasian-origin genetic mark than previously recognised. 'Roughly speaking, the wave of West Eurasian migration back into the Horn of Africa could have been as much as 30 per cent of the population that already lived there – and that, to me, is mind-blowing,' the senior author of the study, Dr Andrea Manica, from the University of Cambridge's Department of Zoology, said. 'The question is: what got them moving all of a sudden?'[28]

While the East African story is interesting, Manica's study exaggerated the impact of this reverse migration on African populations beyond East Africa, as a result of what he described as 'software incompatibility' errors. Pontus Skoglund, a Harvard population geneticist who re-examined the genome data, commented: 'Almost all of us agree there was some back-to-Africa gene flow and it was a pretty big migration into East Africa but it did not reach West and Central Africa, at least not in a detectable way.'[29] One legacy of the migration from the Middle East into East Africa was a 'European' gene, SLC24A5, associated with lighter skin. This is prevalent in Ethiopia and Tanzania but it does not seem to lighten skin colour there to the same extent as in Europe.[30]

Recent research involving full genome DNA sequencing shows how mixed we all are genetically. For all human existence, populations have replaced each other, absorbed each other and blended with each other. Whenever populations come into contact, they end up having sex, often not consensually. Very few populations are anything other than genetic mixes, and most of that mixing has taken place within the last 10,000 years. This applies as much to Europe as it does to Africa. As we shall see in Chapter 6, bearing this in mind, skin colour takes on its appropriate proportions as a minor genetic detail.

4

IS AFRICA REALLY 'BACKWARD'?

Eight years after South Africa became a democracy, an architect working on the renovation of the University of Pretoria came across a locked room. Oddly, no keys were to be found. But he needed access to complete his measurements and so, eventually, he broke the door down. At first its contents – scores of small boxes – seemed unremarkable. But when he opened them, he was astonished. He found figurines made from gold; carved ivory and bone; glass beads; refined iron; copper wire; pottery; and Chinese Celadon ware. One of the items was a beautifully made rhinoceros, its body covered in thin sheets of gold leaf attached to its wooden core with tiny gold nails.

The objects were treasures from the Kingdom of Mapungubwe, a precolonial state in South Africa's Limpopo province that flourished for about 150 years after 1075, part of a network of kingdoms on the gold trading route to Rhapta and Kilwa Kisiwani in present-day Tanzania. The question was, what were they doing in Pretoria, lying in a locked room with no key, and how long had they been there?

MAPUNGUBWE, GREAT ZIMBABWE AND THE GOLD TRADE

The excavations of Mapungubwe, led by members of the University of Pretoria archaeological department, started in 1932. Three years later, Captain Guy Gardner, the chief archaeologist, announced that its builders were of 'pre-Bantu' stock and contained 'a dash of Mongoloid mixed ... with a primitive Hamitic culture',[1] a reference to an abiding, but baseless, belief in a mysterious non-African population that travelled south and did clever things. But just to be sure no one got ahead of themselves, the findings were kept quiet and not publicly released until the early 1990s. And it was only in 2002 that its treasures were rediscovered and publicly displayed.

It emerged that the site was built by an Iron Age Bukalanga kingdom, which covered parts of the current South Africa, Zimbabwe and Botswana. Gold objects were used as burial gifts for its elite. On the top of a hill, the site was a key staging point in the kingdom's vast trading network: the Chinese Celadon ware gives a hint of how far it extended. The site was reserved for the kingdom's ruling class; at its peak, it had about five thousand inhabitants. Its ruins include both stone and wooden enclosures and show evidence of a large-scale pottery industry.

What most people in South Africa did not know, until after apartheid ended, was that the Iron Age in Africa went back 3,000 years. The technique of iron smelting, which involves heating iron ore to 1535°, found its way to Southern Africa about 2,500 years ago and was used very widely on the continent to produce weapons and agricultural tools. By the later Iron Age, sub-Saharan Africans were building in stone and the production of iron had spread to thousands of sites. They were sculpting in gold, and producing ornaments using glass, tin, ivory and copper. Bantu-speaking farmers, who originated in West Africa, arrived in South Africa at least 1,800 years ago. Soon they were clearing land, rotating crops and mining for salt. Later they started mining for gold, tin and copper; 800 to 900 years ago there were several sites devoted to smelting. They were also spinning and weaving, making rope and establishing hundreds of trading centres. The most impressive of these late Iron Age sites is Mapungubwe, closely linked to an even grander site, Great Zimbabwe, which emerged in its full glory soon after Mapungubwe's decline.

Great Zimbabwe is a 722-hectare royal palace, whose construction began about a thousand years ago. Today it remains the most impressive of two hundred Iron Age fortified structures around Southern Africa. When I first visited with my family, in 1973, the grandeur of its huge stone walls, conical towers and giant soapstone birds gave an instant impression of a powerful kingdom from the distant past. I remember asking the Rhodesian guides who had built it and being told it was probably the work of Arabs. It later emerged that Rhodesian bureaucrats had ordered the archaeologists to say it was the work of non-Africans. The state bowdlerised guidebooks, museum displays, school textbooks and other media to maintain this myth. The site's chief archaeologist was instructed that it was 'okay to say the yellow people built it but I wasn't allowed to mention radiocarbon dates'. He said this was the 'first time since Germany in the Thirties that archaeology has been so directly censored'.[2]

While the rewriting of the archaeological record was astonishing, the mythology was nothing new. J. Theodore Bent, an English archaeologist patronised by Cecil Rhodes, wrote, after 'discovering' the ruins in 1891, that they were too sophisticated to be the work of Africans. He surmised that 'a civilised nation must once have lived there' and presented his conclusion: 'I do not think that I am far wrong if I suppose that the ruin on the hill is a copy of Solomon's Temple on Mount Moriah and the building in the plain a copy of the palace where the Queen of Sheba lived during her visit to Solomon.'[3] This bizarre speculation became the favoured explanation in white Rhodesia, with Arab, Semitic, Eastern and 'Hamitic' versions presented as alternatives. Anyone would do, as long as they weren't a sub-Saharan African. The problem was that by the 1950s Rhodesia's archaeologists knew full well that it was, indeed, the work of sub-Saharan Africans, constructed from the eleventh century onwards by the ancestors of the Shona people, serving as a seat of power for monarchs who ruled a late Iron Age kingdom.

The white rulers had a problem: how could this be explained? That 850 years before Rhodes, black Africans built a city with walls 250 metres long, eleven metres high and six metres thick; a city of densely packed housing units and nine-metre-high towers? How could they admit that the soapstone sculptures, iron tools and copper wire, the bronze spearheads and copper ingots, the crucibles and coins, the glass and gold beads and ivory ornaments, were products of an African society that traded up and down the Swahili coast, part of a network that extended all the way to China, India and the Middle East? When Ian Smith's government insisted on its right to rule, to deny the vote to more than 95 per cent of the population and to control the mines and farms, how could it acknowledge that the ancestors of the 95 per cent had extracted twenty million ounces of gold from around the ruins (early Portuguese traders wrote of it being surrounded by gold mines) or engaged in cattle trading and agricultural production on a huge scale?[4]

It might have been even harder to explain to white voters raised on colonial mythology that the area was settled by the ancestors of the Shona in the fourth century CE, 1,500 years before Rhodes cast his imperial eye on its mineral riches, and that soon after their arrival these people were farming the area and mining its iron, let alone the fact that by 1300 CE around eighteen thousand inhabitants lived within the city's walls, and its kings ruled over a huge empire, stretching from the Zambezi to the Limpopo, for more than two hundred years.[5] The coastal cities of this trading kingdom did brisk business with seafaring European and Arab cities and kingdoms, starting with ancient Greece and ending with colonial Portugal.

It began 2,600 years ago, when Africans used fast, light dhows for coastal trade. Later, when the gold trade took off, it expanded into the Arabian Peninsula and Asia. The Greco-Roman epic, *Periplus of the Erythraean Sea*, written in about 60 CE, makes extensive reference to trade on the east coasts of Africa and Ethiopia. The author writes about the 'last port of trade on the coast of Azania, called Rhapta, a name derived from the ... sewn boats, where there are great quantities of ivory and tortoise shell'.[6] He adds that Indian and Arabian traders came to buy these products, including rhinoceros horn, in exchange for glass, spears and daggers. These descriptions led experts to place Rhapta in Tanzania. The archaeologist Felix Chami said he found evidence of its location on the coast near the Rufiji River and Mafia Island, where his team found Greco-Roman and Indian pottery and Persian glass beads that have been carbon dated to 600 BCE. Other archaeologists found Roman coins on nearby Pemba Island. There is also locally produced pottery in the area, some of it 2,000 years old, suggesting it was a key centre of local and international trade, at a time when European and Asian contact with this African civilisation appears to have been reciprocal and perhaps even respectful.

The gold trade came later; the Tanzanian city Kilwa Kisiwani was its key centre. Early Portuguese traders described it as a large city, with houses made of coral stone and a hundred-room palace garnished with gold, silver and jewels. They said their trading partners were 'black Moors', reflecting the conversion of the local population to Islam (one of Kilwa's remaining sites is its eleventh-century Great Mosque). The gold was mined in South Africa and Zimbabwe and carried to the Mozambican trading centre of Manyikeni, before ending up in Kilwa (as shown by a Kilwa-minted copper coin found in Great Zimbabwe).[7]

THE 'PRIMITIVE' AFRICAN AND OTHER MYTHS

This is just one illustration of precolonial industry, trade and construction in Africa. I chose it not only because it throws some light on the 'dark continent' but also because it provides examples of the falsification of archaeological records, illustrating how racist rulers found the idea of black precolonial achievement so threatening. But the other side of this cautionary tale is of a passive acceptance made possible by pre-existing perceptions. These stories illustrate not just the obvious point that those with political agendas tend to distort, invent and ignore history to create narratives that suit their ends, but also that focusing on a narrative of European or American civilisation and ignoring African history can create more insidious, less obvious distortions.

In my South African childhood and later, living in America and Britain, I've periodically been confronted by enquiries that began something like this: 'I'm not a racist but can you just explain to me why Africa has produced nothing of worth? I mean, is it something in their culture or does it go deeper?' Always, the point is that there's something missing in African history that reflects something missing in Africans, exposing the idea that Africans are incapable of invention and innovation or feats of construction, engineering, science or literature, and that the apparent lack of any evidence to the contrary rather proves the point.

I am spoilt for choice when it comes to prominent European thinkers who echo the view of the inherent inferiority of the African brain,[8] from the eighteenth-century Scottish enlightenment thinker David Hume[9] to the twentieth-century English historian Arnold Toynbee.[10] I'll focus on just one, particularly virulent, example: Carl Jung, the German-Swiss founder of analytical psychology, whose views retain their appeal for many 'seekers' despite convincing evidence that he indulged in scientific fraud in a bid to promote his ideas,[11] and his not-exactly contemporary views on women and on homosexuality.[12] Jung was deeply influenced by the writings of the racist German biologist Ernst Haeckel, whose ideas he placed in a mixing bowl with his Lamarckian view of evolution, Aryan mythology, polytheistic religion and beliefs in magic and the occult, all contributing to the race-based edge in his notions of universal archetypes in an inherited collective unconscious. One example: he believed African, Jewish and European mythical themes of sacrifice and death could be explained by inherited distinctions between 'races', justified by his Lamarckian perspective that characteristics acquired in the lifetime of members of one generation could be passed on genetically to members of the next. From Haeckel he drew his 'recapitulationist' view that we carry traces of our primitive past in our psyches. He viewed the adult African as comparable to the European child, prompting his desire to study the 'archetypes' of European children and 'primitive' African adults.[13]

Jung visited Kenya and Uganda in 1925 to carry out a 'scientific inquiry' into 'primitive psychology' and 'primordial darkness', although he admitted his real intent was to ask himself 'the rather embarrassing question: What is going to happen to Jung the psychologist in the wilds of Africa?'[14] He referred to the fact that while in Africa he'd had just one dream featuring a black person; the barber he'd used in Tennessee twelve years earlier. In the dream, the barber held a red-hot iron to Jung's head to style his hair. Jung woke in terror but later decided this dream was a warning from his unconscious that he was in danger of being swallowed by the primitive African world. 'At the time I was obviously all too close to "going black"', he concluded. At times he admired 'primitive' African qualities, such as the 'tender monkey-love of primitives' but warned Europeans against being drawn in to the point where they coexisted with Africans, because they, too, could face the danger of going black. 'Even today the European ... cannot live with impunity among the Negroes in Africa; their psychology gets into him unnoticed and unconsciously he becomes a Negro,' he wrote.[15] He added to this warning by drawing on Haeckel's theories, noting that the 'inferior man has a tremendous pull because he fascinates the inferior layers in the psyche'.[16]

If these Jungian views seem so ridiculous as to not be worth mentioning, we should remember that today's most vocal Jungian is the YouTube darling of the alt-right, Jordan Peterson, who shares one dimension of the master's views on race: that there are innate differences in the brains of different groups. Peterson insists ethnic differences in IQ are biologically based but worries 'this is something you can't say anything about without being killed.'[17] As we shall see, Peterson is wrong about IQ, and wrong about racially based intelligence differences.

To return to Africa; we can see the 'nothing good ever came out of the dark continent' perspective in the writing of a former Oxford biology professor, John Randal Baker. His 1974 book, *Race*, classified human races in the same terms as animal subspecies, and offered twenty-three criteria for determining whether they had reached civilisation. Baker concluded that no indigenous civilisations in Africa had ever been civilised.[18] His work was embraced by the contemporary racist psychologist Richard Lynn, who claimed that Negroid Africans never made the Neolithic transition and 'achieved virtually none of the criteria of civilisation'.[19] In his 2006 book, *Race Differences in Intelligence*, Lynn went further, saying that African intelligence had evolved to a level where they could make 'a little progress in the transition from hunter-gathering to settled agriculture but not sufficient to develop anything that could be called civilisation with a written language and arithmetic, construction of a calendar, cities with substantial stone buildings and other criteria set out by Baker'.[20]

You can see the picture: a pervasive view, held over centuries, that Africans were too primordially primitive to become civilised; a view that persisted in the intellectual mainstream until the horror of its implications emerged in the Holocaust. Since then, you might suppose it was confined to backwaters such as apartheid South Africa, and yet it continues to pop out of the mouths of today's race scientists, psychologists and right-wing politicians. A quick taste: the Australian Gregory Lauder-Frost, vice-president of the Traditional Britain Group and a former Tory Monday Club stalwart, claims: 'The Africans never had it so good as when Britain governed their colonies there ... We owe Africa nothing whatsoever. It owes us eternal gratitude for lifting it out of barbarism.'[21]

AFROCENTRIC MYTHOLOGY

It is not only white racists who indulge. As a child, I read several biographies of Muhammad Ali, and was astonished to discover some of the Nation of Islam nonsense he propagated: Elijah Muhammad's talk of a race of white devils created by an evil scientist, and so on. But ideas not a million miles from these continue to flourish. Afrocentrists range from those re-tilting the colonial balance by emphasising African achievement to those who dip into faux-biology to make claims about the superiority of the African character.

Like white racists, some Afrocentrists portray Africans as a single race, with even ancient Egyptians portrayed as black. Egyptian art showing otherwise is said to be the result of 'symbolic drawing', in which all people were portrayed as having reddish-brown skins.[22] In reality, the ancient Egyptians had a wide ethnic range, from dark-skinned to light, all reflected in Egyptian art. Afrocentrists also make outlandish claims about Africans having been behind every advance in civilisation and white people behind every disaster, such as the claim that HIV/Aids was a Western plot to attack Africans.[23] This mythology also extols the mystical powers of melanin (the pigment that darkens hair and skin), the subject of much discussion on a Web forum, MelaNet. People with high doses (black people) are portrayed as stronger, better coordinated and more athletic, and in some versions, quicker-thinking, more attuned to light and music, more communal and even more psychically attuned. White people are, like albinos, portrayed as suffering from a melanin deficiency, making them physically and morally weaker.

One advocate of this view is Leonard Jeffries, professor of black studies at the City College of New York, who claimed melanin allowed black people to 'negotiate the vibrations of the universe'.[24] He also said it was the source of the higher intelligence and creativity and compassion, as well as the peaceful nature of the 'sun people', whereas melanin shortfalls among the 'ice people' make them cold, greedy, militaristic and authoritarian. Jeffries was fired as chairman of his department after a long legal battle over his claim that Jews financed the slave trade and used Hollywood to hurt black people. In a subsequent interview he was asked what kind of world he would wish for his children. 'A world in which there aren't any white people', he replied.[25]

I have no hesitation in describing Jeffries as a racist, both for his anti-Semitism and his general attitude to white people, which returns me to the question I touched on in Chapter 1: can black people be racist when they don't have power? There, I used the example of anti-Semitism as a form of racism, making the point that both black and white people are capable of being anti-Semitic. But it goes further. By redefining racism (or, indeed, sexism) and coupling it with 'power', it is all too easy to fall into the trap of 'let's change the meaning of words so that only we can use them'. In Rwanda, in 1994, black Hutus committed genocide against black Tutsis, partly motivated by the false belief that the Tutsis had Eurasian 'blood'. Not racism?

Hardly surprisingly, the view that while black people can be prejudiced they can't be racist, because they don't have power, is largely confined to the United States, where by every measure they face prejudice, powerlessness and expressions of racism that many white people struggle to understand. Even the wealthy and well educated are not immune from its sting. When it comes to power, the word 'imbalance' hardly covers it. At an individual level it might be tricky to define and it is easy to get tied in knots ('Surely you aren't saying that unemployed Mr White from the trailer park has more power than Barack Obama?') but pound-for-pound it is clear-cut: a poor white person is likely to have better prospects than a poor black person. Some might argue that the racism of a white person is therefore harder to excuse, particularly if that white person wields real power. But it doesn't mean that attitudes such as those of Jeffries don't count. The point is that racism is all about the false belief in racial attributes. Power is something different.

SOUTH AFRICA: THE HAMITE MYTH DRAWS A VEIL ON THE AFRICAN PAST

Let's return to an example of people with racist beliefs who wielded real power. Nowhere was this belief more systematically advanced than in South Africa, where the obfuscation of the historical record was viewed as imperative, and scientists, psychologists, church ministers and historians contributed.

The Calvinist churchmen who provided the theological underpinning for white rule used the tale of the Tower of Babel to make their point that God's intention was to separate people and create nations by causing them to speak in different languages. The African nations, however, had an irreparable drawback. They were said to be descendants of Ham, Noah's youngest son who, according to Genesis, saw his drunken father naked, prompting Noah's curse that Ham's son should be the 'lowest of slaves' to his brothers.[26] This curse conflated with Joshua's on the Gibeonites: that they should 'never cease being slaves, both hewers of wood and drawers of water'.[27] Apartheid's founders borrowed this biblical gloss from European and American theology to justify policies built on a nationalist version of the Boer experience. But it also needed scientific gravitas, which came from the same well American segregationists drew on: skull measuring, eugenics and IQ testing.

The notion of 'Hamites', variously described as a brown or Mediterranean race, was mobilised in different guises throughout the nineteenth and early twentieth centuries. For example, Professor Augustus H. Keane, of University College London, was more positive about the South African Bantu than the 'Negro', because of the 'leavening' impact of Hamitic blood. In 1899, Keane, a former vice-president of the Royal Anthropological Institute, wrote about the temperament, character and intelligence of different races, and described the Hamite-infused Bantu as 'far more intelligent than the true Negro, equally cruel but less fitful and more trustworthy'.[28] This myth was embraced in South Africa. One of the most influential historical accounts of the black African presence in South Africa came from the gentleman historian, explorer and colonialist Sir Harry Johnston, who explained the Bantu migration to South Africa as the product of a 'trickle of Caucasian blood' through a 'vigorous type of Negro' being 'impregnated' by a 'semi-white race' such as the 'Hamites'.[29]

The implication was that Africans, by themselves, were incapable of exploration but give the sprightlier a few drops of whitish Hamite blood and a bit of whitish Hamite influence and off they would go. The odd idea of the vaguely defined 'Hamites' prompting civilisation in Africa remained popular until the Second World War. The notion that a more intellectually advanced, lighter-skinned people was the impetus for everything they admired in Africa: the grammatical sophistication of the Nguni languages, the advanced military strategies of the Zulus, the construction of Great Zimbabwe and the Buganda kingdom, the Benin bronzes, even the inclination to be converted to Christianity. One widely read British historian, Charles Seligman, writing in 1930, put it like this: 'The Hamites were, in fact, the great civilising force of black Africa.'[30] The Hamites did not, in fact, exist.

When I attended white South African state schools in the 1960s and 1970s, the history textbooks and teachers were largely guilty by omission, but they didn't shy away from outright distortion. The story we heard, year after year, was of the God-fearing Voortrekkers leaving the Western Cape to start their Great Trek, in protest against autocratic British colonial rule and its anti-slavery policies; travelling north and east with their ox wagons, Bibles and rifles, and meeting African tribes travelling in the opposite direction. These competing groups collided in the middle, and the Boers were compelled to defend themselves against the rapacious aggression of the Bantu.

The Afrikaner historian Gustav Preller went further, insisting that the Portuguese explorer Bartolomeu Dias erected his cross on South Africa's south coast ten years before the first 'Bantu migrants' arrived and that white settlers reached the Eastern Cape before the Xhosa. '[S]o far from the white man having been the intruder, who robbed the Bantu of his land,' he wrote, 'it was the Bantu who ... crept into the white man's backyard in the Eastern Province,'[31] And if history weren't enough for Preller, there was always biology. 'Science is now gradually discovering the remarkable physiological differences between the brain of the white man of European descent and that of the Bantu – differences that are innate and constitute the measure of their respective intellectual capacities.'[32]

I was taught that the superior technology (guns), organisation, morality and courage of the outnumbered Voortrekkers secured them divinely blessed victories because, like the Israelites, they were God's chosen. By then, official apartheid propaganda was focusing more on the need for 'separate development' of nations but the classroom story remained one of treacherous, superstitious African tribesmen who, for all their animal cunning, had not advanced beyond spears and shields, and were outwitted by more honourable men who farmed the land and formed the nation. Africans, whose farming methods were illogical (they stupidly ploughed down hills, rather than around them), were pushed by crop failure and pulled by bright lights to work in the white man's farms, factories and mines.

The archaeological record shows that African farmers lived in South Africa for 1,400 years before the Dutch settlers arrived, and that the Xhosa had lived in the Eastern Cape (where they first confronted the trekkers) for about 1,100 years. The claim that they were incompetent farmers is belied by the historical record, which includes evidence of white farmers demanding action against the more efficient African farmers. In fact, one reason why some white trekkers headed north was because of their inefficient subsistence farming, and many English settler farmers were no better. In 1870, a state-appointed statistician in the Eastern Cape noted that 'the native district of Peddie surpasses the European district of Albany in its productive powers.' Ten years later, a prominent white politician and lawyer noted that 'man-for-man the Kaffirs of these parts are better farmers than the Europeans, more careful of their stock, cultivating a larger area of land and working themselves more assiduously.'[33] But it was not to last. By 1913, black Africans were confined to 7.5 per cent of the land that had once all been theirs. The resulting overcrowding, rather than poor farming methods or bright lights, forced them to seek work on the white man's terms.

The problem with the history I was taught at school comes less from its distortion of the historical record than from the stories it chose not to tell. More specifically, from its refusal to impart anything of the record of African achievement before colonisation beyond a none-too-accurate version of the stories of Shaka's rise and fall and Dingane's treachery. I've already touched on the gold trading network of city states from Mapungubwe via Great Zimbabwe to Rhapta and Kilwa Kisiwani in present-day Tanzania. There is no space here for a detailed history but it is worth at least a cursory glance at some other examples of precolonial African development.

PRECOLONIAL AFRICAN KINGDOMS

The most durable was Nubia (in modern Sudan), which lasted for nearly 5,000 years. Ancient Egyptian hieroglyphic scripts suggest it was known for its treasure, dancing girls, wrestlers and slaves, while Egyptian artwork portrays dark-skinned Nubian traders, different from the lighter-skinned Egyptians.

Nubia was a formidable kingdom. It now lies beneath the Sahara Desert but then, it was verdant. Archaeological evidence indicates it was first settled more than 7,000 years ago, and there is evidence of a cattle cult as well as Stonehenge-type astronomical devices that pre-date Stonehenge by 2,000 years. Hundreds of stone gongs, used to communicate between villages and settlements, have been found, which suggests a shift from foraging to farming. A unified Nubian kingdom emerged about 5,300 years ago, interacting closely with Egypt and serving as a trade corridor between Egypt and tropical Africa. This role continued when the first Nubian kingdom declined, about five hundred years later; gold, copper, ivory, incense, ebony and exotic animals were imported through Nubia. In this period, Egypt built fortifications in Nubia to protect its assets but there is also evidence of close interaction between Nubians and Egyptians. During Egypt's Twelfth Dynasty (1991–1786 BCE), several of its pharaohs were black African Nubians.

Nearly 4,000 years ago, the new Nubian Kingdom of Kerma emerged and unified the region. It used slave labour to build monuments, palaces and tombs and traded goods from sub-Saharan Africa to the Red Sea. There are many examples of intricate, glazed pottery but also evidence of a darker side; some thirty thousand graves of men, women and children showing evidence of ritual human sacrifice. Nubia was annexed by Egypt in 1520 BCE and Kerma was destroyed, after which it became the centre of Egypt's gold production. The Egyptians eventually decolonised and a new Nubian kingdom, Kush, became a major trading centre. Archaeological digs revealed evidence of a town surrounded by walls and moats and containing a palace, two hundred houses and a huge religious-ritual platform. Kush invaded and controlled Egypt in the eighth century BCE, after which the Nubian king Taharka became the Egyptian pharaoh. The Nubians colonised Egypt for almost a century before the Assyrians forced them out. The Nubian rulers moved north to Meroe, where the next kingdom emerged on the east bank of the Nile; it lasted for 1,150 years, until 350 CE. Its archaeological sites provide a glimpse of a civilisation that had developed its own writing, moving away from Egyptian hieroglyphics to a twenty-three-sign alphabet, and built nearly two hundred pyramids. Moroe was the military and industrial centre of the region, specialising in iron production. The Greek historian and philosopher Strabo wrote that this kingdom's Nubian archers, led by their one-eyed queen, fought off the invading Romans in 24 BCE.[34] But it eventually fragmented and was conquered by another kingdom to the south; the desert encroached, and Nubia became Christianised. Later, Arab traders introduced Islam, prompting the collapse of the last Nubian kingdom in 1504.

Ethiopia was another durable kingdom, claiming roots in the biblical tale of King Solomon and the Queen of Sheba, a story repeated in a thirteenth-century Ethiopian manuscript, which says the kingdom was founded in 950 BCE by Menelik, son of Solomon and Sheba, but the Ethiopian kingdom itself had deeper roots, going back 2,700 years. This kingdom lasted four hundred years before falling. In about 100 CE, the Aksumite empire, described by a Persian writer as one of the world's great empires (along with Rome, Persia and China) emerged. It became a major international trading power, ruling over an area that extended to Yemen.[35] Ethiopia was the first major power to adopt Christianity (coins minted there in 324 CE bear crosses, whereas earlier coins are decorated with suns and moons) but from at least 200 BCE, a section of the Ethiopian population had converted to Judaism. Later, Muslims and Christians coexisted with minimal conflict, both supporting the state. Although Ethiopian dynasties rose and fell, the country was never colonised. In the seventeenth century, Emperor Fasiladas, in his ornate hilltop castle, ordered that the covetous Portuguese and ambitious Jesuits should be expelled; in 1896, under Menelik II, the Ethiopian army repelled an Italian invasion. But in the second Italian attempt, forty years later, Mussolini's troops used bombs and mustard gas to overcome Haile Selassie's defences, although resistance continued until Allied forces liberated it five years later. Selassie continued to rule until, overcome by hubris and corruption, he was overthrown in 1974.

In 2012, the *Daily Mail* ran a story carrying the headline: 'Meet the 14th century African king who was richest man in the world of all time (adjusted for inflation!)'. It claimed that Mansa Musa I, who was born in 1280 and ruled an empire that covered modern Ghana and Mali, had personal wealth worth the equivalent of US$400 billion in today's money 'by the time of his death in 1331 [*sic*].'[36] This was based on research showing that his empire produced half the world's salt and gold, which he used to build huge mosques that still survive.[37] Musa I did indeed rule an empire of astonishing power and wealth. During his twenty-five-year reign trade revenues trebled and he doubled the size of the empire, making it larger than any of the European kingdoms whose courts his envoys visited. Several cities became trading centres; by the middle of the fourteenth century Timbuktu had become a world trading centre and cultural hub, hosting huge libraries, a lively international book trade, a university and ornate mosques. It became a permanent settlement by about 1100 CE, growing rapidly, as a result of changed regional trading routes, into a centre of international trade for more than four hundred years. Repeated invasions and the encroachment of the desert heralded its decline but in its heyday, between the thirteenth and fifteenth centuries, it traded gold, ivory and slaves, and smelted precious metals, which were mixed with alloys and cast by local craftsmen.[38] The first bronze sculptures in Africa were produced in Timbuktu, using a technique later (from about 1550 to 1700) employed with great sophistication by the artists who produced the Benin bronzes.

Nigeria's Benin Empire, established in the fifteenth century, featured five thousand kilometres of earth walls, some nine metres tall, in and around Benin City, that divided it into forty wards. It traded with the Portuguese from the late 1400s, exchanging ivory, palm oil and pepper for manila and guns; in the early 1500s the two countries exchanged ambassadors. Its metal workers produced bronze sculptures of people and, from the 1700s, weapons, including firearms. As the empire grew, it supplemented its well-drilled forces with mercenaries and made more use of cannon and muskets. It traded with the British in the sixteenth century but later cut its ties, to forestall colonial ambitions, and rebutted persistent British attempts to take control of its trade routes and rubber plantations. The British invaded in 1897, burning and looting Benin City, colonising the country, and seizing the best of the bronzes, which were rehoused in the British Museum, where they remain. After examining these sculptures, the racist biologist John R. Baker speculated in 1974 that the Greeks were behind them, because Arabs were unlikely to have condoned the creating of images of humans and uncivilised Africans couldn't possibly have been up to it.[39]

Finally, two closely related Great Lakes kingdoms. Archaeologists trace Bunyoro's origins to 1000 CE. In its heyday, in the 1500s and 1600s, it controlled the region through its powerful military and an economy that included salt mining, metal-forging and jewellery production. It declined in the eighteenth century, as a result of clan rivalries, but rose again in the nineteenth under King Kabelega, who unified the country and expanded trade networks, buying guns from Zanzibar traders in exchange for ivory. It was crushed in 1899 by a British-led African force; massacre, starvation and disease killed two million people in five years. The Bugandan kingdom emerged in the late 1300s, formed by a prince (Kimera) who escaped the conflict in neighbouring Bunyoro. It expanded after 1700, eclipsing Bunyoro and doubling its territory to become a regional superpower. Its centralised government was based on the *kabaka* (king) who appointed chiefs to ensure that roads, bridges and viaducts were constructed, to enable tax collection. The kingdom had a navy on Lake Victoria, with a fleet of 230 large war canoes to transport commandos to trouble-spots. They traded extensively with Arabs from the Swahili coast, exchanging ivory and slaves for guns and ammunition. In 1875 the explorer Henry Morton Stanley estimated Buganda's army as 125,000 strong. When he visited its capital, he found a well-ordered town of forty thousand people. In its centre, on a hilltop, was the king's palace and compound, surrounded by a four-kilometre-long wall. Stanley called it the 'Pearl of Africa', describing its king as an enlightened despot.[40] But other explorers, unable to understand such an advanced African society, reached for the Hamite legend. The Victorian explorer John Hanning Speke said Buganda's 'barbaric civilisation' must have been a product of the influence of the fair-skinned 'Hamitic' pastoralist race.[41] In the late nineteenth century, as its great rival Bunyoro rose again, Buganda declined. The last of the great *kabakas*, Mwanga II, was defeated by the colonising British in 1897. He was captured but escaped to lead a rebellion, which was crushed.

There were several other significant examples of innovative sub-Saharan kingdoms and centres of power, including the Ashanti confederacy in the eighteenth and nineteenth centuries, which used firearms, sophisticated military organisation, a highly stratified social structure and the slave trade to forge an empire that stretched from its centre in Ghana to the Ivory Coast and Benin. Another is the Zulu kingdom, which used tight military organisation, rigid discipline and new fighting techniques and tactics to unite most of the northern Nguni tribes under a single monarch in the eighteenth and early nineteenth centuries.

The point has been established: the notion that Africa and Africans are incapable of technological, architectural and organisational sophistication – a view held by just about everybody in the nineteenth century, by apartheid South Africa and white-ruled Rhodesia in the twentieth and by the likes of James Watson and Richard Lynn in the twenty-first – is simply wrong. My purpose in these sketches is to counter the flawed premises of an argument that has echoed through the centuries: that absence of evidence of civilisation is evidence of the absence of civilising potential. But really, this should be superfluous. Even if there were no examples of great African kingdoms, it would tell us nothing about the intellectual potential of its people, then or now. Their collective intelligence can't be read from the level of the 'civilisation' or 'advancement' of a particular population at a particular time.

WHY DID AFRICA'S KINGDOMS DIE OUT?

One more question needs an answer: why did these civilisations die out without leaving many traces on modern African society? Why wasn't Africa like Europe, which built on early advances? The answer relates partly to geography but also to what was happening in the rest of the world, which, through slavery and colonisation, had an immense impact on Africa, and to internal factors in the empires. Some lasted longer than others, but they all fell eventually. The factors prompting decline were peculiar to each. In Nubia and Mali, the spread of the desert contributed. In most of Africa the factors included slavery and colonisation, disruption of trade routes and the decline of local industries (not least iron smelting, mainly as a result of cheaper imports under colonial rule).

In Eurasia, population density and favourable transport routes meant that one empire was likely to influence another through trade, migration and conquest. The 'invention' of writing in Mesopotamia had an impact on Egyptian writing, which influenced the Greek and Hebrew alphabets; the Latin alphabet emerged from the Greek, and so on. To take a different example, gunpowder was invented in China in about 800 CE; its first recorded use in weaponry was to defend against the Mongols. It was then used by the Mongols when they invaded Asia and Europe; its use spread across Asia and the Islamic world and through Europe, reaching Britain in the mid-1300s. Similarly, the mathematical knowledge of ancient Greece and Rome was appropriated by Islamic scholars and, through this repository, contributed to the Renaissance in Western Europe.

Wider spaces, lower population density and different climatic and geographical conditions meant there was not the same domino effect between the decline of one African civilisation and the rise of another. I have stressed the impact of trade within and between African kingdoms but, overall, precolonial Africa was more isolated than Europe. The availability of space and game allowed, in some places, for coexistence between hunter-gatherer, agricultural and urban peoples, whereas in Europe population density, conquest, disease and trade put an end to the hunter-gatherer lifestyle about 4,000 years ago.

As I discussed in Chapter 3, some tiny communities in South America and Africa are trying to maintain hunter-gatherer lifestyles. In some countries (such as Botswana) there has been a desire to force them into modernity, born of the misconception that premodern modes of existence are demeaning. Another way to see the choices made by these communities is that they are trying to pursue a way of life that has worked for thousands of years, which might seem to them a better option than the dire fate of other hunter-gatherers who latched on to modernity. Either way, their choices say nothing about brain power.

If you see history as a relentless forward march, and each advance in technology, industry and wealth a sign of progress, then the culture of the Botswana Bushmen, the Congo pygmies or the native

Amazonians might indeed seem 'backward'. Then again, the notion of history as progress doesn't look quite so clever when we consider the human contribution to global warming, the spread of weapons of mass destruction or the coexistence of extreme wealth and extreme poverty. This doesn't imply cultural relativism and absence of value judgements. Some cultures *are* superior to others; not, as the likes of Christopher Hitchens had it, because of their intellectual achievements but rather because they treat people better.

5

WHERE DID 'SCIENTIFIC' RACISM COME FROM?

We live in a world in which racism is widely viewed as abhorrent. Most people would deny they are racist. And yet many find it hard to avoid slipping into a kind of casual stereotyping with a racist tinge: 'He's so meticulous – it's those Germanic genes; those hot Latino emotions are never far from the surface; she's a bit of a mystic – it's in her Irish roots; he loves a good argument – it's in his Jewish blood.' It's all too easy to place the assets of others in an ethnic package but it's no great leap from this anodyne blurring of cultural and racial stereotypes to ascribing unequivocally negative traits. Jews are not just argumentative and clever – they're scheming money-grabbers; blacks are not just great dancers and runners – they're unintelligent and oversexed. Or, conversely, non-Jews are born anti-Semitic; lighter-skinned people always look down on the darker-skinned.

Racism emerges not only from what we hear and see but from what we don't hear or see. If the only black people we come across are in subservient positions, most of us would struggle to imagine a different world. It's the solipsism of the known; the assumption that our *milieu*, our way of doing things, our manner of thinking, is appropriate and natural. What we are used to experiencing is absorbed as the norm; everything outside is abnormal. And most of us lack the imagination to normalise the unfamiliar. George Eliot put it poignantly:

If we had a keen vision and feeling of all ordinary human life, it would be like hearing the grass grow and the squirrel's heart beat and we should die of that roar which lies on the other side of silence. As it is, the quickest of us walk about well wadded with stupidity.[1]

Racism also emerges from what we fear, even in ourselves, which is perhaps why some of its forms, including anti-Semitism, have been so resilient. As the Soviet-era Russian Jewish novelist Vasily Grossman put it: 'Anti-Semitism is always a means rather than an end; it is a measure of the contradictions yet to be resolved. It is a mirror for the failings of individuals, social structures and State systems. Tell me what you accuse the Jews of – I'll tell you what you're guilty of.'[2]

Even the brightest among us are unlikely to challenge racist views unless we are exposed to alternative ways of thinking, or experience life among a subjugated or socially scorned group. Even that is no guarantee. Take the very different attitudes to race held by the two men who independently conceived the theory of evolution by natural selection, Charles Darwin (1809–1882) and Alfred Russel Wallace (1823–1913).

When I first read *The Descent of Man* (written in 1871), I was surprised to discover how overtly racist Darwin was. I knew of his opposition to slavery, despair at the maltreatment of colonised people, and kindness towards individual black people. I also knew he was capable of a humanism lacking in many of his followers, as illustrated by this quote from *The Voyage of the Beagle*: 'If the misery of the poor be caused not by the laws of nature but by our institutions, great is our sin.'[3] But this did not get in the way of his perception that white people were at the apex of the evolutionary pile, and black people some way below. Darwin's prejudice became more virulent as he grew older and absorbed the scientific racism of the late nineteenth century. He recognised humanity's African roots but assumed that the brains of different races had evolved differently. Writing about 'Negroes', Darwin said: 'Their mental characteristics are ... very distinct; chiefly, as it would appear, in their emotional but partly in their intellectual faculties.' He said 'civilised races' were superior to 'savage races' and would 'exterminate and replace' them. And he added: 'The break will then be rendered wider, for it will intervene between man in a more civilised state, as we may hope, than the Caucasian and some ape as low as a baboon, instead of as at present between a negro or Australian and the gorilla.'[4]

It was a staple of nineteenth-century thought that women and black people shared a child-like inability to analyse and calculate. Darwin embraced this prejudice, writing that women differed from men in 'mental disposition' and referring to female 'powers of intuition, of rapid perception and perhaps imitation', faculties he said were 'characteristic of the lower races and therefore of a past and lower state of civilisation'.[5] He was particularly hard on the hunter-gatherers of Tierra del Fuego: 'Viewing such men, one can hardly make oneself believe that they are fellow creatures placed in the same world. I believe if the world was searched, no lower grade of man could be found.'[6]

Darwin's racism reflected prejudices that were close to universal then, but there were exceptions, including the co-founder of the theory of natural selection, Wallace, whose work on evolutionary theory, ecology and anthropology made him one of the most innovative thinkers of his time. Wallace is today known mainly for independently proposing a theory of natural selection, although neither man 'invented' evolutionary theory, which had been bubbling along since the eighteenth century; its theorists included Charles Darwin's grandfather, Erasmus Darwin (1731–1802), who wrote that 'all warm-blooded animals have arisen from one living filament.'[7] The most influential of the early theorists was the French naturalist Jean-Baptiste Lamarck (1744–1829) who wrote that the environment prompted changes in animals but today is best known for his mistaken belief in the inheritance of acquired characteristics; the idea that modifications arising during an animal's life could be passed on to its offspring.

Wallace had been developing his evolutionary ideas for fifteen years before working out how, through natural selection, one species could diverge from another. He wrote a paper entitled 'On the Tendency of Varieties to Depart Indefinitely From the Original Type' and sent it to Darwin, asking him to forward it to the geologist Charles Lyell for publication. Darwin, who had held back from publishing his theories partly because he was worried how Christians might respond, dutifully passed it on, although he feared he would be 'forestalled' (beaten to the punch) and that 'all my originality will be smashed.' His friends helped him to cobble together some of his writings, which were presented alongside Wallace's paper at the Linnean Society of London on 1 July 1858; the first public airing of the theory of natural selection. Had Wallace sent his paper to a different publisher, Darwin might indeed have been 'forestalled'.

Darwin and Wallace supported each other, and agreed on the main contours of natural selection, but placed different emphases on the details. Darwin focused on competition between individual organisms, while Wallace emphasised environmental pressures, a perspective now generally accepted. They both wrote on sexual selection, but Darwin concentrated on females choosing males (he thought female animals had an aesthetic sense) and Wallace on what we'd now call genes for traits (such as bigger horns) that increase mating success in the dominant males; closer to the contemporary view. In explaining speciation, Darwin's 'principle of divergence' focused mainly on competition in the same habitat, while Wallace was more interested in the separation of populations, also now accepted.

On so many issues Wallace was ahead of his time: his environmentalism, support for women's liberation,[8] socialism and fervent opposition to colonialism, militarism and eugenics. But some of his views were very much of his time, particularly, in later life, his spiritualism and belief in a spiritual origin of the higher faculties of human intelligence, which was criticised by Darwin, who, by then, was agnostic, though never an atheist. On race, his perspectives differed most sharply from Darwin's, which might be explained in part by their different social circumstances. Unlike the wealthy, patrician Darwin, Wallace grew up poor, left school at fourteen and for much of his life was short of money. While Darwin conducted his post-*Beagle* primary research in English fields and gardens, Wallace, who questioned the older man's reliance on pigeon breeders, lived among indigenous people in the Amazon and South-East Asia. This experience, combined with his iconoclastic nature, prompted him to see beyond the prejudices of his day – so far beyond that he described these people as 'near perfect' in contrast to the 'barbarism' of Britain. Some of Wallace's ideas of 150 years ago would stand up, with a word change or two, in the most liberal of today's *salons*:

> Now it is very remarkable that among people in a very low stage of civilisation we find some approach to such a perfect social state. I have lived with communities of savages in South America and in the east who have no laws or law courts but the public opinion of the village freely expressed. Each man scrupulously respects the rights of his fellows and any infraction of those rights rarely or never takes place. In such a community, all are nearly equal. ... Our mastery over the forces of nature has led to the rapid growth of population and a vast accumulation of wealth; but these have brought with them such an amount of poverty and crime and have fostered the growth of so much sordid feeling and so many fierce passions, that it may well be questioned whether the mental and moral status of our population has not on the average been lowered.[9]

Wallace's views were out of step with the conventional wisdom of his day. In the nineteenth century there were few alternative glimpses of a world not defined by race, unless you immersed yourself in places beyond the comfort zone of white society, as Wallace did. Even intimate exposure to non-white people was no guarantee, especially if the power relationships were unbalanced. America's third president, Thomas Jefferson, ensured the six children he had with his slave 'mistress' were freed after his death in 1826 but his attitudes on race reflected his time. He wrote that blacks were equal to whites in memory but 'in reason much inferior', adding, 'I advance it, therefore, as a suspicion only, that the blacks, whether originally a distinct race or made distinct by time and circumstance, are inferior to the whites in the endowment both of body and mind.'[10]

On the subject of American presidents, we may think of Abraham Lincoln as the liberal who freed the slaves but his belief in white superiority was barely different from that of the slave-owning Jefferson. In the Douglas debates of 1858 Lincoln said:

> There is a physical difference between the white and black races I believe will forever forbid the two races living together on terms of social and political equality. And ... while they do remain together there must be the position of superior and inferior and I as much as any other man am in favour of having the superior position assigned to the white race.[11]

Today, we live in a world in which views such as Jefferson's, Lincoln's and Darwin's are seen as abhorrent, while Wallace's are unremarkable. As we shall see in more detail in later chapters, most scientists agree that race is no more than a convenient shorthand description, with little scientific value, partly because all population groups are mixed and partly because genetic differences within groups defined as races are far greater than between them. As the British evolutionary biologist Steve Jones put it: 'Individuals and not races and not nations are the repository of our variation in genes whose function is known. The idea that humanity is divided up into a series of distinct groups is quite wrong.'[12]

MORTON AND THE GOULD CONTROVERSY

Race was bound up with religion before evolutionary theory took over. One view, monogenism, was based on the idea that we are all descended from Adam and Eve and that races reflect stages of degeneration since the Garden of Eden, with blacks having fallen furthest (the 'curse of Ham' idea was a variant of monogenism). The alternative, polygenism, viewed Adam and Eve as allegorical and started with the premise that races were different species. The polygenists made most progress in the nineteenth century, particularly in the scientific establishment, through their zeal in measuring skulls and weighing brains.

An early pioneer was a Philadelphia physician, Samuel George

Morton, regarded as one of the giants of American science by the time he died in 1851. He collected six hundred skulls to test his polygenist belief that the races had evolved separately from the start, which, in turn, prompted his hypothesis that you could rank races by brain size. Because brain size and skull size are correlated, he set about measuring skulls, filling them with lead shot to test volumes. He decided whites were at the top of the intellectual pile, Native Americans in the middle and blacks at the bottom. He went further, writing about the character of each race; the Hottentots were at the bottom of the bottom, 'the nearest approximation to the lower animals', he decided, adding that their women were 'even more repulsive in appearance than the men'.[13]

The twentieth-century palaeontologist Stephen Jay Gould re-analysed Morton's data and, in his book on scientific racism, *The Mismeasure of Man*,[14] concluded that although there was no fraud involved, the results were 'a patchwork of fudging and finagling in the clear interest of controlling *a priori* convictions'.[15] However, one aspect of Gould's conclusions was disputed by a team from the University of Pennsylvania, which owned the skull collection.[16] The team found that in only 2 per cent of cases were Morton's measurements significantly wrong and that Gould had made errors in his own calculations, declaring this was 'the stronger example of bias influencing results'. This vehemence was surprising to anyone who'd read Gould's book, because he referred only to 'minor' calculation errors by Morton and emphasised unconscious bias in interpreting the results. An editorial in the scientific journal *Nature* hit back at the Pennsylvania team, noting that the researchers 'have their own motivations' and several 'have an interest in seeing the valuable but understudied skull collection freed from the stigma of bias'.[17]

Gould died in 2002, so he was not around to defend his measurement, but the verdict of the skull-owners, trumpeted in the *New York Times* by the race science author Nicholas Wade,[18] delighted those who supported scientific racism. They used it to accuse Gould of scientific fraud, saying that Morton was proved right and Gould's accusation that he was falsifying data was proven wrong, invalidating Gould's book and Gould himself. The author and YouTube star

Sam Harris was particularly vituperative, claiming 'there's actually evidence that Steven Jay Gould, when he wrote *The Mismeasure of Man*, was consciously lying and was a true bad actor here and actually faked data.'[19] The publisher of *The American Conservative*, Ron Unz, wrote that Gould's 'fraud on brain-size issues' was 'presumably in service of his self-proclaimed Marxist beliefs'.[20] Gould was not, in fact, a Marxist.

Those less invested in rubbishing the reputation of a scientist who had done so much to discredit race science stressed caution. *Nature* noted there had been no post-Gould independent measurement of the skulls and in any event, 'the critique leaves the majority of Gould's work unscathed.'[21] Reading these attacks, one can only conclude either that those making them hadn't bothered to read the nineteen pages devoted to Morton in Gould's 444-page book or were deliberately over-egging the relevance of the measurement dispute. Gould never claimed Morton had deliberately falsified his data. In fact, he stressed the precise opposite: 'I find no evidence of conscious fraud'[22] and later, 'I detect no sign of fraud or conscious manipulation.'[23] Gould emphasised not Morton's 'minor numerical errors' but rather their interpretation. Neither the skulls' owners nor any of Gould's critics dispute that Morton drew some of his conclusions by comparing the skull sizes of 'Hottentot' and 'Negro' women with those of white English men, or that his 'Hindu' skulls came from people with small body size.

BROCA, BEAN AND BRAIN BITS

The 'science' of measuring skulls gained further impetus from the Paris-based professor of surgery Paul Broca. He too was obsessed with brain size, weighing them after performing autopsies. He was forced to concede that some 'lowly races' had big brains, prompting him to conclude that 'if the volume of the brain does not play a decisive role in the intellectual ranking of races, it nevertheless has a very real importance.'[24] Broca decided intelligence was located in the frontal sutures of the brain (now known as 'Broca's area') and said that in black people these closed before the back sutures, making further education ineffective. He also examined the foramen magnum (the hole in the base of the skull through which the spinal cord passes), reasoning that a more posterior foramen magnum was more apelike. When he found whites had a more posterior foramen magnum, he altered his criterion, arguing this was because blacks had smaller front sutures and bigger back ones. 'It is therefore incontestable', he concluded, 'that the conformation of the Negro, in this respect as in many others, tends to approach that of the monkey'.[25] He also felt it was apelike to have relatively longer lower arm bones, compared to those of the upper arm, but abandoned this after finding that, on this score, whites were more apelike than Australian Aboriginals, Eskimos and 'Hottentots'. Broca had no doubt of the correct ranking. As he put it in 1866: 'A prognathous [projecting lower jaw] face, more or less black colour of the skin, woolly hair and intellectual and social inferiority are often associated, while more-or-less white skin, straight hair and an orthognathous [straight-jawed] face are ordinary equipment of the highest group in the human series.'[26]

Gould reviewed Broca's data and found that his methods involved starting with his conclusion (white males on top, blacks below), manipulating facts to fit prejudices, explaining away exceptions and changing criteria along the way.[27] Science has long since discredited Broca's claims. No racial factors are involved in the timing or order of cranial closure, nor in the size of the front and rear sutures, nor in the position of the spinal cord.

Another brain claim was made at the start of the twentieth century by Robert Bennett Bean, a Virginian doctor who studied the corpus callosum (the bundle of nerves that connects the brain's two hemispheres). He decided he could rank races by measuring the genu (front) and the splenium (back) of the corpus callosum. He thought whites had bigger genua and therefore higher intelligence. He added that the genu was bound up with the sense of smell, and everyone knew that blacks, like animals, had a better sense of smell and yet still had smaller genua, meaning that their intelligence must be remarkably low. He also noted a sex-based difference in the corpus callosum, a view revived, and then discredited, in the late twentieth century.[28] Bean used his data to claim that blacks were an intermediate form 'between man and the ourang-outang [*sic*]'[29] and his claims prompted the journal *American Medicine* to publish an editorial calling on politicians to 'remove a menace to our prosperity – a large electorate without brains'.[30] But his claims came unstuck when his mentor at Johns Hopkins University repeated his research without prior knowledge of which of the brains were black or white and found no difference between the size of their genua or splenia.

The weighing of brains and measuring of skulls and brains lost ground in the twentieth century but it hasn't died completely: over the past fifty years brain size claims have continued to pop up, including one from Richard Lynn, who still insists brain size causally correlates with intelligence: 'To accommodate this enhanced intelligence,' he said of Europeans, 'they evolved larger brains ...'.[31] His views on the relevance of brain size proliferate in alt-right corners of the Web but are not taken seriously by neurologists or anyone else dealing with the brain, for the simple reason that big people have big brains and small people have small brains, which tells us nothing about their smarts.

SOCIAL DARWINISM AND EUGENICS

By the time *On the Origin of Species* was published, and natural selection entered the intellectual mainstream, the creationist version of monogenism was in retreat but there was no backing away from its race-based precepts. Rather, they were given fresh impetus by some of Darwin's disciples, whose racism was unleavened by their hero's humanism. 'Darwin's bulldog', Thomas Huxley, had no doubt: 'No rational man, cognisant of the facts, believes that the average Negro is the equal, still less the superior, of the average white man,' he wrote, adding that for this reason the 'Negro' would be unable to compete 'with his bigger-brained and smaller-jawed rival, in a contest which is to be carried on by thoughts and not bites'.[32]

Ernst Haeckel, the German biologist, who was strongly influenced by Darwin, noted that the 'immense superiority which the white race has won over the other races in the struggle for existence is due to Natural Selection'. He predicted it would become 'more marked in the future so that still fewer races of man will be able ... to contend with the white in the struggle for existence' and that it would be 'the more perfect, the nobler man, that triumphs over his fellows ...'.[33] The British philosopher and economist Herbert Spencer embraced some of Haeckel's ideas, including his recapitulation theory, which he explained thus: 'The intellectual traits of the uncivilised ... are traits recurring in the children of the civilised.'[34]

This became part of the package of social Darwinism, an extension of evolutionary thought that started with the error of viewing human behaviour as a direct consequence of natural selection. It was amplified by 'survival of the fittest', a term coined by Spencer but embraced by Darwin because he was worried 'natural selection' could suggest conscious choice. What Darwin didn't know, because he was unaware of Mendel's discovery of genetics, was just how unconscious a mechanism it was. Spencer went further, applying his term to culture, politics and economics, although there is no good reason to equate biological evolution with how people organise their affairs. Darwin despaired of Spencer's desire to turn natural selection into a theory of everything, but social Darwinism flourished, prompting the 'science' of eugenics; the belief that humanity could be improved by breeding out undesirable traits by sterilising or eliminating those who had them. Its most vociferous advocate was Darwin's cousin, Sir Francis Galton, who wrote of the European superiority over 'lower races', describing black people as being 'so childish, stupid and simpleton-like as frequently to make me ashamed of my own species'. Colonisation would lead to their replacement by betters because, he said, they would fail to submit to the needs of a 'superior civilisation'.[35]

Again, Wallace stood out as one of the few intellectuals to row against the tide, arguing, in 1890, that society was too unjust and corrupt to decide who was fit or unfit. He added: 'Those who succeed in the race for wealth are by no means the best or the most intelligent.'[36] But Wallace's warnings were drowned out by the clamour for eugenics, which attracted enthusiastic support from people with power. Winston Churchill was a big fan, as were Teddy Roosevelt, Woodrow Wilson and, of course, Adolf Hitler, whose policies on the extermination of Jews, Gypsies and disabled people were influenced by Haeckel and, specifically, by eugenics. Several leading biologists followed suit, including the Austrian biologist Konrad Lorenz, co-founder of ethology (the biological study of animal behaviour) and Nobel Prize laureate, who joined the Nazi Party and supported its policy of seeking 'racial hygiene'. Eugenics, married with IQ scores, was embraced in the USA, Sweden, Norway, Denmark, Iceland, Finland, Switzerland, Japan, Canada, France, Belgium, Brazil, Estonia and Australia. It prompted mass sterilisations in a vain bid to improve 'genetic stocks', and selective immigration policies, often motivated explicitly by racism.

Before World War II, public declarations of racism were routine; sometimes based on race science, sometimes on IQ scores, and often expressing a 'gut-racism' that was unafraid to speak its name. These views went right to the top. In 1902, Winston Churchill boasted of killing Sudanese 'savages' and called for further colonial conquests because he believed 'the Aryan stock is bound to triumph'.[37] In 1919, he argued: 'I am strongly in favour of using poisoned [sic] gas against uncivilised tribes,' adding it would 'spread a lively terror'.[38] Later, during an Indian rebellion, he said: 'I hate Indians. They are a beastly people with a beastly religion.'[39] Later still, when the Kenyans rebelled, he described them as 'brutish children'[40] while the Palestinians were 'barbaric hordes who ate little but camel dung'.[41]

Churchill and Gandhi were enemies in India but they shared similar attitudes to black South Africans when they both lived there at the turn of the nineteenth century. 'We believe as much in the purity of races as we think they [whites] do,' Gandhi wrote, early in his South African sojourn. Later, he was more explicit: 'About the mixing of the Kaffirs with the Indians, I must confess I feel most strongly.'[42] In 1908, he complained that 'Kaffirs are as a rule uncivilised – the convicts more so. They are troublesome, very dirty and live almost like animals.'[43] Unlike Churchill, he at least half-changed his mind: 'However much one may sympathise with the Bantus,' he wrote in 1939 '... Indians cannot make common cause with them.'[44]

The Holocaust revealed the full implications of race science and a fresh consensus emerged, its tone set by the United Nations Educational, Scientific and Cultural Organization (UNESCO) whose initial statement on race noted that scientists agreed that 'all men belong to the same species, *Homo sapiens*.' This did not mean that race science had died, rather that its proponents were less voluble. A few broke ranks, such as Ronald Fisher, the British eugenicist dubbed by Richard Dawkins as 'the greatest biologist since Darwin'. He proposed that science showed that 'the groups of mankind differ in their innate capacity for intellectual and emotional development, seeing that such groups do differ undoubtedly in a very large number of genes.'[45] Fisher's advice was rejected, and UNESCO opted for a toned-down version of the original. 'Scientists are generally agreed that all men living today belong to a single species,' it began before noting that 'within different populations ... one will find approximately the same range of temperament and intelligence.'[46]

SOCIOBIOLOGY AND THE REVIVAL OF BIOLOGICAL DETERMINISM

The determination to keep race out of the conversation held for a while and was accompanied by a shift away from biological determinism. But this consensus began to fracture, on the one hand by psychologists using IQ scores to revive race science and on the other by sociobiologists embracing the biological determinism of the likes of Spencer, with the added gloss of genetics.

Several popular science books revived the idea that we are no more than products of our biological inheritance. First came Konrad Lorenz, who explained in his 1963 book *On Aggression* that his subject 'is *aggression*, that is to say, the fighting instinct in beast and man which is directed *against* members of the same species',[47] behaviour programmed by natural selection. Despite (or perhaps because of) his Nazi past, which involved support for racial purity, he avoided referring to race but his writing contributed to a genetic determinism that viewed social divisions as natural. In 1973, Lorenz was, with

Timbergen and von Frisch, the joint winner of the Nobel Prize for Physiology or Medicine.[48]

Next came Robert Ardrey, a failed playwright whose 1966 bestseller, *The Territorial Imperative*, explained the world as a fight over real estate. Our ancestors were 'killer apes', characterised by heightened male aggression, and war over territory and interpersonal aggression drove evolution. Ardrey, a supporter of apartheid South Africa who went to live and die there, had none of Lorenz's racial reticence. 'Today there is not a black African state which ... does not stagger along on one side or the other of the narrow line between order and chaos, solvency and bankruptcy, peace and blood,' he wrote, adding that 'the pariah state of South Africa is attaining peaks of affluence, order, security and internal solidarity rivalled by few long-established nations.' Applying his ideas about how we evolved to secure defence of territory, he concluded that the impact of international opposition had been the opposite to that intended, making South Africa 'a stable, united, incredibly prospering nation in which the threat of racial explosion is almost non-existent'.[49]

Desmond Morris's *The Naked Ape* was an even bigger popular success, translated into twenty-seven languages and made into a dire feature film. Morris, a former curator of mammals at London Zoo, stressed our similarity with the great apes and said human nature evolved to meet the challenges of the savannah: 'Despite our grandiose ideas and our lofty self-conceits, we are still humble animals, subject to all the basic laws of animal behaviour.'[50] It was pointless to mess with these, because 'the naked ape's old impulses have been with him for millions of years, his new ones for a few thousand at the most and there is no hope of quickly shrugging off the accumulated genetic legacy of his whole evolutionary past.'[51] Morris combined despair about the implications of modernity with disdain for 'simple tribal groups'. Overpopulation and nuclear weapons mean '[t]here is a strong chance that we shall have exterminated ourselves by the end of the century,' he wrote in 1967.[52] His solution was either 'massive depopulation or the rapid spread of the species to other planets',[53] yet he was contemptuous of groups who resisted modernity, and of those who studied them, writing of 'cultural backwaters so untypical and unsuccessful that they are nearly extinct' and referring to anthropologists' inclination to 'return with startling facts about the bizarre mating customs, strange kinship systems or weird ritual procedures of these tribes as if they were of central importance to the behaviour of our species as a whole', arguing the focus should be on 'the ordinary successful members of major cultures – the mainstream specimens who represent the vast majority'. He reiterated: 'The simple tribal groups that are living today are not primitive, they are stultified ... The naked ape is essentially an exploratory species and any society that has failed to advance has in some sense failed, "gone wrong".'[54] But despite the retention of the word 'negro' (already dubious in 1967) in the book's 2005 edition,[55] he steered clear of any discussion of race, talking instead about evolved 'social in-groups, which 'could never be eradicated without a major genetical change in our makeup and one which would automatically cause our complex social structure to disintegrate'.[56] Desmond Morris became the Barbara Cartland of sociobiology, pumping out several books a year, most having 'naked' or '-watching' (man-watching, body-watching, and so on) in their titles. His reputation suffered from the law of diminishing returns and perhaps also from the bizarre quirkiness of his public output. My personal favourite is his claim, made on television, that lipstick was an evolved product of the male attraction for the inflamed bottoms of female apes on heat.

An ant-obsessed biologist steered these ideas back into the realm of academia. In 1975 Edward Osborne (E. O.) Wilson's seminal work, *Sociobiology: The New Synthesis*,[57] made it to the covers of *Time* and the *New York Times*, prompting sales that were impressive for a 697-page tome, which is now in its twenty-fifth edition. He followed this with further books on the subject,[58] promoting his ideas with a missionary zeal redolent of his Southern Baptist upbringing in Alabama, particularly in his portrayal of the relationship between immortal DNA and the mortal body. When he spoke of moments of 'climactic and exhilarating'[59] discovery he made them sound like divine revelation. Some of the writing of this biologist, who calls himself a 'provisional deist', makes evolution sound like Calvinistic predestination: people are 'adaptation executors,'[60] free will is an illusion and there is a 'genetic leash'.[61] For Wilson, all human behaviour is governed by the need to preserve genes and pass them on. He aimed to sublimate everything under one crusading banner: sociobiology was the 'systematic study of the biological basis of all social behaviour'. His project involved a Cartesian ambition to 'synthesise' biology and the humanities, prompting genetic explanations for racism, sexuality, altruism, war, religion, tribalism and entrepreneurship.

Fellow Harvard biologists Stephen Jay Gould and Richard Lewontin castigated E. O. Wilson for social Darwinism and portraying a 'deterministic view of human society and human action'.[62] Accusations of ethnocentrism, racism and sexism made him incendiary on US campuses: the ethnocentric charge drawn from his limited choice of examples of 'universal' human behaviour; the racism charge from the implications of his views and from his specific claim that racism itself had a genetic base. The International Committee Against Racism called *Sociobiology* 'dangerously racist' because it encouraged 'biological and genetic explanations for racism, war and genocide'. Wilson, a self-described 'social conservative' has spoken of his despair at such accusations. 'I came from the Old South. I was raised as a racist,' he said. 'I mean, we all were. It was only in my teens that I began to change. But here, if you were called racist – well, in the Seventies – it was like a death sentence.'[63] Despite this protestation, Wilson was not beyond giving open support to race science. In 1994, he backed the work of the overtly racist psychologist J. Philippe Rushton, who had written a book arguing that blacks were inherently less intelligent.[64] In 2014, Wilson provided a gushing back cover endorsement for a book by the popular science author Nicholas Wade, who explained just about every cultural stereotype, including Jews being good capitalists and Africans being tribalists, in terms of natural selection. Wilson called it 'truth without fear'.[65]

DAWKINS RESCUES SOCIOBIOLOGY FROM ITS RACIST LABEL

It took an Oxford-trained Englishman to decontaminate sociobiology and build a bridge to evolutionary psychologists, the next breed of neo-Darwinians, serving as their high priest. In his pomp Richard Dawkins was free of any whiff of racism, although the virulence of his regular late-life rants against Islam changed his reputation.[66] Islam aside, the politics associated with social Darwinism frustrated him. 'I believe in the survival of the fittest as an explanation for the evolution of life,' he said:

> ... but there have been people who have advocated the survival of the fittest as a kind of political creed, where they will justify a form of right-wing politics or economics on the grounds that it conforms to the laws of nature. And that I do object to, as indeed so does any other modern Darwinian. We don't want to see Darwinism being used to justify things like Fascism, which it has been.[67]

The young Dawkins was an unlikely fit for his future role as 'Darwin's Rottweiler'.[68] The son of a colonial official, he spent his childhood in Africa before being packed off to boarding school in Britain, where he embraced Christianity with the kind of fervour that would become familiar territory to Dawkins watchers. At thirteen, he had 'a fairly active fantasy life about a relationship with God', imagining 'creeping down to the chapel in the middle of the night and having a sort of blinding vision'.[69] A year later, after hearing Elvis Presley was a Christian, he said he 'felt Elvis calling me to be a messenger from God'.[70] By sixteen he still had a sense of a creator along the lines of 'There must be SOME sort of designer, some sort of spirit, designed the universe and designed life'. Then he read about Darwinism and had a second Road to Damascus moment. '(I)t was when I understood Darwin that I saw how totally wrong that point of view was; that, rather suddenly, scales fell from my eyes.'[71]

Destiny took a while longer to reveal itself. He was an indifferent pupil, with little aptitude for biology, although he followed the family tradition of going to Balliol College, Oxford. Next came 'a bit of a drift, really' until 'more or less by default' he discovered zoology.[72] Despite his second-class degree, he was permitted to study further, and passion took over. He completed his PhD, chose the fledgling discipline of ethology, and for a while worked in relative obscurity. This changed in 1976, with the publication of *The Selfish Gene*,[73] which had huge impact as a work of popular science. Eight more books and numerous articles, television programmes and radio interviews followed, and Dawkins became known more for popularising science and atheism than for any of his own theorising, let alone field research – so much so that E. O. Wilson, when asked about his various disputes with Dawkins, replied rather cattily: 'I have no dispute with Richard Dawkins. My disputes are with other scientists and Richard Dawkins is a journalist.'[74] Journalists, he added, merely 'report' what real scientists do.

The Selfish Gene was original not so much for its ideas but for its knack in expressing them. The theory is drawn from Dawkins's mentor, W. D. Hamilton, whose notion of kin selection he swallowed whole, although he steered clear of Hamilton's support for eugenics.[75] Where Dawkins excelled was in his knack for creating catchy terms that entered the public consciousness. He presents a picture of selfish genes struggling to survive in passive host organisms, which in turn are controlled by the genes (or 'replicators' as he sometimes calls them). The relationship among alleles (alternative forms of the same gene), genes and host is mutually beneficial. Genes that help an organism reproduce will also help themselves reproduce. The term 'selfish gene' suggests genes have a will of their own but what Dawkins meant was that the surviving genes were those whose *consequences* continued to be replicated. In this sense, the term is metaphorical, and metaphors are part of what makes his book so compelling. However, in his hands, the metaphorical selfishness of the gene and the innate selfishness of the individual are blurred. As he put it in the introduction:

> Be warned that if you wish, as I do, to build a society in which individuals cooperate generously and unselfishly towards a common good, you can expect little help from biological nature. Let us try to teach generosity and altruism, because we are born selfish. Let us understand what our own selfish genes are up to, because we may then at least have the chance

to upset their designs, something no other species has ever aspired to.[76]

Dawkins focuses on culture as a product of natural selection, requiring its own metaphor. To describe how culture 'evolves' he invented the word 'meme', which he felt captured the idea of a 'unit of cultural transmission or a unit of imitation'.[77] These units 'evolve' over time, in a way comparable to natural selection. However, the catchiness of the term disguises the flaw in the analogy: genes exist, memes don't. Even as a metaphor, memes barely work, because culture does not come in particles and is not subject to laws comparable to those of natural selection. The philosopher Mary Midgley memorably compared thought and culture to patterns of traffic flow that could not usefully be studied under a microscope. Of memes, Midgley said: 'Since such patterns are not composed of distinct and lasting units at all, it is not much use trying to understand them by tracing reproductive interactions among those units.'[78]

ANOTHER VIEW: COOPERATIVE GENES AND EVOLUTIONARY SPANDRELS

The idea of a selfish gene leading its recipient by the nose in its march for self-reproduction and unleashing cultural memes along the way is frequently ridiculed. Dawkins acknowledged he should have taken his publisher's advice and called it 'The Immortal Gene'.[79] E. O. Wilson, whose sociobiology was often compared with Dawkins's, now dismisses the selfish gene notion. 'There might be a few scientists who still hold to it but they have largely been silenced,'[80] he said. A prominent critic is the Stanford evolutionary biologist Paul Ehrlich, who says that to talk of genes as 'self-replicating' or 'selfish' is misleading. 'Genes cannot be reproduced except by being imbedded in a complex cellular mechanism – they're about as self-replicating as a printed page lying in a copying machine,' he wrote, adding, 'It is critical that genes act in concert and not interfere with the functioning of other genes. If one is silly enough to assign human-like motivation to them, it would be as sensible to describe them as "cooperative" as to call them "selfish".[81]

Another criticism of the gene-centric theory is that it focuses too narrowly on natural selection, suggesting that evolution is a one-trick pony. There are also non-adaptive evolutionary factors, the most significant being genetic drift, which refers to changes in the frequency of an allele (gene variant) in a population over time due to chance. Such changes may be good, bad or indifferent, and usually occur in isolated populations or population bottlenecks, where mutations for a particular trait have a higher chance of taking root. Once that happens, and the group is reintegrated into a wider population, the trait can spread. The pioneering Japanese geneticist Motoo Kimura argued that *most* evolutionary change at the level of the molecule is prompted by random genetic drift.[82] Other non-adaptive evolutionary forces include chance events (such as the meteor-strike that hit the earth sixty million years ago, creating conditions that dinosaurs could not survive but in which mammals thrived); structural constraints on the body plan of an organism; and selection at the level of the group or species, in which DNA sequences become fixed and spread because they benefit the group, so the group acts as the vehicle of selection.

Perhaps the most persistent critic of biological determinism was the palaeontologist, geologist and evolutionary biologist Stephen Jay Gould, whom we met earlier in this chapter in the Morton skull controversy. Gould was born the same year as Dawkins (1941), but unlike Dawkins had an early sense of vocation: after seeing a *Tyrannosaurus rex* skeleton when he was five years old, he decided he wanted to be a palaeontologist. He went on to research for a doctorate in geology at Columbia University before becoming professor of geology at Harvard and professor of biology at New York University in the last six years of his life.[83]

Gould persisted with field research throughout his life, focusing on snails. These studies prompted his most significant contribution to evolutionary theory: the theory of punctuated equilibrium, which he developed with his fellow palaeontologist Niles Eldredge. This theory challenged the premise that 'life's major transformations can be explained by adding up, through the immensity of geological time, the successive tiny changes produced generation after generation by natural selection.'[84] Instead, fossil records show millions of years of stasis, followed by brief periods of rapid change; Gould and Eldredge concluded that punctuated equilibrium was the 'dominant pattern of evolutionary change'.[85] This theory gained ground partly because of evidence from the fossil record. The idea that the human brain might not have evolved significantly for 100,000 years therefore does not go against the evolutionary grain: it is what we might expect.

Another challenge to genetic determinism came from a further evolutionary departure by Gould, this time with his Harvard colleague Richard Lewontin: the notion of non-adaptive side-consequences of evolution, which they called 'spandrels'.[86] They invented the term 'to expose one of the fallacies so commonly made in evolutionary argument: the misuse of a *current utility* to infer an adaptive *origin*'.[87] Spandrels (more formally called exaptations) were key features of human evolution, they said, particularly when it came to cognition. For instance, the human brain reached its current complexity tens of thousands of years before the emergence of reading and writing, so the mental machinery for these capacities must have originated before the need for them. But they went further, saying that *most* mental capacities were spandrels. As Gould put it: 'In pure numbers, the spandrels overwhelm the adaptations.'[88] This allows room for culture and environmental influences and avoids the assumption that just because behaviour such as racism is widespread, it is primarily genetically rooted. The implication is that we can't assume that behaviour differences between population groups are 'hardwired'.

Contemporary neurological research has reinforced this direction of thinking by focusing on the 'plasticity' (flexibility) of the human brain: the brain's development is constantly mediated, and even sculpted, by its environment. This plasticity is most marked during childhood but is not confined to the early years; the adult brain also changes constantly in response to environmental stimulus. This was aptly expressed by the cognitive psychologist Geoff Bunn: 'It seems the cerebral cortex is hardwired to be culturally constructed,' he said, adding, 'If there is one organ of the body that demonstrates that biology is not destiny, it is the sentient brain.'[89]

WHY SOME PEOPLE ARE RACIST AND OTHERS AREN'T

If racial distinctions go beyond skin colour and relate to brain type, it would follow that we would be naturally conscious of them. Such was the view of the historians who provided the ballast for apartheid, including Gustav Preller, who wrote in 1938:

> Science is only now gradually discovering the remarkable physiological differences between the brain of the white man of European descent and that of the Bantu – differences which are innate and constitute the measure of their respective intellectual capacities; but it is a striking fact that the Boers of a hundred years ago were aware of these natural differences.

He added that the 'ganglion cells of the native's brain remain undeveloped' and his intellectual progress ceased at puberty 'as if it were not capable of further development'.[90]

No one talks of 'ganglion cells' in relation to race any more but as seen from the example of E. O. Wilson, genetic determinists tend to view racism in biological terms. Dawkins rejects race as a scientific concept but has a different take on racism. In *The Selfish Gene*, he borrowed the kin selection theories of his hero, W. D. Hamilton, in proposing that 'racial prejudice could be interpreted as an irrational generalisation of a kin-selected tendency to identify with individuals physically resembling oneself and to be nasty to individuals different in appearance.'[91] He argues we evolved to notice physical difference: '[R]acial classification is not totally uninformative but what does it inform about? About things like eye shape and hair curliness. For some reason it seems to be the superficial, external, trivial characteristics that are correlated with race – perhaps especially facial characteristics.'[92] Like Morris and Wilson, Dawkins adds that we are genetically predisposed to make distinctions between in-group and out-group: 'Perhaps there are special reasons for a disproportion-ate amount of variation in those very genes that make it easy for us to notice variation and to distinguish our kind from the other.'[93] In other words, we evolved to be aware of racial difference and to act on it. Dawkins's error lies in inventing a biological case for something that does not require it. People are not racist because of a genetic propensity to notice eye shape or hair curliness, nor because they have an innate tendency to be nasty to people who don't look like them. They may, however, be racist because they ascribe certain connotations to eye shape or hair or skin colour; connotations that are absorbed from the environment they move in.

We are capable of picking up racial cues even before we are capable of speech. One study found that while newborn babies had no preference for looking at black or white faces, by the age of three months they were more inclined to look at same-race faces than at other-race faces.[94] This implies nothing; it relates to babies becoming familiar with the faces they see most often. Another study found that three-to-four-month-olds who spent more time with their mothers preferred to look at female faces, while those who spent more time with their fathers preferred male faces.[95] In other words, babies are attracted to faces that are familiar. It's no more profound than that.

This consciousness is consolidated in childhood; observation of implicit adult responses is more significant than explicit responses. In one study, preschool children were shown short films featuring a white actor ('Gaspare') interacting with a black actor ('Abdul'). One shows Gaspare formally shaking Abdul's hand while saying he's not prejudiced and would be happy for Abdul to live in his city, but Gaspare avoids eye contact and sits two seats away from Abdul. In another film, Gaspare talks only about his work but looks in Abdul's eyes, shakes his hand enthusiastically and leans towards him. The children were asked how they felt about Abdul. Those who observed the positive body language were far more likely to have positive responses.[96] Reinforcing this, a study of mothers and young children, using what is known as the Implicit Association Test, showed that the stronger the mother's implicit negative responses to black people, the less likely it was that their child would choose a black child to play with and to consider a black child in a positive light. Consciously expressed parental attitudes to race made less difference.[97]

Children pick up cues, starting with faces and going on to absorb the implicit attitudes of their parents. These can be hard to dislodge, because even when explicit attitudes change, implicit attitudes can be passed on to succeeding generations by more subtle signals. I have noticed the marked distinction in the way white South Africans now talk about race compared with talk in the apartheid era. In private, as in public, racist attitudes are more muted, and many white people have had profound changes of mind. But it is also notable that South Africans, both black and white, use more racial adjectives than people in, say, Britain. A person is not simply a person but is a black person, white person, Coloured person, Indian person. Hardly surprisingly, a heightened colour consciousness persists in what was once euphemistically called 'The Rainbow Nation'. This may take a while longer to shift.

But just as Alfred Russel Wallace rose above the assumptions of his era by immersing himself in indigenous cultures, so it has been with many South Africans who were breast-fed on racism. Through working with black people, being exposed to non-racial ideas and seeing black people in positions of power and prestige, most South Africans have changed both their use of language and their attitude. It is possible that as a result of early childhood experience, subliminal over-consciousness of race persists but we need not despair. Attitudes that are culturally rooted, rather than fixed by biology, can be changed.

6

ARE RACE GROUPS REAL?

In 1955, in the dark days of apartheid, in the ultra-conservative South African town of Piet Retief, a daughter was born to the white Laing family. After the initial moment of delight there was a terrible shock for the new parents: the baby they named Sandra was what was called a 'throwback': a brown-skinned, kinky-haired girl.

South Africa's 1927 Immorality Act had made it a crime for black people to have sex with white people. After coming to power in 1948 the apartheid government tightened the law, making it a crime for any non-white person to have sex with any white person. So, cases such as Sandra's were becoming rare by the time she was born. It was therefore assumed that the real reason for Sandra's colour was that her mother, Sannie, had slept with a black man. The locals muttered and shunned, causing huge anguish in the Laing family because, they insisted, there had been no straying with anyone at all, let alone with one of *them*.

Sandra's apartheid-backing, church-going Afrikaner parents, grandparents, great-grandparents and older brother were clearly white, so there was no indication of where this 'throwback' started. In their desperation to prove it must be so, her mum and dad signed an affidavit affirming that they were her natural parents. When that didn't cut it, her father, Abraham, took a paternity test to show he was indeed Sandra's biological father. Waving the proof, he pleaded: 'If her appearance is due to some coloured blood in either of us ... then it must be very far back among our forebears and neither of us is aware of it.' This was not good enough for a community that knew a *Kleurling* (Coloured) when it saw one. Sandra was horribly bullied and told she was 'dirty'. Other children refused to drink from the water fountain when she'd used it. Her parents were shunned, because of the stubborn belief that whatever the paternity test said, Sannie *must* have had sex with a *kaffir*.

When Sandra was ten, she was given the 'pencil test' at school by two police officers. This involved sliding a pencil into her hair. If it fell, she would pass, and could be classified as European; if it stuck, the hair was held to be kinky and she would be defined as 'Coloured'. It stuck. She was immediately expelled from school, escorted by two policemen. She was redefined as 'Coloured'; her parents' legal bid to overturn this ruling failed.

The only children who'd play with Sandra were the sons and daughters of black workers. This didn't change even when, as a result of all the publicity, the Minister of Home Affairs intervened to reclassify her again, as 'white'. Despite this, nine white schools refused to have her, so she was sent to a Coloured convent boarding school. She sought affection wherever she could find it; when she was sixteen, she eloped to Swaziland with Petrus, a Zulu-speaking vegetable vendor, but was caught and jailed. After her release, her father threatened to kill her and to shoot Petrus on sight, so she fled again. She went on to have six children; to keep them, she had to be reclassified as 'Coloured'. Sandra lost contact with her family and although she was eventually reconciled with her mother, her brothers cut her out.[1]

In the South Africa of Sandra Laing's first thirty-nine years, the idea that race wasn't real, or merely reflected superficial differences, would have seemed ridiculous to most whites and many blacks too. Your position in life was defined by your racial classification, which could be decided by criteria as random as the fate of a pencil, and further refined by ethnic, linguistic or tribal subdivisions. These definitions governed your existence: where you lived, how you lived, who you lived with, your human rights, your friends, your neighbours, your teachers, your leisure opportunities, your job prospects, your access to public health, your pension, your life expectancy, the range of laws you had to obey and the punishments meted out to you if you broke those laws; longer sentences, worse food, more prison overcrowding and more whippings and hangings for blacks. The implications of racial definition were huge.

Apartheid South Africa was not alone in enforcing deep racial divisions. Throughout Africa colonial powers parcelled land, power and wealth on a racial basis, with the colonists taking the best bits for themselves. The relationship between white settlers and black indigenous populations in colonial Kenya or Rhodesia was not a long way from that in South Africa, including the propensity to brutality when the rulers were challenged: the British were torturing and castrating Kenyan Mau Mau prisoners as late as 1960. The gap between wealth and poverty, land and landlessness, power and powerlessness, was similar. In slightly different forms, the same could be said of French, Portuguese, Belgian and German colonies, which were often even more barbaric for the colonised.

Colonialism was all about the desire for land, labour, produce and raw materials but the business of subjugating colonised people required a particular mindset. Part of its rationale – the conviction that Africans and Asians evolved to have inferior brains – came from race science imported from Europe and America; behind this colonial and apartheid mentality lay a belief that race was far more than skin-deep, and that natural selection had endowed different races with different levels of intelligence and different characters.

WHAT IS RACE?

Today, most biologists and anthropologists regard race as mainly a social construct. By this they mean that what we define as 'races' reflect the criteria people consider significant in distinguishing between populations, rather than profound biological distinctions. Those criteria usually relate to things we can spot immediately – skin colour, facial features, eyelid form and hair type – but they are not differences that predict character, intelligence or any other abilities and liabilities. At the turn of the millennium, the geneticist Steve Jones wrote, hopefully: 'At last there is a real understanding of race and the ancient and disreputable idea that the peoples of the world are divided into biologically distinct units has gone forever.'[2] Perhaps gone forever among academic experts but as we shall see later this disreputable idea has made a concerted comeback among the bloggers and vloggers of the alt-right, and among the psychologists and journalists who feed them.

The decoding of the human genome, the complete set of genes that hold all a person's genetic information, undermined the idea that the world is divided into biologically distinct races. We now know that just about every population group has mixed with others throughout recorded history, and unrecorded prehistory; very few have been cut off from this gene flow for very long. While we might still use the word 'race' as a linguistic resort, it is not much more. Geneticists prefer to focus on ancestry and to talk of population groups, which are smaller units that cut across traditional notions of race. This view was bolstered by research conducted by the Human Genome Diversity Project on fifty-three genetically distinct indigenous population groups. It emerged that one of the remarkable conclusions is how little genetic variation there is in humans, compared to other species of mammal, partly because of our relatively recent evolution. As Richard Dawkins puts it: 'We are indeed a very uniform species if you count the totality of genes or you take a truly random sample of genes.'[3]

THREE PRIMARY RACE GROUPS?

Some divide the world into three primary races, reflecting the separation of populations thousands of years ago: Negroid, Mongoloid and Caucasoid. In his race science book, *A Troublesome Inheritance*, the journalist Nicholas Wade argued that these groups had a firm biological basis. The geneticist Mark Stoneking, from Leipzig's Max Planck Institute, disputed this:

> How to define the concept of race biologically is not easy but to me one prediction is that not only should one be able to define discrete clusters of people that correspond to races,

there should be distinct boundaries between them – and if you look at patterns of genetic variation in human populations, you find they are distributed along geographic 'clines' with no distinct boundaries.[4]

Race groups have traditionally been categorised in terms of appearances but even these distinctions look shaky under interrogation, because the boundaries between where one starts and another ends are blurred. Even if we restrict ourselves to the superficial signifiers of skin colour, eyelid type and nose width, we run into problems. One is that, as the evolutionary biologist Paul Ehrlich puts it, 'patterns in one characteristic (say, skin colour) are not congruent with those in another (say, nose width).'[5] For example, start with what would seem to be an easy distinction: the epicanthic fold that is a feature of East Asian eyes. That might seem like a clear defining feature until you consider it is also common among the San people of Southern Africa and some Europeans.

When racists talk about race, they invariably have skin colour in mind: blacks, whites, 'yellow people' and even 'red Indians'. But colour is hardly the most significant genetic difference between human populations. Even in terms of appearance it is a fuzzy signifier: Caucasoids range from pale, freckled Scots to dark-skinned Indians, Negroids from the yellowish-brown skins of the San people to the nearly black skins of some West Africans. Even within West Africa there is a significant range, from the dark-skinned Nigerian Yoruba people to the light-skinned Igbos. Mongoloids range from the lighter-skinned Chinese and Japanese to darker-skinned Polynesians and Filipinos. Categorising races by skin colour becomes even more complicated when we factor in culturally based definitional differences. In the United States, anyone with even a little bit of West African genetic inheritance tends to be defined as black or African-American. This, like so many things, is a residue of slavery. President Jefferson's 'mistress' Sally Hemings was a slave, and therefore defined as black, even though 75 per cent of her genetic inheritance came from European slave-owners. This 'touch of the tar brush' definition persisted after the abolition of slavery and was absorbed by African Americans.

The West African genetic inheritance of African Americans is said to average between 70 and 80 per cent but it varies from state to state and includes people whose African genetic heritage is less than Sally Hemings's 25 per cent. Some are people who might look white and be defined as such in other countries. It includes others who might be defined as 'mixed-race' in Britain or 'mulatto' in the former Spanish and Portuguese colonies, where only dark-skinned people are called black. If 'black' people can be light brown and 'Caucasian' people can be dark brown, skin colour hardly offers a suitable basis for definition.

There might be some purpose to definition when it comes to policing, and to state statistical data and the social policies those data might inform. Population breakdowns on racial lines can be useful in defining disadvantage: why, for example, do white working-class boys in the UK consistently achieve the lowest exam grades compared to the performance of children in other ethnic groups who also receive free school meals (a marker of low family income in the UK)? Why is the arrest and imprisonment rate for African-American men so much higher than for Caucasian Americans?

But such racial categories are not intended to be genetically cohesive. The American racial category 'Hispanic' includes people with widely varying proportions of Native American, European and African genes; some light, some dark. 'Hispanic' may be a convenient social category but it tells us very little about the ancestry of Hispanics.

Ehrlich produced four world maps showing the distribution of skin colour, hair structure, average height and head shape. What is remarkable is how little they correlate. 'Races in people', he says, 'are arbitrarily defined entities'. He adds that they are particularly arbitrary when it comes to skin colour: 'Attempts to treat divisions of humanity based primarily on skin colour as natural evolutionary units have always been and still are, nonsensical.'[6]

WHY DID PALE SKIN, HAIR AND EYES EVOLVE?

If we want to explain the colour range of modern humanity the obvious starting point is gene swapping. But if we go further back, to the first 200,000-plus years of human history, when we all lived in Africa, was everyone black? Until recently this was assumed to be the case. Paler skin absorbs more of the energy from sunlight, which supports the synthesis of vitamin D by the body. When humans settled in cold European regions, with less sunlight, those with darker skins were more prone to diseases caused by low vitamin D, such as rickets, and therefore in prehistoric times were less likely to produce lots of children. Blond hair and blue eyes were thought to have similar origins, although some have suggested these were selected for because they were innately attractive. But a combination of archaeology and genetics has rubbished the attractiveness case and made the explanation for the evolution of lighter skin colour significantly more complex.

The first challenge to the prevailing theory came in 2006, when scientists analysed the DNA of a man who lived 7,000 years ago in the mountains of north-west Spain. They found he had blue eyes, dark hair and significantly darker skin than modern Europeans. DNA from the skeletons of other Western European hunter-gatherers, dating from between 5,000 and 9,000 years ago, produced similar results. 'Prehistoric Europeans in the region we studied would have been consistently darker than their descendants today,' said the palaeo-geneticist Sandra Wilde, the first author of a paper on the subject.[7] A revised theory emerged: that in cold northern climates there was enough sunlight to support sufficient synthesis of vitamin D if combined with the food sources of a hunter-gatherer lifestyle, including regular consumption of fish, meaning there were no pressures for the evolution of lighter skin. Only when agriculture was embraced came the adaptation towards lighter skin: a less fishy diet required greater synthesis of vitamin D.

However, a recent analysis of the DNA of 1,570 people from ten ethnic groups in Ethiopia, Tanzania and Botswana, carried out by a team headed by Sarah Tishkoff,[8] a geneticist from the University of Pennsylvania, suggested the explanation is not quite so simple. The team measured the amount of dark melanin on the inner arms of these African volunteers and then analysed more than four million spots on their genomes to find variations linked to skin colour. They homed in on six genes implicated in skin colour that accounted for 29 per cent of the variation among the volunteers. They found that most of the variants, for both light and dark skin, came from an ancient African past, all the way back to *Homo erectus*. And in some cases, the older variant was for paler, rather than darker skin.

One theory is that our earliest ancestors had pale skin, like most primates, but as they moved out of the forests and lost their hair, they evolved darker skin. Complicating the picture is the variation of skin colour in sub-Saharan Africa. Tishkoff's team found that the Nilotic people in East Africa had some of the darkest skins, while the Southern African San had some of the lightest, because they lacked a dark skin gene called MFSD12, yet both groups had lived in hot climates for many generations. They also found that MC1R, one of the light skin colour genes that spread among European populations, was common in Africa. What clearly emerges is that the genes for light skin colour originated in Africa, even if, in the cold of Europe, they were more widely selected for over time.[9]

The recent decoding of the whole genome of the skeleton of 'Cheddar Man' raised the 'why' question once again. Cheddar Man was an ancient Briton who met a violent end more than 9,000 years ago; his remains were found in cave in Cheddar Gorge, Somerset in 1903. He too was found to have had brown skin, dark curly hair and blue or green eyes. He was also lactose-intolerant. Why did a man who had to survive cold winters have a brown skin? And why did Brits evolve from having brown skins to having fair skins?

The answer to the first question may relate less to the vitamin-D-synthesising plus hunter-gatherer diet, and more to the way natural selection works. Natural selection is not a teleological process; it has no set direction or destiny. The environment can't conjure up genes that don't already exist in a population; natural selection can only work on what is available. It is possible that the dark-skinned Western European hunter-gatherer population did not contain the alleles for lighter skin colour. And mutation occurs randomly; with respect to skin colour, mutation cannot be prompted by cold, heat or anything else.

The answer to the second question is that Cheddar Man's descendants didn't evolve to have lighter skin. Rather, the genetic line of Cheddar Man, and of the similarly dark-skinned continental European hunter-gatherers, fizzled out and was replaced by that of the light-skinned descendants of the people of the Russian steppe around 4,500 years ago. Less than 10 per cent of the DNA of modern Europeans comes from the hunter-gatherers of that time. Why the new lot had light skin might relate to selection for genes that allow vitamin D synthesis in cold climates, but it might well have been natural selection acting on existing genes; in all probability, genes originating in Africa. It is worth remembering that natural selection is not the only evolutionary force. There is also genetic drift; 'neutral' evolution that usually occurs in isolated communities. This could be another reason why the people from the Russian steppe had light skin and brown eyes, while Western European hunter-gatherers had dark skin and blue eyes.

When it comes to skin colour, full genome DNA analysis suggests there were at least three variants in the Europe of 5,000-plus years ago: the brown skins, dark curly hair and blue or green eyes of the Western European hunter-gatherers such as Cheddar Man, the lighter skins, brown eyes and dark hair of the first farmers who migrated from Anatolia, and the pale skins, brown eyes and mainly dark (but also blond) hair of the pastoralists from the Russian steppe. The latter two groups largely replaced the dark-skinned hunter-gatherers, although blue eyes might be one hunter-gatherer genetic trait that has survived. The reason so many Europeans today have light skin therefore relates to the extent of the migration and swamping of darker people bearing ancient north Eurasian ancestry.[10]

The discovery that Cheddar Man and other European hunter-gatherers had blue eyes raises doubts of previous claims that this eye colour came from a single European ancestor who lived between 6,000 and 10,000 years ago; rather, it suggests an older origin. While one of the genes relating to eye colour is associated with melanin-producing cells, there is no clear evolutionary reason why it evolved among dark-skinned European hunter-gatherers and not among light-skinned European farmers and herders. This raises the possibility that it evolved through genetic drift rather than through natural selection.

Blond and red hair are closely associated with light skin, and therefore might have been selected for in Europe for the same reason that light skin, low in melanin, allows for more efficient synthesis of vitamin D. This would have given blond or red-haired people, who also had light skins, a selective advantage in cold conditions, and might help explain why genes for blond hair are most common in colder regions such as Scandinavia. Japanese research in 2006 found that the genetic mutation that prompted the evolution of blond hair dates to the ice age that happened around 11,000 years ago. Since then, the 17,000-year-old remains of a blond-haired North Eurasian hunter-gatherer have been found in eastern Siberia, suggesting an earlier origin. And we now know that there are several genes associated with blond hair, which means it could have evolved in different places independently, perhaps related to genetic drift. This may be why between five and 10 per cent of the dark-skinned people in the Solomon Islands, east of Papua New Guinea, are blond; it's all due to a single gene, one not shared by European-origin blonds.

Red hair is thought to have evolved at least 20,000 years ago. It was also found among the Neanderthals, although prompted by a different gene. It is most prevalent today in Ireland, where around 35 per cent of people carry the red hair allele and 10 per cent actually have red hair. It is also widespread in Scotland, among European Jews, in other clusters on the European continent and in parts of the world where Europeans migrated, with little pockets elsewhere, including in China. This might suggest that it took root in small founder communities and spread (in other words, genetic drift). It might also suggest that red hair, like blond hair, has different genetic roots in different areas. But whatever the evolutionary causes of blond and red hair, their spread in Europe had little to do with their possible innate attractiveness and much to do with the success of the all-conquering herders from the steppes who carried these genes.

FROM RACISM TO COLOURISM

Lighter skin, hair and eyes have something to do with cold climates but the relationship is less direct than is often assumed. Had it not been for the success of Anatolian farmers who spread north, and the migrations of the warlike herders of the Yamnaya, Corded Ware and Bell Beaker cultures from the Russian steppe, Western Europeans could still be dark-skinned. We have only to look at the Inuit people, who have lived in the Arctic for 5,000 years yet retain brown skins, to appreciate that there's nothing inevitable about a link between cold climates and fair skin. And as the desert people of North Africa and the Southern African San illustrate, there's also nothing inevitable about hot climates and dark skin. Yet skin colour has taken on massive proportions in the history of the last few thousand years, with lighter people invariably higher in the social and economic pecking order than darker people.

Racial prejudice is routinely blamed on colonialism and slavery. When people of one colour conquer those of another, they invariably see the conquered as lesser beings, and skin colour is part of that package of perceived inferiority. When they enslave people of another skin colour, they tend to regard the enslaved as less than human, and any differences in skin colour are sure to be part of this perception, although it is worth noting that slavery did not only involve lighter-skinned people enslaving darker-skinned people.[11]

The decoding of DNA from the bones and mummified bodies of ancient Eurasians suggests that colour prejudice may have deeper roots. We've already seen how lighter-skinned European farmers and later, even paler-skinned herders, saw off the descendants of dark-skinned hunter-gatherers throughout Western Europe. A similar process happened in India, where the indigenous people of northern India were first swamped by the mass migration of farmers from Iran from about 9,000 years ago, and then by the arrival of the Yamnaya culture from the Russian steppe, 5,000 years ago. These groups introduced Indo-European languages and also the caste system, although there is a view that this was only systematically practised after Britain colonised India three hundred years ago. The castes correspond to distinct genetic patterns that have persisted over time because of limited interbreeding, with the elite, priestly Brahmins having higher levels of ancestry from the light-skinned people of the Russian steppe.[12] It's likely that people with light-skinned genetic heritage looked down on those from lower castes who had the darker skins that were a legacy from deep in India's past, and that these prejudices were exacerbated by British colonialists, who probably couldn't believe their luck when they arrived and set about strengthening and reinforcing the caste system.

In both Europe and India, prejudices perpetuated by colonialism might well have had deeper cultural roots. Certainly, these colour-based hierarchies persisted after slavery was abolished in the nineteenth century and after colonialism ended late in the twentieth, and were absorbed by the colonised and by former slaves, because this reflected their life experience. Under apartheid in South Africa, and under segregation in the United States, it was not uncommon for lighter-skinned people to 'try for white'. More generally, light-skinned people, or 'mulattos' in the former Spanish colonies, had more privileges and faced less hostility.

In much of the world 'colourism' has persisted, often assuming a sex-based complexion. The idea often reflected in nineteenth-century literature, of the dark and handsome man pursuing a fair, pale-skinned beauty, was a resilient staple of the colonial era. This too has deep roots. Male colonisers and slave plantation owners were ruddy outdoor types while, until the popularisation of sunbathing in the 1950s, the ideal 'lady' retained an alabaster skin unkissed by the sun. Unfortunately, the pursuit of white skin, which until the early twentieth century often included the use of lead-based cosmetics, killed some women, shortened others' lives and caused all sorts of ailments.[13]

In many parts of the world this feminine ideal was accepted by the victims of this prejudices. Ifemelu, the blogging USA-based Nigerian heroine of Chimamanda Ngozi Adichie's novel *Americanah*, complains that successful American black men have white wives, and those who don't have light-skinned black wives. 'And this is the reason dark women love Barack Obama,' she writes. 'He broke the mold! He married one of their own.'[14] She also writes of the pressure on African-American women to have 'relaxed', hanging hair,[15] a pressure not faced by African-American men. Long, straight, hanging hair, is, of course, a white feminine ideal. Even today, most black female film stars do not have 'natural' hair and are light-skinned. This is even more pronounced in Bollywood, where the idea that men should seek out lighter-skinned women stubbornly persists and is seldom challenged. I could cite similar examples from Britain, Japan and Korea and many African countries, where a brisk business in skin-lightening creams, mainly used by women, continues.

Mutations involving a handful of genes have assumed an overwhelming significance in the human history of the past few thousand years, and particularly the past few hundred. These mutations led to traits far less significant than others that separated populations, such as susceptibility to certain diseases. And they are recent developments, at least in their European form, perhaps representing not much more than 3 or 4 per cent of modern human history. Skin colour has minimal intrinsic significance but has taken on immense cultural significance and will continue to do so for a long time to come.

Sarah Tishkoff, the American geneticist who led the team that analysed African DNA for skin colour variation, emphasised the anti-racist implications of their results. 'One of the traits that most people would associate with race – skin colour – is a terrible classifier,' she said after their pioneering paper was published. 'The study really discredits the idea of a biological construct of race. There are no discrete boundaries between groups that are consistent with biological markers.'[16]

DIFFERENCES WITHIN AND BETWEEN POPULATION GROUPS

The range of skin colour in sub-Saharan Africa illustrates the point that genetic variation within Africa, and particularly within long-settled African communities, is greater than that between the different population groups that are sometimes defined as races. The San people have a greater genetic range than, say, that between Germans and Italians. If we restrict ourselves to the most visible differences, Africa has the shortest and tallest populations in the world, the biggest and smallest heads, the thickest and thinnest lips and the widest and narrowest noses.[17]

An early stab at how this relates to genetics came from the Harvard population geneticist Richard Lewontin, who in 1972 published a paper on population-based variation in blood type. He looked at seven regional populations: Africans, West Europeans, East Asians, South Asians, Aboriginal Australians, Native Americans and Oceanians, and found that 85 per cent of the variation was found within those populations and just 15 per cent between them. 'Races and populations are remarkably similar to each other', he concluded, 'with the largest part by far of human variation being accounted for by differences between individuals.' He went on to say that racial classification had 'virtually no genetic or taxonomic significance'.[18]

Lewontin's conclusions are not in doubt. They were replicated by subsequent studies and led to a consolidated perspective that for most traits, there is so much overlap in human populations that it is impossible to find a single trait that definitely distinguishes from which group people come.[19] In other words, it is not useful to try to define 'race' in this way. Today, scientists therefore focus on ancestry and how that relates to population groups. A recent study analysed 377 variable positions on the human genome in relation to six American population groups. The conclusions were consistent with Lewontin's; the data showed significantly more variation within the groups than between them. But it also showed clusters of mutations that correlated to these population groups, helping to distinguish them.[20]

This is now the direction of travel for population geneticists. They acknowledge that most variation takes place within populations and that people carrying most traits can be found in any population, but they also look for areas of difference relating to genetic ancestry between the genomes of population groups. This information is useful in tracing the course of history and prehistory, and in providing a sharper picture of the distribution of human assets and liabilities. The old idea of three primary races is too broad a brush to cope with the amount of genetic variation in those groups and of genetic overlap between them. However, this doesn't close the case, because the fact that some populations were separated for tens of thousands of years means there are indeed genetic differences between them. And it turns out that these ancestral distinctions can sometimes be significant.

Some relate to adaptations to cope with particular climates or geographic surroundings. People native to Tibet have a genetic advantage over others when it comes to coping with altitude; an adaptation inherited from the Denisovans. Likewise, a recent study conducted among the Bajau tribe in Indonesia, famed for its members' ability to free-dive to depths of more than sixty metres for up to thirteen minutes while spear-fishing, found that as a result of a genetic adaptation relating to thyroid hormone levels their spleens were 50 per cent larger than those of their land-dwelling neighbours, the Saluan tribe. The Bajau's larger spleens help them stay underwater for longer.[21] Another adaptation is one that enables adults to digest milk (lactose tolerance). Ninety per cent of people in northern Europe have this adaptation but fewer than 5 per cent of people in parts of Africa and Asia. Genes have also been isolated that help explain why, on average, northern Europeans are taller than southern Europeans.

More important for humanity as a whole are differences in disease prevalence between groups, due to population genetics. Some distinctions relate to natural selection. People whose ancestral origin is from regions where malaria is prevalent, particularly sub-Saharan Africa, are more likely to have sickle cell anaemia. Inheriting one allele for this syndrome protects against malaria but having two causes sickle cell disease. Because malaria killed so many people, there was selection for the protection-against-malaria trait, even though there was a risk of sickle cell disease if both parents passed on the allele.

However, it is worth emphasising that natural selection was not a factor in the spread of most ethnic diseases. Many evolved through genetic drift,[22] taking root during population bottlenecks and then spreading more widely.[23] This 'founder effect', sometimes traced to a single person, has been cited for the higher than average prevalence of a number of ethnically influenced diseases, including Tay–Sachs and Gaucher's disease among Ashkenazi Jews, cystic fibrosis in northern Europe, type 1 diabetes in Sardinia, asthma in Tristan da Cunha, Meckel syndrome in Finland[24] and schizophrenia in Palau, Micronesia.[25]

In some of these diseases, there are significant differences in prevalence among the population as a whole. Genetic research conducted by David Reich's laboratory has shown that self-identified African Americans are 1.7 times more likely to have prostate cancer than European Americans,[26] and they are also more likely to have end-stage kidney disease.[27] On the other hand, European Americans have a higher risk of developing multiple sclerosis than Asian Americans or Hispanic Americans.[28] People of Caribbean and African descent have higher genetically prompted rates of hypertension and stroke than Caucasians, but Caucasians have higher rates of coronary heart disease.[29]

Where does that leave us on the question I posed at the beginning of this chapter: are race groups real? Skin colour, hair type, nose shape or eyelid form might help us to identify people, and they help states make social classifications. They are also real from a negative perspective; if people believe in the significance of race and act on those beliefs, this will have consequences for people, such as Sandra Laing, who are at the wrong end of those beliefs. But the broad categories of race don't tell us very much else that is important.

Even if we move away from race and narrow our gaze to smaller population groups, this will not be enough to tell us whether someone has sickle cell anaemia (one in 365 African Americans) or Gaucher's disease (one in 450 Ashkenazi Jews), or whether they are susceptible to high blood pressure, prostate cancer, multiple sclerosis, heart disease or various other diseases that show ethnic differences in prevalence. But then again, if I were a doctor carrying out tests for certain symptoms, I might be interested in my patient's ethnic origins, because it relates to a higher or lower risk factor for some diseases that can affect their lifestyles, longevity and mental health.

In trying to avoid the obvious errors of over-interpreting or stereotyping by racial definitions, we should not fall into the opposite pit of pretending that no relevant biological distinctions exist between populations, or that it's dangerous to allude to them and discussion should be silenced. To do so would be pointless, because genetic science relating to ancestry and human populations is moving at a very rapid clip and is not likely to be stalled by such sensibilities. Analysing the real differences between population groups regarding certain physical abilities and certain diseases that affect the body, and may also affect the mind, is really not a slippery slope. It is a vital asset to medical science and to discovering more about humanity.

7

CAN WHITE MEN JUMP AND
BLACK MEN SWIM?

Let's begin with nation X. It's not a country that devotes billions of dollars to sporting success, nor is it one with disproportionately drug-fuelled question marks. But, not long in the distant past, it emerged as an astonishing powerhouse of middle- and long-distance running. It started in the mid-1970s, when one of its men broke world records in the 2,000 and 3,000 metres, another shattered the 10,000-metre world record, a third broke the 5,000-metre world record and a fourth the world marathon record. Two of its men took turns in breaking the 800, 1,500 metres and mile world records, and gathered five Olympic medals. They were overtaken by an even faster fellow, who smashed world records in the 1,500 metres, 2,000 metres and the mile in a nineteen-day spell, as well as winning world championship gold. And by another who, after winning world and Olympic medals, was the highest-ranked 800-metre runner in the world.

There was a gap before nation X's women caught up but one of its runners went from winning the 10,000-metre Olympic silver to world championship gold, along with winning the world half-marathon title and three big city marathons. She was overtaken by the greatest-ever female long-distance runner, who twice won the world cross-country championships, the gold medal for the marathon in the world championships, won the world half-marathon championships three times, won seven big city marathons, and obliterated the world record twice. She still holds that record. Nation X's female middle-distance athletes kept pace; one doubled up, winning Olympic gold in both the 800 and 1,500 metres.

What was it about nation X that gave it the talent to have such consistently astonishing success? Why did its middle- and long-distance runners do so much better than other nations'? What made this group so fast, holding twenty-eight world records between them?[1] Surely, something in their genes? No one asks these questions, because nation X is the United Kingdom. Churchill, who was obsessed with race, called its citizens the Island Race but really, Brits are Europe's mongrels; the words 'race' or 'ethnic group' simply don't fit.

I asked my running club chairman, Jerry Odlin, a contemporary of Sebastian Coe and Steve Ovett in the 1980s, why British middle- and long-distance runners excelled in his era. He mentioned youth cross-country racing, very competitive then, less so now. 'That created a critical mass, so that we were all competing with each other and pushing each other,' said Jerry, who ran the 10,000 metres in twenty-eight minutes in the 1980s. 'If I did that time today, I'd be number eight in Britain. Back then I was number fifteen.'[2] The change, he suggested, came from the decline of competitive sport in state schools, a more sedentary culture in which fewer people walk and more play computer games, the increased availability of alternative sporting options and gym membership. Football has grown from dominant to hegemonic, and participation in smaller sports has grown, including some such as cycling, BMX, triathlon and boxing that require similar cardiovascular fitness to distance running.

More Brits are running than ever before. The growth has been most rapid among women generally and the over-thirty-fives specifically, but the elite standard has declined, even if a few outliers occasionally break records. Steve Jones's 2:07:13 was the UK marathon record for thirty-three years, until finally broken in 2018 by the Somali-born Londoner Mo Farah; Steve Cram's British 1,500-metre record stood for twenty-eight years, finally broken in 2013, also by Farah. Just as outstanding performance is fostered by cultural developments that encourage participation, together with elite training programmes – the rise of British cycling and rowing are other examples – so negative cultural developments contribute to decline. In neither case is it wise to assume that genetics plays an overwhelming role.

KENYAN DISTANCE RUNNERS AND THE NANDI PHENOMENON

This takes us to Kenya. It is widely assumed that its runners have genetic advantages. Attention focuses on the Kalenjin tribe and particularly a small area, the Nandi Hills in the Great Rift Valley, where fewer than 2 per cent of the population live. According to the South African sports medicine specialist Tim Noakes, a Nandi runner has, on average, twenty-three times more chance of succeeding than his average Kenyan counterpart.[3] Or, as the author Matthew Syed puts it:

> Far from being a 'black' phenomenon or an East African phenomenon or even a Kenyan phenomenon, distance running is actually a Nandi phenomenon ... Or, to put it another way, much of the 'black' distance running success is focused on the tiniest of pinpricks on the map of Africa, with the vast majority of the continent underrepresented.

The 'pinprick' has fascinated running enthusiasts for decades. There are few more adamant advocates of the biological case than the Kenya-based American journalist and runner, John H. Manners, who claimed that natural selection endowed the Kalenjin tribe, and the Nandi clan within it, with the perfect raw material. The Kalenjin represent less than one eighth of the population but comprise three quarters of the country's running elite. 'I contend that this record marks the greatest geographic concentration of achievement in the annals of sport,' he wrote.[4] How did this happen? The livelihood of the tribe was once bound up with long-distance cattle rustling, he noted. The quickest runners brought home the cows, had more wives and sired more children; the slowest were caught and killed. In this way, the genetic predilection to be able to run quickly was born. For good measure, he also threw in an odd nurture element; stoicism in the face of pain is highly valued, as in Kalenjin men's circumcision, which had to be endured without flinching. This helped them to develop their courageous attitude to running through pain barriers.

These ideas are entertaining but, really, quite silly. The fact that someone can persuade stolen cows to run faster doesn't imply a speedy genetic trait that can be passed on. Natural selection doesn't work this way. It's usually a slower process because there can only be selection for randomly produced mutations, which require relative genetic isolation and often take thousands of years to gel in a population. Also, the relationship between keeping still during the pain of adult circumcision (which is by no means unique to the Great Rift Valley) and maintaining your form when lactic acid is building up is rather obscure.

Noakes also veers towards the biological. He notes his research on black South African distance runners has prompted the conclusion that their superiority 'must be due to at least some genetic determinants'.[5] He speculates: 'This is perhaps due to an ability to sustain levels of skeletal muscle recruitment before developing fatigue. Black distance runners also seem to have more Type II (white) muscle fibres than their white counterparts.' Intriguingly, however, the distance runner who for him 'embodies athletic perfection'[6] is the white South African Bruce Fordyce (who won the 87-kilometre Comrades Marathon nine times in the 1980s, held the world record for the 100 kilometres and still holds it for fifty miles), both because he was 'genetically superior'[7] and because he trained properly and was strong-minded. Noakes also considered the explanations for Kenyan success – long thin legs, stretchable hearts, different muscle fibres and cattle-raiding genes – and decided that 'the ultimate explanation might be that all of these factors are contributory but all are under the control of a single regulatory mechanism, the brain.'[8] Later he dropped the heart explanation, noting there was no evidence that the Kenyans had either a superior cardiovascular function or a greater capacity to transport oxygen to their muscles. Instead, he said, it 'seems' their success is due to a 'multitude of physiological factors' that allow their brains to 'drive them to higher levels of skeletal muscle recruitment'. In future he would focus on legs: 'To run so fast, the Kenyans' muscles must be incredibly powerful – a point that has been fastidiously ignored.'[9]

It is hard to know what to make of all this because the evidence is lacking. Noakes's use of qualifying words such as 'seems', 'seem to', 'must be', 'perhaps' for his claims suggests a gulf between speculation and evidence and his point about leg strength seems, well, rather obvious, which is no doubt why it is ignored. Is he really suggesting Kenya's elite has something extra in the leg department that Mo Farah and Steve Jones lack?

The first blow to the genetic view came on the podium: the Ethiopians arrived. No Kenyans could quite match the record-breaking heroics of Haile Gebrselassie and Kenenisa Bekele on the track. They were further challenged by Eritreans, Ugandans and Tanzanians in long-distance events, and by Moroccans and Algerians over the middle distances. The era of Kenya-worship seemed to have passed in the 10,000-metre final in the 2012 Olympics, when a Somali-born Brit took gold and was followed in by a white American and an Ethiopian. Or perhaps it ended when the Brit, Farah, went on to win the 5,000 metres, and repeated the double win at the 2013 and 2015 World Championships and 2016 Olympics. However, the heroics of the marathon world record holder, Eliud Kipchoge, helped restore the balance.

The decisive blow came from the research conducted by Yannis Pitsiladis, a Greek academic based at the Institute of Cardiovascular and Medical Sciences, University of Glasgow. He patrolled the Great Rift Valley and Ethiopian highlands with his test tubes and swabs, collecting DNA samples from leading athletes past and present. His findings kicked away the underpinnings of the Nandi-biological explanation. This required the Nandi to be genetically isolated; without that, they couldn't develop the distinctive features that gave them the edge in running. Pitsiladis's research showed the Nandi population was genetically diverse and had never been isolated. His analysis of Ethiopian DNA showed the same thing: great genetic diversity. He also showed that the Ethiopians and Kenyans were far apart in terms of key markers, scotching the idea of a shared East African genetic inheritance for running. 'The more we have studied the phenomenon, the more we have realised that the patterns of success are not genetic, despite being specific to certain populations,'

Pitsiladis said. '[W]e can already say with reasonable confidence that social and economic factors are the primary factors driving the success of Kenyan distance running.'[10]

Even before the decline of the biological case, attention was shifting to environmental considerations: the impact on distance runners of living at the perfect altitude (2,100–2,400 metres), the active lifestyle of young people, many of whom run to and from school, the perception, ever since Kip Keino won Olympic gold in 1968 and 1972, that running is a way out of poverty and a source of adulation, the critical mass of top-level runners, whose talents are forged by training groups that work at high altitude and by academies that transform them into world-class athletes. Several of these factors, including altitude, also apply to the Ethiopians.

And, of course, there are the drugs. Until very recently, there was minimal testing in Kenya and evidence of considerable corruption in the system, which meant that Kenyan runners could dope with impunity. In 2015, the World Anti-Doping Agency (WADA), the international body tasked with rooting out drugs cheating, declared Kenya 'non-compliant'. Athletics' governing body, the International Association of Athletics Federations (IAAF), took action, suspending officials and demanding compliance. Threats and pressure finally brought a modicum of order to Kenya's anti-doping regime, after which several international athletes who had used the country as a training base moved to Ethiopia, which was even more lax than Kenya. The suspicion remains that at least some of the outstanding success of Kenyan and Ethiopian runners is drug-fuelled.

This doesn't imply that genetics plays no role at the population level. Slow-twitch muscle fibres, longer legs and lighter bones provide a significant advantage, and these are in higher supply in Kenya, Ethiopia and Morocco than in, say, Samoa, Nigeria and Jamaica. And when these long legs and light bones combine with high altitude and a running-to-school lifestyle, potential kicks in. A Swedish exercise physiologist, Bengt Saltin, compared Swedish distance athletes with highland Kenyan runners in the 1990s and found they had identical muscle make-up. It also emerged that sedentary Kenyans had no aerobic capacity advantage over sedentary Swedes. But when considering why Kenyan teenagers were consistently leaving his elite Swedish adult runners behind (he estimated at least five hundred Kenyan schoolboys could outpace his men), he came up with an intriguing insight: Kenyan runners had more blood capillaries surrounding their muscle fibres and these contained more mitochondria, which help to generate energy. And after intense exercise sessions, the Kenyans' muscles contained less ammonium lactate (which is associated with fatigue), meaning their bodies were more efficient in using fat as fuel. However, Saltin suspected this wasn't genetic, suggesting it was a result of running at altitude throughout their lives and training more intensively. In support of this view, he found that Kenyan children who ran to school had a VO_2 max (maximum oxygen intake) 30 per cent higher than those who did not.[11]

The epidemiologist Robert Scott, of the University of Cambridge, who has studied the DNA of Kenyan runners since 2004, also stresses environmental factors but raises an additional biological consideration: genetic variation among East Africans is far greater than among white Europeans.[12] This relates to a point I made in earlier chapters: there is far more genetic variability in sub-Saharan Africa than in the rest of the world, because only small groups migrated from Africa tens of thousands of years ago. In any sub-Saharan African population, the genetic range will therefore be wider than among people with European ancestral heritage. A population with greater genetic diversity would be more likely to have genes favourable for distance running; more people with slight physiques, light bones and long legs. And also more with physiques wholly ill-suited to distance running. Then again, the body types needed for record-breaking are not confined to East Africa, as Ovett, Coe, Cram, Jones and the rest showed in the 1980s and Paula Radcliffe twenty years later.

WHITE MEN CAN'T SPRINT?

Since the inaugural event in 1983, every 100-metre gold medallist at the World Athletics Championship has been black. More than 93 per cent of the occasions that sprinters have dipped below the ten-second mark have involved athletes of West African genetic descent, and they are also responsible for 83 per cent of the 200-metre times below twenty seconds.

People with West African ancestry have, on average, bigger gluteus muscles, heavier bones (which are less vulnerable to injury) and less fat beneath the skin, increasing the potential for more of the fast-twitch muscle fibre that is essential for sprinting. Referring to fast-twitch fibres and world-class sprinters, the Swedish exercise physiologist Bengt Saltin said: 'If you don't have at least 70 to 80 per cent fast-twitch muscle fibres, I'd say it's unlikely you could be among them. But if you have that kind of level, you could probably do well – and if you have 80 to 90 per cent that's even better'.[13]

A study of sprinters by the Neuromuscular Research Centre at the University of Sydney took this a step further in 2003, isolating a gene, ACTN3, that promoted fast-twitch muscle development. They also discovered the 'sprint version' of this gene was more common in people of West African origin; 98 per cent of black Jamaicans had at least one copy of ACTN3 compared with 82 per cent of people of European origin. If the two populations were the same size, the curve of their performance potential would be heavily skewed in favour of Jamaicans but population statistics redress the balance, because there are 742 million Europeans and just 2.9 million Jamaicans. Subsequent research found that a slightly higher proportion of Kenyans than Jamaicans had the sprint version of ACTN3 and yet the Kenyans have had no international success in sprinting.

Nevertheless, this news prompted one of the greatest sprinters of all time, the American Michael Johnson, to propose that Jamaican and American dominance reflected the fact that their sprinters were the descendants of slaves, who had inherited this 'superior athletics gene'. The hardiest survived hazardous voyages, and slave owners bred slaves for strength. 'It is a taboo subject in the States,' said Johnson, 'but it is what it is. Why shouldn't we discuss it?' The Jamaican Usain Bolt, the fastest sprinter of all time, agreed that this brutal heritage was relevant. 'It's a background from slavery,' he said.

'The guys back in the day were so strong from physical work ... The genes are really strong.'[14]

Bolt's Lamarckian argument implies that hard work helps create strong genes, which is not how evolution works. Johnson's contention is barely more realistic, because it implies there was selection or breeding for the ability to run fast, which affected the entire gene pool. Slave owners might indeed have picked strong-looking slaves but that's not the same as selecting fast-twitch muscles. Being big doesn't improve your chances of surviving for months aboard slave ships, and it doesn't make you fast. Elite male sprinters include slight men, such as the former 100-metre World Champion Kim Collins (1.77 metres, 77 kilograms), bulky men such as the former 200-metre World Indoor Champion John Regis (1.8 metres, 98 kilograms) and the long, lean Bolt (1.95 metres, 94 kilograms). Elite women range from the 1.52 metre-tall Shelly-Ann Fraser-Pryce, former world and Olympic 100-metre and 200-metre champion, to the 1.78 metre-tall Marion Jones, the former Olympic champion, who was stripped of her medals after admitting drug use.

When the Cambridge epidemiologist Robert Scott studied Jamaican athletes, he found that ACTN3 was absent in two of the country's leading sprint stars.[15] Daniel MacArthur, one of the Australian geneticists who found the ACTN3 gene in 2003, poured further cold water on the science behind Michael Johnson's claims, saying that ACTN3 played a 'pretty small role' and that there was no clear relationship between its frequency and Jamaican sprinting success. 'It is almost certainly true that Usain Bolt carries at least one of the "sprint" variants of the ACTN3 gene,' he wrote. 'But then so do I – along with around five billion other humans world-wide. That doesn't mean you'll see me in the 100-metre final ... Unfortunately for me, it takes a lot more than one lucky gene to create an Olympian.'[16]

The contemporary focus is therefore on individual genomes rather than whole populations. Dr Ken van Someren, former director of sports science at the English Institute of Sport, noted that the closer individuals come to the outer limits of their potential, the more likely it is that their personal genetic make-up will create 'some sort of glass ceiling'. He added that no single gene accounted for speed and power or for sprinting, instead describing a 'complex interaction' of multiple genes. 'So it's impossible to say there's a West African genotype for sprinting or an East African genotype for endurance running.'[17]

Let's stick with the Jamaicans. Peruse the all-time sprinting records and you find an even mix of African Americans and Jamaicans but over the last decade Jamaicans had the edge. In the 2012 Olympics, they won five of the six available medals in the men's 100 metres and 200 metres, plus the 100-metre relay; in 2016, they won men's and women's gold in the 100 and 200 metres. If West African genes are the main explanation, where were the West Africans? If it's all about slavery, why are the elite of a population of 2.5 million descendants of slaves outpacing the elite of a population of 45 million descendants of slaves?

The answer relates mainly to the high status of sprinting in Jamaican culture. Speedy African Americans are likely to end up in basketball or American football. Speedy Jamaicans sprint. As with British distance runners in the 1980s, or Kenyan and Ethiopian distance runners today, specific running cultures have created a critical mass vying to be the best of the best. It starts with athletics programmes at primary school level and goes on to scholarships to elite athletics high schools. Even the Junior National Championship draws crowds of twenty thousand. In recent years the Jamaican Amateur Athletic Association, backed by the state, has fine-tuned elite training at its High-Performance Training Centre. A further reason might relate to evidence of widespread use of performance-enhancing drugs and lax testing.

To return to the question of whether black people are naturally better sprinters ... put like that, the answer is a tentative 'no'. East Africans or Southern Africans seldom break records. Nor, for that matter, do Australian Aboriginals, Central African Bambuti (Pygmies) or the San (Bushmen). However, there would appear to be a slight genetic advantage in some people of West African origin, magnified by the running cultures of North America and the Caribbean, particularly Jamaica. Without that strong, well-organised, well-funded sprinting culture, genes are not enough. Which is why, of the sixteen West African countries, only Nigeria has enjoyed consistent international sprinting success.

Also, despite the domination of people of West African ancestry, it is not a complete shut-out. Now and then, someone breaks through to remind us that ACTN3 and fast-twitch muscles and sprint-suitable bodies are not restricted to a single ethnic origin. One such was the white South African sprinter Paul Nash, who equalled the 100-metre world record four times in 1968, running ten seconds flat. Another was Frankie Fredericks, a mixed-race Namibian who won four Olympic medals and World Championship golds in the 1980s and 1990s and broke the twenty-second mark for the 200 metres twenty-four times and the ten-second mark in the 100 metres twenty-seven times, putting him fourth on the all-time record list. The stand-out non-West African woman is Dafne Schippers, from the Netherlands, the 2015 and 2017 world 200-metre gold medallist and the third fastest woman of all time at the distance. Above all is the 'Coloured' South African Wayde van Niekerk, the world 400-metre record holder, world and Olympic champion and the only man in sprinting history to have run under ten seconds for the 100 metres (9.94), under twenty for the 200 (19.84) and under forty-four for the 400 (43.03).

However, it's hard to ignore the preponderance of sprinters with West African ancestry. ACTN3 is not the only factor. Other considerations – heavier bones, larger gluteus muscles, less fat beneath the skin – also play a role. Here we return to the bell curve; as the geneticist David Reich shows, a small genetic advantage in the average sprinting ability (say a 0.8-standard deviation advantage over average innate European sprinting ability) 'would lead to a hundredfold enrichment in the proportion of people above the 99.9999999th percentage point'.[18]

There is another possibility, one already suggested in relation to East African distance running ability: the far higher genetic variability in sub-Saharan Africa. This could mean there is more variation in innate sprinting ability in people with West African ancestry; more people with naturally high abilities and more with naturally low abilities. As Reich points out, there is a 33 per cent higher genetic variability in West Africans than Europeans.[19]

FIGHTING GENES?

When I lived in South Africa in the 1980s, I had a journalistic side-line reporting on professional boxing. At first some white trainers assumed that because I was white, I shared their prejudices, including the conviction that the abilities of fighting men were racially implanted. One trainer took me aside, telling me that while 'darkies' had the 'fancy stuff' they also had a 'yellow streak', which meant they couldn't take it to the body. This myth was immune to logic, because whenever a black man wilted from a body blow, it seemed to be confirmed, whereas white men who struggled with body blows or black boxers who withstood them weren't noticed; a classic example of confirmation bias.

To any objective observer, differences in style between black and white boxers related to how these young men were taught in the gyms, and to the *milieux* of the township and the town. But the best learned from each other, worked hard and ignored mythology about racial genes. The finest of the lot was a fiercely driven little fellow from a broken family in the rougher white suburbs. Brian Mitchell did not possess the flair of several of his countrymen – he was a so-so amateur – but he had a work ethic and capacity to learn equal to none, and an ability to focus. He put in the hours, lived the life and forged his skills by fighting in black townships, which few other white boxers were prepared to do. He got better and better in defence and attack. He went on to become South Africa's greatest-ever boxer, winning two versions of the World Super Featherweight title and fifteen world title fights during a five-year reign. He ended his career with just one loss (avenged three times) in forty-nine fights and a forty-two-fight unbeaten streak. He is the only South African boxer to have been elected to the International Boxing Hall of Fame.

Mitchell is an example for those holding to the '10,000-hours rule', who believe that application is more important than innate talent. This idea started with a paper written in 1993 by Anders Ericsson, of the University of Colorado, and others, entitled 'The Role of Deliberate Practice in the Acquisition of Expert Performance'. It referred to a Berlin study of violin students, showing that those who reached the top did not seem to possess superior natural gifts. Ericsson argued that many characteristics thought to reflect innate talent were really the result of intense practice for ten years or more, averaging out at about ten thousand hours. He added that the 'differences between expert performers and normal adults reflect a life-long period of deliberate effort to improve performance in a specific domain'.[20]

This idea was boosted by Malcolm Gladwell's bestselling book, *Outliers*,[21] which cited Ericsson's research and in turn, associated his name with the phrase. Ericsson complained that the '10,000-hour rule' was a Gladwell invention, and that 10,000 was simply an average. Many of the best musicians put in 'substantially fewer' hours to reach the top.[22] He also complained that Gladwell failed to mention deliberate practice. Gladwell hit back that he disagreed with Ericsson's view that deliberate practice was enough, saying he believed that elite performance also requires 'a big healthy dose of natural talent'.[23]

Yet when we look at those who reach the top, we often find other boys and girls who initially seem more gifted but lack the drive to push themselves single-mindedly to excel with a focus that allows no compromises; missing the compulsion to train harder and more consistently than anyone else and to let nothing get in the way. In several of the elite-of-the-elite I've interviewed over the years, it seemed that some childhood deprivation or damage, or some personality defect, prompted them to sacrifice everything that interfered, including family and friends. But their solipsism often came at a cost; they were seldom in the habit of interrogating their inner lives and when they tried, their inner reach didn't extend far. Their lurking demons so often caught up with them when they retired, all too often resulting in alcoholism, drug addiction and other forms of self-harm.

In stressing the psychology of extraordinary sporting achievement, I do not pretend that genetic inheritance is irrelevant. Some people are born with quicker reflexes, more fast-twitch or slow-twitch muscles, better hand-eye coordination, lateral vision and balance, greater strength relative to body weight, a naturally lower resting pulse rate or more appropriate bone structure. No one reaches the pinnacle without huge dollops of ingrained individual talent. I'm also not disputing that some populations, because of shared genetic inheritance, are more likely to succeed in certain sporting disciplines. The point is that when considering success, either at the individual or population level, it's all too easy to put it down to favourable genes. The answer often lies elsewhere, and boxing provides a good example.

For the first half of the twentieth century professional boxing was dominated by Americans of Irish, Italian and Jewish heritage. The 'fighting Irish', in particular, were seen as innately scrappy. When Billy Conn was knocked out by Joe Louis in the thirteenth round of the World Heavyweight Championship in 1941, having decided to go for glory after leading on points, he explained his hot-headed error to be a product of his Irish blood. 'What's the use of being Irish if you can't be thick?' he said. Black boxers, on the other hand, were viewed as having fewer noble qualities. When the former World Heavyweight Champion Jim Jeffries returned to the ring in 1910 to fulfil his destiny as 'the chosen representative of the white race' (as the novelist Jack London called him), he was the favourite in the betting against the first-ever black Heavyweight Champion, Jack Johnson, because it was believed the black man had a latent 'yellow streak'. Johnson dished out a one-sided beating, which prompted race riots in twenty-five states, leaving eight black people and five white people dead.

This gave promoters an excuse to bar black men from fighting for the heavyweight crown for twenty-two years, until Louis arrived to rule for more than a decade. The last white American World Heavyweight Champion was Rocky Marciano. He retired in 1955 and would have fought at light-heavyweight if competing today. After Marciano, white heavyweights were derided as 'hopeless white hopers', so much so that none embraced the label. In 1969, when the promising Irish-American heavyweight Jerry Quarry was asked whether he thought of himself as a Great White Hope, he replied: 'Screw White Hopes. I'm a fighter.' That he was, and a good one too, but not quite good enough or large enough to beat the best black heavyweights such as Muhammad Ali, Joe Frazier and Ken Norton.

Black Americans were assumed to be naturally bigger, stronger, more athletic; men whose genes gave them a huge edge. White heavyweight contenders were rare beasts, until the fall of the Berlin Wall released boxers from the Soviet sphere into the professional game. The Ukrainian Klitschko brothers, Vitali and Wladimir, took over, until one retired and the other was beaten by a white Briton. A fractious supporting cast of Eastern Europeans took their turns, along with the occasional black American, Briton or Cuban, while the elite at the lighter weights included Mexicans, Puerto Ricans, Kazakhs, Japanese, Thais and Filipinos.

JUMPING GENES?

The 1992 film *White Men Can't Jump* tells the tale of the white hustler Billy, who can sink a shot from beyond the half-court line but is useless at slam-dunking. In the end, through persistence, Billy wins the crucial game with a perfect slam dunk, refuting the assertion. The film's title was drawn from the higher proportion of black jumpers than white in the National Basketball Association (NBA), which in turn had something to do with the fact that African Americans comprise 78 per cent of NBA squads. And this had something to do with the reasons why basketball took off in poorer black areas: it is a sport that can be played on small concrete courts rather than needing big open fields.

After the release of the film, several in the game felt inclined to go with the title rather than the message. 'It's a physiological fact that black players tend to be faster, more flexible and better jumpers,' Mark Hannen of the English Basketball Association told me.[24] Alex Anzelmo, chief physiotherapist of the British and Irish basketball associations, added that players of West African genetic origin had bigger gluteus muscles and longer gastrocnemius muscles and Achilles tendons, all of which help jumping. However, they had less developed quadriceps and hamstrings, increasing knee injuries.

He also acknowledged 'socioeconomic influences' in black slam-dunking, 'which are important in the same way that the lifestyle of the Kenyan distance runners is in aiding their performance'.[25]

The picture is clear: black players have a genetic edge, magnified by 'socio-economic' factors. Or is it? Surely a better source of jumping data would be the high jump. The men's world record has long been held by the black Cuban Javier Sotomayor. But the gold medals in the last five Olympics were won by white men. Over the last six Olympics, black jumpers have won only five of the twenty men's medals and two of the nineteen women's medals. It would seem that the business of jumping high might have more to do with the culture of the sports concerned than with any ethnic genetic advantage.

WHITE CYCLISTS AND BLACK LINEMEN

This brings us to professional cycling, a sport that includes sprinters, with chunky pedal-churning thighs, and long-distance specialists who need the lean, slow-twitch muscles and recovery powers of a marathon runner. All the leading competitors on track and road, male and female, are white; so much so that when the British commentator Phil Liggett spotted a black Frenchman, Yohann Gène, Team Europcar's talented *domestique* (a rider who works for the benefit of the team) in the 2013 Tour de France, he was compelled to put his foot in his mouth, referring to him as a 'coloured cyclist'. No one would wonder aloud whether the shortage of black cycling stars was the result of a genetic shortfall, because cycling is so obviously a sport that black people have not taken to in large numbers. There are indeed more black cyclists than ever before, and some, such as Gène, are very good but they make up such a small proportion of the total that there are unlikely to be many on the medal podiums for a few years to come.

On the other hand, African Americans, who comprise 14 per cent of the American population, dominate NBA and National Football League (NFL) fixtures; just 37 per cent of NFL players are white. What's the reason for this dominance? Those West African genes again? A higher proportion of black Americans see sport as a potential profession, and it is not hard to understand why. The black unemployment rate is nearly double that of the white; the picture is even more skewed in youth unemployment, even though the education gap is closing.[26] Young black men still face people who avoid looking at them when they pass on the pavement, being refused job applications because their names sound black, and being suspected of criminal behaviour because of the colour of their skins. And they have a significantly higher chance of ending up behind bars.[27] Sport is seen as a way out and up, offering income and status not available elsewhere. Success breeds success, encouraging further participation, not least because of the mythologies of genetic superiority. If you believe you're likely to do well because you've heard you have the genes for success, you're more likely to sign up. Conversely, if you're a white sprinter, a non-African distance runner or a black swimmer, you might believe you'll never excel and follow a different path.

African-American sporting participation is therefore more concentrated, focusing on American football and basketball. In other sports, black participation is on a downward trajectory. In 2017, just 7.1 per cent of Major League baseball players on opening-day rosters were black, compared to 19 per cent in 1995. Black participation has also fallen in professional boxing, where Hispanics have become more prominent. And despite the triumphs of Tiger Woods and the Williams sisters, there has been no flood of black players into tennis or golf. Sporty whites, on the other hand, spread their bets. Their participation in American football is falling, as it is in basketball and sprinting, but it is rapidly increasing in soccer and remains unchallenged in baseball, motor sports, tennis, golf, swimming, road running, triathlon, ice hockey, lacrosse, cycling, mountain biking, martial arts, horse racing and other equestrian sports, squash, handball, surfing, rowing and other water sports, and a wide range of esoteric individual sports. If more black Americans are entering professional sport and concentrating on fewer sports, while white Americans are diluting their sporting focus, it's no great surprise that a few major professional sports are black-dominated.

BLACK MEN CAN'T SWIM?

In 1980, I overheard a nineteen-year-old student at the University of Cape Town discussing the evils of apartheid. One of his friends agreed, remarking how terrible it was that black people couldn't use white facilities such as beaches and swimming pools. The nineteen-year-old nodded earnestly but added that perhaps this was not so much of an issue as it was a 'known fact' that blacks couldn't swim because their bodies couldn't float. How did he draw this astounding conclusion, his friend asked? He replied that he'd 'heard it somewhere' and, in any event, he'd 'never seen one' in the water.

Beliefs along these lines have persisted in the swimming world and have found their way into some academic studies focusing on the 'lower buoyancy' of black Americans. The swimming genes debate parallels the sprinting genes debate, with the race bias reversed. The same is true of participation. Visit any running track and you'll find most of the sprinters are black. Go to a swimming club and the picture will be reversed. Swimming is perceived as a white sport, just as sprinting is perceived to be a black one. But is the racial culture of the sport the main reason why there are so few top black swimmers, or is this culture at least in part a reflection of innate ability?

First, the genetic argument: are black people really less buoyant? Among people of recent African descent there is more variation than among other population groups, so there are people at both the upper and lower ends of the bone mass and bone density scales. However, those of West African descent tend, on average, to have slightly higher bone density and mass. All things being equal, a lower bone density might aid buoyancy but even among top swimmers there is a significant range.

There are other, more important, assets that help make a good swimmer, including height, arm length and foot size. Michael Phelps, the most decorated Olympian of all time, stands 1.93 metres tall, has a long upper body and long arms (a 203cm span), which help the thrust from his arms' propulsion, and size 14 feet that can bend at an angle fifteen degrees greater than the average. But Phelps's main advantage is his outstanding stroke mechanics, which come from expert tuition and thousands of hours of practice. The former USA team physician H. Richard Weiner, put it like this:

> I guess it's hard for people just to believe that it can just be stroke mechanics for Phelps ... I'm sure if we could measure Phelps as much as we would like, we would find attributes better than average for swimming but I don't think we would find any glaring abnormalities. I suspect if we could comprehensively measure all Olympians in finals, we would see significant differences but we would not see them having freakish things like 200 per cent more lung capacity or muscles that can contract at twice the force of a normal human muscle. I mean, come on.[28]

Part of the argument against the buoyancy idea comes from the success of elite black swimmers, mostly Americans of West African heritage, over the last thirty years. By 2016 black swimmers had won eighteen Olympic medals (six gold) and had set several world and Olympic records.[29] Their example shows that ethnic heritage and skin colour are not inbuilt impediments to international success in the pool. But just like Frankie Fredericks, Dafne Schippers and Wayde van Niekerk in sprinting, these exceptions only prove that the rule does not exist; they do not explain the imbalance. For that, we need to look at the culture of American swimming. Fewer African Americans can swim than the US population average. According to a study conducted by USA Swimming, only 31 per cent of black children have learnt to swim reasonably well, meaning 69 per cent either can't swim or have low swimming ability. And most of the children who can't swim have parents who can't swim. This contributes to a fatal drowning rate for black children three times that of white.[30]

One explanation relates to slavery; fear of them escaping meant slaves were forbidden from teaching their children to swim. The more recent history relates to the lack of public pools in poorer areas and lack of money for swimming lessons. In the UK, learning to swim is part of the school curriculum but not in America, where policies vary from state to state. In the 1920s and 1930s many white Americans learned to swim in two thousand new municipal swimming pools, from which black people were usually barred. This colour bar was lifted at most pools by the 1950s, so many whites joined whites-only private clubs and some had garden pools installed. Swimming thereby gained a reputation as a country club, big-garden thing.[31] According to USA Swimming's research, this is exacerbated by the fear of non-swimming parents that if their children learn to swim, they would want to swim and so open the possibility of drowning. Another reason given by African-American women is that the chlorinated water would damage their 'relaxed' hair.[32] It would seem that the reason black people are under-represented in Olympic pools has everything to do with the culture of the sport and nothing to do with population genetics.

More generally, claims made about biological reasons for sporting success, particularly when they attach themselves to ethnicity, should be treated sceptically. It might be true that genes boosting fast-twitch muscles and gluteus muscles are slightly more prevalent among people of West African descent, but as we see from the lack of sprint medals from West Africa, genes are never enough. It is certainly true that some populations grow taller and have greater muscle mass and higher bone density than others, although the reasons may relate not just to genetics but also to epigenetics, diet and other environmental factors. Increased protein consumption has consistently prompted the growth of Chinese and Japanese people over the past two generations and South Koreans are, on average, more than five centimetres taller than their North Korean neighbours because of their better diet.

One sport does provide a pointer to the spread of talent: football. Because of its ubiquity – more people follow and play it than any other team sport and there are few countries that don't share the passion – it provides a more appropriate picture of population-by-population prowess. Yet in the 'beautiful game', it is hard to discern a pattern of ethnic dominance. When fans debate who was the greatest-ever player, they might opt for the black Brazilian Pele or the white Argentinean Maradona. Among today's stars they might discuss whether the black Frenchman Kylian Mbappé is set to surpass the white Argentinean Lionel Messi or the white Portuguese player Ronaldo. Colour is not a major consideration when it comes to assessing talent.

APPLES AND PEARS, FISH AND BICYCLES

Even when sporting genes are coupled with ethnicity, these are not *racial* genes; they are genes that are more prevalent in a particular population. Variation within and between populations can relate to anything from bone density, muscularity and fast-twitch muscles to disease patterns. But the fact that there has been population-based evolution in such areas does not imply that there has been continued and varied population-based evolution in something as complex as intelligence.

It is said you can't compare apples and pears. But you can: 'apples are not as sweet as pears; pears are squishier than apples.' You could even compare a fish and a bicycle: 'I'd rather buy a cheap bicycle than an expensive fish.' You can't, however, get away with: 'because apples are crunchy, so must pears be' or 'fish can swim; therefore, so can bicycles'. The former examples use valid points of comparison. The latter are based on the fallacy that apples and pears, or fish and bicycles, are equivalent. So it is with the comparison between the evolution of fast-twitch muscle and intelligence. A valid point of comparison would be: 'it's easy to find single genes for fast-twitch muscles but difficult to find single genes that have a significant bearing on intelligence,' or 'the relationship between natural selection and lactose tolerance is clear but its relationship with intelligence is opaque.' In these examples, the valid points of comparison are gene identification and the role played by natural selection. But it is not valid to say: 'because lactose tolerance evolved over a few thousand years through natural selection, intelligence therefore evolved the same way,' or: 'because we can identify single fast-twitch muscle genes, we can therefore do likewise with intelligence genes.' The reason is that intelligence and muscles are not the same, nor even similar, things. It's pure fish and bicycles.

8

ARE THERE RACE-BASED INTELLIGENCE GENES?

Andrew Sullivan, super-blogger and race science advocate, once wrote a column in the *Sunday Times* in which he anticipated that neuroscience and genetics would help 'find out for a fact whether there are any measurable genetic differences in IQ between different ethnic groups'.[1] None have emerged but biologists on the make continue with the quest for the holy grail of genetic determinism: a racially distinct intelligence gene.

The first big stab came from Bruce Lahn, a Chicago-based Chinese biologist, who made the bold claim that he had indeed found the golden ticket. Lahn, a professor of human genetics at the University of Chicago, left China for America in the late 1980s, after taking part in pro-democracy protests. He sparked a firestorm in 2005, when he claimed in two papers in the journal *Science* that there had been recent genetic changes, linked to brain size and intelligence, among non-Africans.[2] A *Science* report on his research had the headline: 'Evolution: are human brains still evolving? Brain genes show signs of selection.' It was illustrated with a picture of a muscular white man carrying an impressively large shaved head, echoing the pose of Rodin's *The Thinker*. The caption read: 'Big thinker? Certain forms of two brain genes may confer a selective advantage.'[3] The journal honoured Lahn's research as 'Breakthrough of the Year' and ran a glowing profile.

This delighted the far right, which viewed it as proof that non-black people were more intelligent than black people. The white supremacist publication *American Renaissance* featured a story by the overtly racist British psychologist Christopher Brand, in which he wrote that Lahn had 'found that sub-Saharan blacks were the most distinct of the racial groups they studied, in that they had a markedly lower frequency of both variants. This is consistent with the distinct black African profile of smaller brains and lower IQ.'[4]

When I reviewed the evidence a few years ago, a quick Google search found websites with headlines such as: 'DR BRUCE LAHN, U of Chicago Genetics, proves NIGGERS not fully HUMAN!'[5] Sections of the mainstream right also seemed overjoyed. John Derbyshire wrote in *The National Review* that Lahn's 'bombshell papers' showed that geneticists 'have been lying through their teeth about the sup-posed genetic similarity of all races'. And he added: '[I]f different human groups, of different common ancestry, have different fre-quencies of genes influencing things like, for goodness' sake, brain development, then our cherished national dream of a well-mixed and harmonious meritocracy with all groups equally represented in all niches, at all levels, may be unattainable.'[6]

Lahn was obsessed with genetic links between race and charac-ter. He once proposed an article on 'why Chinese are boring' and wondered whether there was 'some selection' against rebellious individuals in imperial China.[7] He insisted evolution of the human brain was continuing, which meant some populations might have higher intelligence than others,[8] and thought he might have turned some of these opinions into fact. He researched two genes (ASPM and microcephalin) that are associated with microcephaly, a condi-tion in which there is a reduction in brain growth. He claimed differ-ent variants of these genes would be associated with positive brain growth, which would correlate with intelligence: 'We're seeing two examples of such a spread in progress. In each case, it's a spread of a new genetic variant in a gene that controls brain size. This variant is clearly favoured by natural selection.'[9]

Lahn's team examined the DNA of 1,184 people from fifty-nine populations and found that the new mutations spread more quickly beyond sub-Saharan Africa, particularly among European and Middle Eastern groups and that they were 'probably' associated with 'higher IQ'. Lahn acknowledged the evidence was 'iffy' but none the less, speculated that a mutation of microcephalin emerged 37,000 years ago and spread to 70 per cent of humanity, coinciding with 'the introduction of anatomically modern humans into Europe, as well as the dramatic shift towards modern human behaviour such as art and the use of symbolism'.[10] Lahn guessed that interbreeding with Neanderthals might have prompted this genetic change and that this exchange made the Eurasian portion of the human population smarter. One reason he offered for his Neanderthal theory was that the new gene variant was more common in Eurasia than sub-Saharan Africa, 'and we know that Neanderthals evolved outside Africa', although he admitted that '[b]y no means do these findings constitute definitive proof'.[11]

The modern version of the other gene (ASPN) emerged 5,800 years ago, Lahn claimed, and spread to 30 per cent of humanity, again mainly in Eurasia. This, he said, coincided with the 'development of cities and written language', suggesting a causal link.[12] Lahn never explained precisely why evolution for higher intelligence would emerge from very limited interbreeding with Neanderthals who, after all, most experts believe were probably not quite the equal of, let alone superior to, humans when it came to creative cognitive capacity. It is also worth remembering that the Neanderthals died out a couple of thousand years before Lahn's start date for microcephalin and the interbreeding with Neanderthals that left a genetic trace came at least 12,000 years earlier.

Another problem was that the achievements Lahn picked seemed all too convenient. We could just as easily point to the cultural leaps of symbolic art and self-adornment made by the sub-Saharan Africans 77,000 years ago, the construction of Göbekli Tepe in Anatolia 13,000 years ago or the first urban centres 7,000 years ago. But this would not fit with his schema. Pointing to several historical errors, the Harvard historian of science Sarah Richardson noted that Lahn's account was poorly grounded in history, geography and demography, relying instead on loose and sweeping generalisations about human cultural history.[13]

The most damning attacks came from other scientists, who savaged Lahn's premises and conclusions. Geneticists at the Broad

Institute in Cambridge, Massachusetts reanalysed his data and disputed his view that natural selection acted to change the ASPN gene. Geneticists from the University of California, Los Angeles tested these gene variants in 120 people and found they had no impact on brain size; a conclusion shared by the results of five other studies, including one from Lahn's own laboratory. Other scientists found these gene variants were probably far older than he suggested. Sarah Richardson concluded there was 'no evidence of an association between the alleles and either IQ or brain size' and that the idea that larger brains led to higher intelligence was 'also not grounded in empirical evidence'.[14]

The political right continued to regard Lahn as a hero for defying 'political correctness' but the response from the scientific world was very different. Following the demolition of his research by fellow geneticists, the two co-authors of his original papers in *Science* distanced themselves from the papers' claims,[15] while his university withdrew a patent application to use his work to develop a DNA-based intelligence test. Lahn defended himself, saying his work was being held to a higher burden of proof than others' but as his claims unravelled he retreated from the debate, saying it was 'too controversial'.

In the wake of Lahn's research, several public figures had their DNA tested to find out which versions of the two genes they had, including the BBC's notably cerebral political commentator Andrew Marr, who was amused to discover that he had the sub-Saharan variant. Even Lahn admitted he once tested himself to find whether he had the Eurasian or the sub-Saharan variant and although the results were indecisive, 'it wasn't looking good'.[16]

Other claims seemed to be drawn from Lahn's faulty research. Steven Pinker wrote in 2009 that he had had his IQ tested and that the results were 'above average'. He then had his genome sequenced, which prompted him to ask: 'who wouldn't be flattered to learn that he has two genes associated with higher IQ?'[17] At that stage and subsequently, the only claims that had been made about single or double intelligence genes were Lahn's, so one can only presume that Pinker had failed to read the many academic critiques of Lahn's work that had been published by 2009.

Lahn's failed bid to find racially linked intelligence genes was hardly the final word. There is prestige and money in genetics, which can attach itself to the scientists making the boldest claims. Lahn retreated with his tail between his legs but his case illustrated the need to be cautious about claims relating to the behaviour of any genes or combinations of genes, because the headlines prompted by publicity-hungry researchers do not always match the peer-reviewed reality.

THE HUMAN GENE SHORTFALL

Before delving deeper into the relationship between genes and environment for intelligence, we must understand the implications of decoding the human genome. In previous chapters I've stressed how full genome analysis has opened the way to discovering a great deal we never knew about ourselves but it is time to explain what genetics cannot do.

The first decoding of the full human genome in 2001 prompted a spate of wide-eyed reports based on the perception that the hard-wired approach to understanding human behaviour had prevailed. Anyone whose knowledge of genetics was restricted to the news media might have decided that nature had trumped environment and culture. There were gushing tales about the discovery of genes 'for' traits ranging from television-watching to alcoholism. In every case, the predictions were later scotched. Claims of single genes 'for' a behavioural trait showed nothing more than the ignorance of the authors. An isolated gene can be no more than one element in the biomechanical pathway implicated in a trait. The environment, and the way it is experienced, has huge implications for the way those genes express themselves. Yet this kind of nuance was trampled over in the early excitement, even by those who should have known better. Daniel Kosman, editor of *Science*, was so thrilled that he claimed the nature-nurture debate was over, and that nature had won.[18]

Those at the core of the human genome project issued warnings against claims of finding genes for any behaviour, stressing that anyone hoping to find single genes for crime, sporting success or intelligence would be disappointed. Craig Venter, the scientist who led the private sector effort to decode the human genome, was particularly forthright: 'In everyday language the talk is about a gene for this and a gene for that. We are now finding that that is rarely so. The number of genes that work in that way can almost be counted on your fingers, because we are just not hardwired in that way.'[19]

He stressed one point, that the gene-centric approach to human behaviour was simply wrong:

> There are two fallacies to be avoided: determinism, the idea that all characteristics of a person are 'hardwired' by the genome; and reductionism, that now that the human sequence is completely known, it is just a matter of time before our understanding of gene functions and interactions will provide a complete causal description of human variability.[20]

The project produced some surprises along these lines. It emerged that the human genome contained around nineteen to twenty thousand protein-coding genes, one-fifth the anticipated number. In fact, not many more than a fruit fly and fewer than a mouse. The Stanford University biologist Paul Ehrlich noted that the human genome did not have anywhere near enough genes to programme the connections in our brain that control behaviour:

> Our 'gene shortage' is one reason human infants and young children are so helpless. Their helplessness allows the physical and cultural environments to do the brain programming that our hereditary endowment couldn't manage. It's that environmental input that gives us the adaptability that is the hallmark of humanity. We could never have evolved as genetically controlled robots.[21]

Like Venter, Ehrlich expressed despair at the trend towards genetic determinism, particularly from evolutionary psychologists:

There is an unhappy predilection, especially in the United States, not only to overrate the effect of genetic evolution but also to underrate the effect of cultural evolution. Uniquely in our species, changes in culture have been fully as important in producing our natures as have changes in the hereditary information passed on by our ancestors.[22]

EPIGENETICS AND THE 'NEW LAMARCKISM'

The decoding of the human genome coincided with fresh evidence, from the field of epigenetics, that genes weren't the sum total of destiny. To paraphrase the Bible, epigenetics is all about the sins of the father or mother being visited on their children and their grandchildren. It refers to anything that can alter the impact of a gene but have no impact on the DNA sequence itself. Put differently, it is the study of heritable changes to the *phenotype*[23] of an organism that don't involve genetic changes. It therefore refers not to a change in genes but to whether a given gene is 'switched on' by environmental factors.

Epigenetics has prompted a huge volume of scientific research over the past two decades, and also produced its fair share of quackery and pseudoscience. Since the early 2000s, research has shown that certain traits can be passed on to the fourth generation. A study of male rats exposed to cancer-prompting insecticide found that while it produced no genetic changes in their offspring, the risks were nevertheless carried through four generations.[24] Another showed that cocaine-sniffing male mice passed poor memories on to their offspring with no evidence of DNA damage, and a third showed pregnant rats exposed to nicotine passed on asthma to their offspring and to their grandchildren.[25]

How much or little we eat, what we eat, the alcohol we consume, even the levels of stress or isolation we face; all can affect the expression of our genes and those of our children and their children. This area of study has been dubbed 'The New Lamarckism' because it rehabilitated the discredited idea that environmental characteristics acquired during a lifetime could be passed on.[26] Several studies have shown that fear and anxiety can be epigenetically transmitted to offspring; experiments on mice showed they could be trained to associate particular smells with fearful memories and that these could be passed on to later generations.[27] Professor Kerry Ressler, a neurobiologist and psychiatrist at Emory School of Medicine in Atlanta, Georgia, who led a research project on the epigenetic transmission of fear in mice, said similar transmission could take place in people, with the fearful experience of a parent being transmitted to subsequent generations, perhaps through chemical changes in the sperm and the eggs acting as a 'mechanism of conserving as much information as possible from a previous generation'.[28]

Several recent studies have indeed shown a transgenerational epigenetic inheritance of fear and anxiety among people, and also of addictive behaviour and a range of physical outcomes.[29] Swedish researchers found that paternal-line grandsons of men who had been exposed to famine during adolescence were less likely to die of heart disease, while those whose paternal-line grandfathers had plenty of food were more likely to die of diabetes, and paternal granddaughters of women who had been exposed to famine while in the womb died at a younger age.[30]

Epigenetics also covers the more immediate impact of the environment on gene expression. A study of children conceived during the 1944–45 famine in the Netherlands found that 'early-life environmental conditions can cause epigenetic changes in humans that persist throughout life'[31]; sixty years on, these people had an increased risk of diabetes, heart disease and other conditions because of the epigenetic impact of these diseases.

A particularly intriguing study involved 153 healthy people in their fifties and sixties. University of Chicago researchers identified the most socially secure, and the loneliest, and examined their DNA. They found that 1 per cent of the participants' DNA (209 genes) responded differently in the two groups, particularly the inflammatory immune response. This study has since been replicated, confirming that social isolation was even more critical than stress as a disease risk factor.[32] It has been further backed up by studies finding that the insecurity and isolation that often emerge from poverty seemed to exacerbate the genetic expression of those experiencing it, weakening their immune systems and making them more susceptible to several diseases. Steve Cole, the UCLA scientist who co-led the Chicago study, explained: 'You can't change your genes but ... you can change the way your genes behave,' or as he put it in a lecture: 'Your experiences today will influence the molecular composition of your body for the next two to three months or perhaps, for the rest of your life. Plan your day accordingly.'[33]

Epigenetics is not about evolution. Epigenetic traits always fade. Rather, it presents an alternative to the binary question of whether a particular trait is genetically innate or environmental. Until recently, it was assumed that some health problems – depression, obesity, heart disease, cancer, schizophrenia – can only be explained by a combination of social conditions and genetic inheritance. Now there's a third option: the environment can leave a biological imprint that may linger. Our experiences, and those of our parents and grandparents, might not only change our perceptions of the world; they might also change the response of our genes to this world. Performance in IQ tests is likely to be affected by the kinds of poverty-related gene expression – anything from susceptibility to asthma to anxiety levels to depression – that these studies have highlighted.

THE QUEST FOR IQ GENES

There have been several recent attempts to discover which combinations of genes and parts of the brain are implicated in intelligence, some more successful than Bruce Lahn's. In 2010, a team of British psychologists and scientists researching the academic performance of four thousand British schoolchildren said they thought that perhaps two hundred genes could be directly implicated. However, their general point was that this form of intelligence was governed by a network involving thousands of genes, each making a tiny contribution, rather than by a handful of powerful genes, as once thought.

'Of the gene variants we looked at, a couple of hundred are emerging which seem to have a small but significant relationship with ability in maths and English,' said Robert Plomin, the King's College London psychologist who led the research. Plomin, a hereditarian who signed a contentious race-based public statement on IQ,[34] acknowledged that this said nothing about how the genes themselves might work, adding: 'It seems that no single gene has a really big effect ... [C]ognitive powers depend on lots of little effects from lots of genes.'[35]

Elsewhere, Plomin and the geneticist Oliver Davis cast doubt on other genetic presumptions, noting that 'for most complex traits and common disorders the genetic effects are much smaller than previously considered: the largest effects account for only 1 per cent of the variance of quantitative traits.'[36] This caution was backed in 2009 by Ian Deary, head of the University of Edinburgh's Centre for Cognitive Ageing and Cognitive Epidemiology. He acknowledged that one of the toughest jobs in genetics was trying to find genes linked to intelligence. A multiplicity of genes might affect intelligence indirectly, he speculated, by changing the way the brain grows. But, he added, 'It is difficult to name even one gene that is reliably associated with normal intelligence in young, healthy adults.'[37]

If the idea of finding a single significant intelligence gene has faded, several research projects claim to have found combinations of hundreds of genes implicated in cognition. A research team from Harvard University and the Universities of Edinburgh and Southampton, including Ian Deary, used data from two genetic studies to compare the variation in DNA in more than 248,000 people, to isolate genes associated with intelligence, which they conflated with IQ. The researchers located 187 regions in the human genome, in the brain and pituitary gland, that they said were linked to intelligence (and also to longevity) and isolated 538 genes that played some role. They applied this finding to a smaller group, after which they claimed to be able to predict, based on individual DNA, nearly 7 per cent of IQ differences.[38] The first author, David Hill, said they were also able to identify 'some of the biological processes that genetic variation appears to influence to produce such differences in intelligence',[39] while Deary added that the study suggested that 'health and intelligence are related in part because some of the same genes influence them.'[40]

The Dutch statistical geneticist Danielle Posthuma has been involved with studies of the relationship between genes and intelligence for several years. The largest and most significant was an international meta-analytical study of almost 270,000 people of European ancestry. This study, published in 2018, claimed to have discovered 205 DNA regions and 1,016 brain genes linked to intelligence, mostly located in the brain's basal ganglia, where learning and emotion are processed. The subjects' DNA analysis was married to their IQ scores to find the links. The researchers also found that higher scores correlated with lower odds of suffering from schizophrenia, attention deficit hyperactivity disorder and Alzheimer's disease but higher odds of autism. An interesting aspect of the study was that all but seventy-seven of the intelligence-related genes had not previously been identified as such.[41]

Both studies found hundreds of genes that correlated positively with aspects of cognition; they also seemed to affect aspects of health and longevity. But neither study found – or looked for – variations in the presence of these genes in different population groups. The closest we seem to have come to a single brain gene that varies from population to population is the KL-VS variant of the *klotho* gene, which modulates ageing. However, this has not been proved to relate to IQ and the research has been conducted on older people, so it is not yet clear how this allele would affect younger subjects. It therefore doesn't cut it as the 'Holy Grail' of genes sought by Bruce Lahn and his ilk.

Research conducted at the University of California, San Francisco, published in 2014, involved an analysis of KL-VS, which, the researchers said, as well as being associated with longevity, might also improve the brain's ability to perform certain everyday mental tasks, particularly relating to memory, by increasing the strength of the connections between nerve cells in the brain. 'Our findings suggest that the KL-VS variant promotes cognition by increasing levels of secreted *klotho*,'[42] the paper's authors wrote, in reference to a protein produced by KL-VS. One of the co-authors, Lennart Mucke, added that the allele, which was present in about 20 per cent of the population they researched, may 'increase the brain's capacity to perform everyday intellectual tasks'.[43] They have since synthesised the protein and given it to mice; they conclude that it 'also enhanced cognition' and that this synthetic version might be useful in countering dementia. The team, which was studying age-related cognitive decline, found that those with the KL-VS variant tended to live longer and have a lower risk of stroke and age-related heart disease. After analysing data on three groups of ageing white Americans, they also found that the one fifth of volunteers who had this variant tended to perform better in cognitive tests, leading to speculation that it could prompt IQ differences of 'up to' 6 per cent.

There are several reasons why these tentative conclusions should prompt more caution than that shown by journalists' talk of an 'IQ-boosting gene'.[44] First, while the team tested memory, attention, visuo-spatial awareness and language, they did not test IQ. The suggestion that this allele could be implicated in higher IQ must therefore be seen as speculative, and some of the media claims even more so. *The Economist* wrote that the KL gene 'could account for as much as 3 per cent of the variation of IQ in the general population ...'.[45] Phrases such as 'up to' and 'as much as' should cause further caution, because when they are attached to statistical data they divest the numbers of real meaning: 'up to 6 per cent' could mean 6 per cent or it could mean 0.00001 per cent.

Bold claims resulting from a single paper should be handled gingerly, particularly when the data are drawn from a narrow population range. Almost all those tested were white Americans and the authors acknowledged the 'potential limitations in extrapolating our data worldwide', adding '[I]t is possible that more diverse genetic or environmental influences could alter or mask the effect in other populations.'[46] But the sampling problems go beyond its narrow ethnic base. All the volunteers were aged between fifty-two and eighty-five. The authors suggest cognitive impact is not age-related, but we really don't know. If they had tested volunteers aged between twelve and eighty-five, would the results have been the same? Or would a 'longevity gene' have more impact on the cognition of ageing people? A 2018 paper dealing with this gene variant noted that it was associated with the right dorsolateral prefrontal cortex, a part of the brain involved with planning and decision making, 'which is especially susceptible to shrinkage with age'.[47] This might suggest its impact is indeed age-related.

Are there potential implications for race and intelligence? Let's begin at the extremes; assume that, on average, people with this allele have higher IQs and that the estimate of 'up to 6 per cent' is 5 per cent, rather than 0.00005 per cent. Let's also assume that this allele is unevenly distributed among world populations. If 25 per cent of population A carries this allele and 5 per cent of population B has it, all other things being equal, the average IQ of population A should be 1 per cent higher than that of population B. This, however, would be impossible to assess. As we shall see in future chapters, comparing IQ averages of different populations is fruitless because of different cultural, educational and other environmental influences. And there have also not been large cross-population studies of the presence of KL-VS.

The most we can say is that, given the limited ethnic data on KL-VS, it does not correlate with the dubious ethnic data on average IQ scores. A study of 107 Iranians found that none carried KL-VS, leading researchers to conclude that this allele 'seems to be [scarcely found] in the Iranian population'.[48] Does the nation that produced the wonders of ancient Persia have a reputation for low IQ? I searched for reports of Iranian IQ studies and found several studies showing that Iranians were average; even the much-derided 'IQ of nations' chart produced by Richard Lynn put Iranians in the middle.[49] Studies of 723 Caucasian and 242 African-American elderly heart patients in Baltimore showed that almost double the proportion of black patients, compared to white patients, carried the KL-VS allele so it would seem that the relationship between KL-VS and ethnicity is not what knee-jerk racists would anticipate or hope for.[50] Even if subsequent studies confirm that people with this gene variant have, on average, fractionally higher IQs, this is a long way from saying it evolved for intelligence.

DO DIFFERENT POPULATION GROUPS HAVE DIFFERENT GENETICALLY PROGRAMMED IQS?

There are three broad perspectives on the contribution of population genetics to intelligence. What is sometimes self-described as the 'human biodiversity' view or 'racial realism' stresses that intelligence-related genes differ substantially among populations and races, as demonstrated by the disparity in IQ scores. The currently most popular example is the higher than average IQ scores of Ashkenazi Jews.

Other than the contrarian 91-year-old James Watson, no other prominent living geneticist has lent their name to this view. The podcaster Sam Harris claimed, in a debate with *Slate*'s editor-at-large, Ezra Klein, that several had 'privately' expressed their backing for this idea to him but were afraid to come out because of a climate of political correctness. However, at the loud and proud level, its most prominent backers are vloggers and YouTube stars such as Harris, Jordan Peterson and Stefan Molyneux, journalists such as Andrew Sullivan and Nicholas Wade, evolutionary psychologists such as Steven Pinker, Richard Lynn and Satoshi Kanazawa, and disparate academics, such as the political scientist Charles Murray, the anthropologist Henry Harpending and the educational psychologist Linda Gottredson.

A very different view is prominently associated with the Harvard geneticist David Reich, who is a passionate opponent of racist science and has sunk his teeth into Watson, Wade and Harpending, accusing them of conflating knowledge about average genetic differences between populations with guesswork that 'has no merit' and no base in serious scholarship.[51] He says their claims are racist, have no scientific authority and their speculations 'correspond to long-standing popular stereotypes – a conviction that is essentially guaranteed to be wrong'.[52] Reich's perspective is that even though the differences within human populations are significantly greater than those among populations, all genetic traits, including those that affect cognition, might differ between populations. However, he adds, we have 'no idea right now what the nature or direction of genetically encoded differences among populations will be'.[53] He disputes the idea of a qualitative distinction between the evolution of physical traits such as skin colour, which involve a handful of genes, and those relating to the brain, which has more than ten thousand genes. Reich points out that some physical traits that have evolved differently between populations also involve a multiplicity of mutations; height is a prime example. He writes: '[I]f natural selection has exerted different pressures on two populations since they separated, traits influenced by many mutations are just as capable of achieving large average difference across populations as traits influenced by a few mutations'.[54]

However, Reich emphasises that the differences that emerge will not follow the boundaries of the traditional social categories of race. We can be sure of this because recent full genome studies of ancient DNA show that the current populations of the world are 'mixtures of highly divergent populations that no longer exist in unmixed form' and are therefore 'not exclusive descendants of populations that lived in the same locations ten thousand years ago'. This knowledge, he says, should warn those who think that the 'true nature of population differences will correspond to racial stereotypes'[55] He adds that the offence of racism is to judge individuals by a stereotype of their group, which is sure to be misleading. 'Statements such as "You are black, you must be musical" or "You are Jewish, you must be smart" are unquestionably very harmful.'[56] The example he offers to show where surprises might be found relates to the greater genetic diversity among sub-Saharan Africans. He refers to the 33 per cent higher genetic diversity in West Africans, compared to Europeans, and proposes this might be a reason why they produce faster sprinters. He generalises from this point, suggesting it might also translate into a higher proportion of sub-Saharan Africans with other extreme genetic abilities 'including cognitive ones'.[57] In other words, it is possible that the range of cognitive potential in Africa, at both ends of the spectrum, is greater than elsewhere.

A third view – held by most of those working in this area – is that significant genetically based intelligence differences between populations are unlikely. Among this view's most prominent advocates are the evolutionary biologists and geneticists Richard Lewontin, Steven Rose, Steve Jones, Paul Ehrlich, Kevin Mitchell and the late Stephen Jay Gould, IQ theorists such as Richard Nisbett, Eric Turkheimer and the late Leon Kamin and the paleoanthropologists Ian Tattersall and Agustin Fuentis. The strand of their case relating to population genetics has recently been argued by Mitchell, a geneticist from Trinity College, Dublin. Like Reich, he uses height as an analogy, noting that some gene variants make people a bit shorter and others a bit taller, on average. However, the balance has been maintained by natural selection, because growing ever-taller has no benefit. On the other hand, evolutionary forces 'drove intelligence in one direction only in our ancient ancestors', as it was our 'defining characteristic and our only real advantage over other animals' and was amplified through culture and language. The selective advantage of ever-greater intelligence prompted a 'snowball effect' that was probably only stopped by limits imposed by the size of the birth canal and the 'metabolic demands of a large brain'.[58] This is the evolutionary reason for the huge cognitive potential of our complex brains. But while random mutations will affect all genetic programmes, with intelligence they are likely to affect it negatively, because of its genetic complexity. '[O]nce that complex system was in place,' Mitchell said, 'the main variation would be in the load of mutations that impair it, which will likely have effects on many traits and impair fitness generally.'[59] But in any population general fitness will always be selected, 'meaning that intelligence will get a free ride – it will be subject to stabilising selection, whether or not it is the thing being selected for'.[60]

Mitchell draws another analogy, this time with a Formula 1 racing car: intelligence, like the car's performance, is 'an emergent property of the whole system'. Thousands of brain genes affect intelligence; there is no dedicated module on which natural selection can act without affecting other traits.[61] And just as random tinkering is unlikely to improve the car's performance, so intelligence won't be affected by a balance of IQ-boosting mutations and IQ-harming mutations. Instead, genetic differences in intelligence may 'largely reflect the burden of mutations that drag it down', which is why evolution will tend to select against them.[62] Mitchell dismisses *The Bell Curve*'s idea that genes have at least something to do with racial differences in intelligence, saying that not only is there no evidence for this but that it is 'inherently *unlikely*',[63] because the genetic forces acting against population-level variation in intelligence are far stronger than those supporting it.

Because so many genes are implicated in the brain's development, there will inevitably be some variation in the mutations affecting cognition among populations, clans, families and individual people but this 'constant churn of genetic variation works against any long-term rise or fall in intelligence'.[64] At the population level, natural selection will remove mutations with large effects; those with minor effects might linger. But significant intelligence differences are highly unlikely because this would require 'enormous' and persistent differences in selective forces among groups, which would have to act across huge areas, in wildly different environments, and persist over tens of thousands of years of cultural change, which 'is inherently and deeply implausible'. Instead, what might once have appeared as distinct races, in fact consist of numerous interbreeding population groups with arbitrary divisions 'and the larger and more ancient, the greater diversity there will be *within* that group'.[65] Mitchell concludes that while genetic variation helps explain why one person is more intelligent than another, there are unlikely to be 'stable and systematic' genetic differences that make one population smarter than another.[66]

Both Mitchell and Reich recognise that most cognition-related genetic differences are between individual people rather than populations. But unlike Reich, Mitchell says that population-level differences are unlikely, and if they do exist, they will be minor. He stresses evolutionary factors acting against significant population differences in intelligence. Reich emphasises the possibility of significant differences in cognition genes among populations but says they won't follow the lines of traditional race groups. The form they will take is unknown, but he offers the example that greater genetic variation in sub-Saharan Africa may lead to more variation in cognition genes.

The view of Jim Flynn, widely recognised as the world's leading IQ theorist, falls between these two stools. He applies this to possible differences between male and female IQ scores as well as to the white-black gap in the USA. In the twentieth century, on average, adult men marginally outperformed adult women in IQ tests, but in recent decades women's average IQ scores have overtaken those of men. The reasons relate to changes in the workplace and education. 'This improvement is more marked for women than for men because they have been more disadvantaged in the past,' Flynn says. 'The full effect of modernity on women is only just emerging.'[67] He made a similar point on racial IQ differences in 2018: 'I think it is more probable than not that the IQ difference between black and white Americans is environmental. As a social scientist, I cannot be sure if they [African Americans] have a genetic advantage or disadvantage.'[68] He noted it was possible that the two groups' ten-point difference in IQ reflected a twelve-point environmental difference, which would mean that black Americans had a two-point innate edge.

BANNON, BREITBART AND THE 'EXTREME WARRIOR' GENE

Claims relating to brain genes do not only concern intelligence. They can also relate to behaviour. Some of these claims have tipped over into the realm of racism. In July 2016, when the *Breitbart* website boss Steve Bannon was about to become the chief of Donald Trump's presidential campaign, he wrote a piece on the shooting of black men by white policemen entitled 'Black Lives Matter is a left-wing conspiracy'. His money quote: 'Here's a thought: What if the people getting shot by the cops did things to deserve it? There are, after all, in this world, some people who are naturally aggressive and violent.' Bannon implied that the disproportionate number of black men killed by the police was a result of the 'naturally' disproportionate aggression and violence of the black victims; probably a reference to an alt-right trope that circulates on websites such as *Breitbart* and other race-obsessed sites of 'the psycho gene' (also called 'the extreme warrior gene') that predisposes its black bearers towards violence. Richard Lynn cited this gene as part of his claim that black people have evolved to have 'greater psychopathic tendencies'[69]. Research on this allele was devoured with particular relish by Nicholas Wade, whose favourite example of genetic differences among populations involves the MAOA-2R allele, which is more common among African Americans; one in 21 carry it, compared to one in 200 Caucasian-American men and one in 150,000 Asian-American men.

The MAOA case indeed illustrates that genes relating to mental well-being can vary among populations. Other examples I touched on in Chapter 6, in relation to genetic drift, include schizophrenia and multiple sclerosis but any of the ethnically varying diseases, including hypertension, prostate cancer, Gaucher's, cystic fibrosis and stroke can have an impact on mental well-being. However, when we take a closer look at MAOA it emerges that the claims of the alt-right are not borne out. It does not make a significant contribution to increasing rates of violence among any population, and contrary to some reports does not appear to have been selected for this reason, or indeed selected at all. A study that dealt with MAOA frequency among South Africa's white Afrikaners noted the role played by genetic drift.[70]

The place to start is with an earlier recipient of the 'warrior gene' label, the far more common MOAO-3R variant, which causes lower levels of production of a protein needed to break down old serotonin in the brain. It has therefore been associated with risk-taking, depression, aggression, violence and anti-social behaviour. The variant is found in 56 per cent of Maori men and 48 per cent of African-American men, compared to 34 per cent of European men. It was used to explain higher rates of Maori criminality, but further studies produced a surprise: its prevalence among Chinese men was the same as among Maoris (56%) but the highest prevalence was in Taiwanese men (61%). Neither Chinese nor Taiwanese men have high rates of criminality, violence or anti-social behaviour.

Research on the 'extreme warrior' 2R variant suggests its association with criminal violence has mainly been confined to men who were abused as children. Poverty, maltreatment during childhood, interrupted education, low IQ, being beaten as a child, as well as high testosterone levels, have been cited as contributory factors.[71]

Other influences include the *in utero* environment of the foetus, and the amount of attention given to children. It is also likely that epigenetics plays a role in the expression of this allele.[72] One study linked its epigenetic expression to women's alcohol and nicotine dependence during pregnancy. In other words, if you carry 2R, and your mother smoked and drank during pregnancy, you're more likely to be aggressive.[73] Dire social conditions are therefore most likely to significantly raise the odds of being agitated, aggressive, impulsive or depressed, at least among African-American men. Caucasian men carrying this gene have higher rates of anti-social behaviour even if they have relatively normal upbringing.

Different environments can make profound differences to how genes are expressed. In one study, scientists took one thousand baby bees from their hives; five hundred aggressive African 'killer bees' and five hundred easy-going European bees. They swapped them, putting the African bees in the European hive and *vice versa*. The impact on gene expression was dramatic. It was not only that the bees behaved differently if placed in a foster hive; their genes behaved differently, even though the genome itself was unaltered. The way the genome behaved was far more socially fluid than previously realised.[74]

Some human genes ('plasticity genes') are particularly susceptible to environmental influence; MAOA is one. A study on the impact of physical discipline conducted by a team led by the University of Pittsburgh developmental psychologist Daniel Choe found that being hit as a young child was causally associated with anti-social behaviour in adolescents and young men with the 2R and 3R variants.[75] Another study showed that 'delinquency risk for the 2R allele is buffered for males close to their biological or social father.'[76] The low-stress uterine environment of a mother who didn't smoke or drink during pregnancy, supportive, non-smacking parents and, particularly, attentive fathers, seems to mitigate against 2R and 3R prompting anti-social behaviour, at least among African-American men.

More than 95 per cent of black American men do not carry MAOA-2R. Kevin Beaver, a Florida State University biosocial criminologist who has studied this allele, is sceptical of claims that it is a candidate for explaining violence and criminality in black men, because it's simply 'not common enough in African Americans'.[77] Speculating loosely, let's say one in twenty members of population X carry 2R, and that 20 per cent of that group suffered the kind of abusive or chaotic upbringing that would make them prone to significantly higher rates of criminal violence, and among those abused carriers, the rate of criminal violence is double that of non-carriers. This would mean that the additional contribution to population X's proportion of criminal violence made by those with 2R would be 1 per cent.

In reality, it's more complicated. Different populations appear to respond differently to what is known as 'low-yield MOAO' (particularly 2R). One study of Caucasian and African Americans found that when all those who'd experienced childhood abuse were excluded, the presence of low-yield MOAO did not significantly raise the odds of anti-social behaviour among African Americans but did raise it by 41 per cent among Caucasian Americans, compared with those with high-yield MOAO. In other words, among African Americans the key factor was not the 2R gene variant alone but rather the combination of 2R and the abuse suffered in childhood, while among Caucasians 2R seemed to raise the risk of anti-social behaviour regardless of childhood abuse.[78] Put differently, there might be proportionately fewer Caucasians with 2R but its negative impact is more severe for them.

The researchers noted there may be both environmental and genetic reasons for this. The genetic reasons may relate to the fact that aggression and violence is influenced by many genes, some of which have yet to be identified, and some of which may be more prevalent among Caucasians. This is a point acknowledged by Wade, who uses the example of the presence in Finns of the HTR2B-Q20 allele, which, as he puts it, 'predisposes the carrier to impulsive and violent crimes when under the influence of alcohol'.[79] This allele is carried by more than 100,000 Finns (2.2 per cent of the population) and although it is mainly associated with drunken violence in men, because of a 'tendency to lose behavioural control while under the influence of alcohol',[80] the results of a recent study show that men with this mutation are also more impulsive 'even when sober and they are more likely to struggle with self-control or mood disorders',[81] said its first author, Roope Tikkanen. Finally, a study on the impact of low-yield MAOA on women found it had the opposite impact to that on men, relating to 'greater happiness'.[82] This contradicts earlier studies that found it was associated with depression in both men and women,[83] illustrating the point that when it comes to research on the impact of genes, it is wise to treat them with caution.

Reich notes that the history of science has revealed over and over again 'the danger of trusting one's instincts or being led astray by one's biases – of being too convinced that one knows the truth'. This, he warns, can lead to racist stereotypes, of the kind peddled by the likes of Nicholas Wade and James Watson, that are 'essentially guaranteed to be wrong'. He adds: 'We truly have no idea right now what the nature or direction of genetically encoded differences among population will be.'[84] This is a debate that will run and run. Claims will be made about 'IQ genes' and about how their distribution differs among populations. These claims will inevitably be over-cooked or misinterpreted by people who want to stress race-based difference, until the logic and the research is scrutinised by peer review, as happened for Bruce Lahn. There's a long way to go on this one.

9

WHAT IS IQ AND
IS IT HARDWIRED?

I was nine years old when I endured my first IQ test. My class was herded into the gym to complete an hour of tests. The visiting tester explained the procedure but I have always been useless at following instructions – I glaze over as soon as they start – so when he shouted 'Start!' I had no clue what to do. I raised my hand. The irritated tester came over and whispered the very simple instructions. A few minutes had lapsed by the time I got going so I didn't complete the first batch of questions, which I presume would have affected my score. It made me wonder at the time that if such a small thing could influence an IQ score, was it really a reflection of how smart I was? Perhaps following instructions is part of smartness but it sure as hell wasn't the part that the test was meant to be assessing.

My next test came when I was starting out at high school as a twelve-year-old. This time I knew the ropes. There were 150 of us, ticking boxes as quickly as we could, to test our supposedly innate ability in verbal and non-verbal logic. Those of us in the top 20 per cent were placed in a special class, A1, in which we were hothoused. There was no attempt to obscure this intent; although we were never formally informed of our IQ test scores, some of our teachers were all too willing to offer explicit hints, sometimes to admonish us for not living up to our supposedly natural intelligence. Today, it seems like a dubious act of discrimination, giving special treatment to the select few on the basis of something they were supposedly born with, while the other 80 per cent felt crappy.

My father, an IQ enthusiast, had no reticence about informing my brother and me of his and our IQ scores. Nor had he any doubt that these magical numbers were set in concrete. But as my schooling progressed, I began to wonder. Most boys who ended up in A1 did well but some faltered. When we got to the end of the year, several in A2, 3, 4 and 5 outshone A1 boys. Others from the non-A1 classes seemed sharper, more streetwise and more creative. I had one friend who was caustically witty, had a way with words, an ability to mimic others to perfection and was intellectually astute. He seemed brighter than most in A1. This made me wonder what IQ was measuring.

Then I read that my hero, Muhammad Ali, had an IQ of 78; on the borderline of 'moronic', in the language of that time. I'd read that those with IQs below 80 went to special schools and could only manage the most basic conversations. But Ali had a wonderful way with words, his doggerel was funny and clever, his magic tricks sharp and he could more than hold his own in debates with university students and talk show hosts, charming them or putting them down with witty repartee and bursts of caustic invective. I began to query whether IQ really reflected something inbuilt. I had yet to do any reading on the history of IQ but my observations congealed into two doubts: first, that IQ really was hardwired and second, that the logic it measured was the same thing as intelligence.

My final IQ test came when I applied for my first job as a journalist. The newspaper decided that the magic number delivered by such tests was just what they needed to decide whether I and the other hopefuls were the right fit for their newsroom and the three-month course that preceded it. It seemed odd then and odder later. I discovered that being a good journalist required skill and cleverness, but not the kinds of skill and cleverness bound up in that number. IQ tests do not assess whether you're good at doggedly pursuing a story, asking the right questions, ingratiating yourself to get what you need, reading people well, building a fat contacts book or writing coherently in the style required and meeting your deadline.

THE INVENTION OF IQ, AND HOW IT LATCHED ONTO RACE

Scientists in the nineteenth century were obsessed with measuring brains to discern the intelligence of different races. After that, the focus changed to measuring what the brain produced. IQ emerged as the chosen method, but it didn't start as that. Until the twentieth century, the only move towards intelligence testing came from Darwin's eugenics-promoting cousin, Francis Galton, who firmly believed intelligence was entirely a product of natural selection and was unevenly distributed among the races. He used tests of reaction time and psychological records to try to estimate individual people's intelligence, but his methods were unsuccessful in predicting educational performance.

Alfred Binet, the Frenchman who invented IQ, had more egalitarian ambitions. He spent several futile years in the skull-measuring tradition but by the end of the nineteenth century he despaired of the idea that brain volume was an accurate way of measuring intelligence. In 1904 Binet, by then head of psychology at the Sorbonne, was commissioned by his government to find students who might benefit from special education. He combined a number of minitests, designed to find children's ability in ordering, comprehension, invention and correction (rather than learnt skills, such as reading), into one test, and reduced the results to a single number. He refined his methodology in 1908, by assigning an appropriate age level to each task, with the tasks getting progressively harder. The age of the last task the children could complete was their 'mental age'. Binet subtracted this from their real age; if the mental age was well below the real age, the child was chosen for special education. Soon after, a German psychologist, William Stern, proposed that it was better to divide the mental age by the real age, creating a 'Quotient', in which a score of 100 indicated that mental and chronological age were the same. The median IQ score is always 100.

Binet was a modest man. It was not his idea to call this number the 'intelligence' quotient; he recognised intelligence was too complex to be expressed in a single number. He also avoided making claims about whether below-par performance was acquired or 'congenital'.

All the number expressed was the average result for the mini-tests. As he put it: 'The scale ... does not permit the measure of intelligence because intellectual qualities ... cannot be measured as linear surfaces are measured.'[1] He feared that, if vested with too much significance, it 'may give place to illusions'.[2]

Binet's fears were realised after his death. Henry Goddard, research director at the Vineland Training School for Feeble Minded Girls and Boys, in New Jersey, introduced Binet's scale to the USA by translating his articles and applying his tests. His goal, unlike Binet's, was eugenically inspired: to identify 'morons', segregate them, stop them breeding, and prevent them from immigrating. He regarded test results as accurate reflections of 'inborn' intelligence that could not be changed, and believed inherited intelligence correlated with the natural order in life: criminals, prostitutes and alcoholics were morons; then came the toiling masses, and so on all the way to the top, where intelligent men deserved their large homes.[3]

As well as isolating American 'morons' in institutions, Goddard's squad were dispatched to Ellis Island to sniff out feeble-minded immigrants. Their tests, conducted in the home language of the immigrants, produced results showing that 87 per cent of Russians, 83 per cent of Jews, 80 per cent of Hungarians and 79 per cent of Italians were morons. 'We cannot escape the general conclusion that these immigrants were of surprisingly low intelligence,' he said, adding that America was 'getting the poorest of each race'.[4] After this, deportations increased dramatically (by 350% in 1913 and 570% in 1914).

IQ received its next big boost through the efforts of Professor Lewis M. Terman, of Stanford University, who introduced mass testing. He extended Binet's methods to include 'superior adults' and increased the number of tasks, calling it the 'Stanford–Binet' scale. He was most concerned about weeding out the feeble-minded who scored in the 70 to 85 range, and would therefore require vocational training, without which they could 'drift easily into the ranks of the anti-social or join the army of Bolshevik discontents'.[5] Terman shared Goddard's view that class differentiation was caused by inherited intelligence, because he found that the children of well-off parents had higher scores on average. He also applied this to race:

the poverty of the 'inferior races' was due to their low IQ and not their lack of opportunity to learn. IQs of between 70 and 80 were 'very, very common among Spanish-Indian and Mexican families of the Southwest and also among negroes', he noted, adding that this 'dullness seems to be racial'.[6] He predicted 'enormously significant racial differences in general intelligences, differences which cannot be wiped out by any scheme of mental culture' and suggested the segregation of these children because they were incapable of mastering abstractions.[7]

Terman even applied his methods to famous dead men. In 1917 he decided his hero, Francis Galton, had an IQ of 200. He went on to encourage an associate to study 282 other past lives, matching childhood precociousness with adult achievement. I remember reading these posthumous scores in one of my father's books, published by *Life* in the early 1970s. Goethe was top dog with an IQ of 204. Cervantes, Bunyan and Copernicus only managed 105, mainly because little was known of their childhood. Galton's cousin, Darwin, was accorded 135. Today, most consider this venture laughable but a few continued to take it seriously, including the man most responsible for the revival of racist IQ theory, Arthur Jensen, who called the Great Man scores a 'reliable estimate'.[8] And he added: 'The average estimated IQ of three hundred historical persons ... was 155.'[9]

The next and biggest IQ-popularising wave came under the rigorous direction of Robert Yerkes, who persuaded the US army to allow him to test 1.75 million recruits during World War I. He and his team (including Terman and Goddard) devised the tests: Army Alpha for literates; Army Beta for both illiterates and those who failed Army Alpha. Each test took less than an hour; the testers would quickly grade them and use the scores to suggest appropriate military placement. The army made scant use of these grades among the ranks, partly because of hostility to Colonel Yerkes's clipboard-wielding imposters, but they influenced officer recruitment and had a huge impact in spreading the popularity of IQ.

One of Yerkes's aides, bearing the tremendous name of Captain E. G. Boring, used 160,000 test results to devise racial averages. The average mental age for whites was 13, just above 'moron' mark.

Thirty-seven per cent of whites and 89 per cent of blacks were in the moron range. Boring concluded that Eastern Europeans and the darker races of southern Europeans were less intelligent than fair-skinned recruits from Western and northern Europe, with several having moronic average ages (Russians 11.34, Italians 11.1, Poles 10.74) but all were above 'Negroes' at 10.41. 'All officers without exception agree that the negro lacks initiative, displays little or no leadership and cannot accept responsibility,' he wrote. 'Some point out that these defects are greater in the southern negro.'[10] Yerkes assumed low scores reflected 'native intellectual ability' and ignored the fact that a large proportion could not speak English fluently, were unfamiliar with American ways, had not gone far in school, that many were unwell or that black recruits had been born only a generation after slavery ended. He even claimed the failure of blacks to attend school was itself a reflection of low intelligence.

Stephen Jay Gould, who studied the tests, found that Army Alpha was overloaded with detail, and that many recruits couldn't understand the questions and wrote nothing. They therefore recorded zero scores, which were included in the averages. He also noted it was really a test of literacy and the results reflected the fact that education levels differed widely, as did access to American cultural symbols. One question asked: 'Washington is to Adams as first is to ...?' Others included: 'Crisco is a: patent medicine, disinfectant, toothpaste, food product?', 'Christy Mathewson is famous as a: writer, artist, baseball player, comedian?' or 'Velvet Joe appears in advertisements of ... tooth powder, dry goods, tobacco, soap?' But for me the most astonishing question was: 'The number of a Kaffir's legs is: 2, 4, 6, 8?'[11]

Army Beta was hardly more encouraging for recruits not well versed in white American culture. Items they needed to know about included a rivet, a pocket knife, the filament of a light bulb, a tennis court net, the horn on a record player, a playing card, a violin and a bowling ball. They were asked about sports and film stars, commercial products and the primary industries of states and cities. Army Beta also required a knowledge of numbers and symbols and how to write them, which was beyond the ken of many recruits.

In addition, the administration was often chaotic. Sessions ran out of paper, took place in overcrowded rooms where recruits couldn't hear properly, and had inadequate time for the instruction. There were also varying standards when it came to selecting candidates for Army Beta; many black and immigrant illiterate or semi-literate recruits took Army Alpha and scored poorly, only to be denied the chance to take Army Beta, thereby reducing their group's average score.[12]

Yerkes's disciple Carl Brigham based his 1923 book, *A Study of American Intelligence*, on these results, claiming they were scientific and provided the 'first real contribution to the study of race differences in mental traits'.[13] He offered two reasons why northern blacks scored higher than southern: only the brightest were smart enough to head north and there was a 'greater admixture of "white blood"' among northern blacks,[14] while ignoring the fact that northern blacks had significantly more education. Discussing why earlier immigrants tended to score higher than those who arrived later, he again dismissed environmental influence and instead put it down to 'a gradual deterioration in the class of immigrants ... in each succeeding five-year period since 1902'.[15] He explained that 'lower and lower representatives of each race'[16] were arriving and that newer immigrants included duller Mediterraneans, whereas the earlier ones were smarter Nordics.[17]

Brigham's book, hailed by Yerkes for expressing 'not theories or opinions but facts', was used to promote immigration restrictions from low IQ countries. The 1924 Immigration Restriction Act set immigration limits at 2 per cent of the population of that group living in the USA but, crucially, based this on the 1890 census (before the arrival of most southern and Eastern Europeans). As a result, Jewish refugees escaping Nazi rule in the 1930s were refused entry, as were millions of Eastern and southern Europeans.

Most of these IQ pioneers – all but Yerkes – realised the error of their ways. Goddard said he had 'gone over to the enemy', admitting he'd over-cooked the results when it came to morons and was mistaken in believing they could not lead productive lives. He concluded that feeble-mindedness was 'not incurable' and became an advocate for 'moron education'.[18] Terman also had a rethink, noting the huge variability in IQ scores within classes and population groups and the overlap between them. He wrote in 1937 that the data did not 'offer any conclusive evidence of the relative contributions of genetic and environmental factors in determining the mean differences observed'.[19] In 1962 Boring described Yerkes's hereditarian conclusions about the army tests as 'preposterous'.[20] The most complete reversal came from Brigham, who rejected his past faith that a single IQ score could reflect intelligence and recognised that the army tests measured familiarity with American culture and language rather than anything innate. He also recognised that Army Alpha and Beta tested different things, which meant they couldn't be used to produce racial averages, and that the tests were internally inconsistent. 'Comparative studies of various national and racial groups may not be made with existing tests,' he concluded, adding: 'One of the most pretentious of these comparative racial studies – the writer's own – was without foundation.'[21]

In each case the *volte face* came too late: too late for Jewish and other European refugees, too late for the thousands of American women who were forcibly sterilised on the basis of their low IQ scores, and too late to prevent such ideas about measuring innate intelligence spreading all over the world.

THE *G*-SPOT, GENETICS AND RACE

One key element was missing in the early American approach to IQ: a scientific method of correlating test results into a common denominator for 'general intelligence'. This came through factor analysis, a mathematical technique invented by the English mathematician Karl Pearson and pioneered by the English psychologist Charles Spearman, who coined the name. Factor analysis is used to determine whether there is a high or low correlation between different matrices of data; if there is a positive correlation, factor analysis will generate a 'general factor' (or 'principle component') between them. IQ testing involves batteries of verbal and arithmetical tests, and factor analysis provided a way of correlating matrices from each of these mini-tests. The general factor between them Spearman called g, while there are also specific factors, which he called s. The more a test is 'g-loaded', the more it is seen as measuring general intelligence.

However, correlation tells us nothing about causation. There might be a causal connection between all the components but the maths won't tell us that. We need additional knowledge. There's a physical connection in the positive correlation between the growth of legs, arms, height, head size and body weight of a child. We know this from medical science, not from factor analysis. But the opposite could also be true, and it would still reveal a common principle component. A factor analysis of the rate of increase in my age, the price of my house and the growth rate of my garden tree might correlate positively but that doesn't imply my tree is making me older or raising the price of my house.

That said, the correlation among the IQ mini-tests is high because those who do well in one kind of test often do well in others. There is also a fairly high correlation between IQ tests and other kinds of academic work, because they measure similar things. The cause of these correlations may be genetic (some people do well because they're born clever). It may be environmental (some people do well because they're exposed to abstract logic from an early age, are taught to read early, are well educated, not under stress and enter the room expecting to excel). It could also be a combination of the two. Spearman was convinced g was entirely inherited. As he put it, 'g is ... determined innately; a person can no more be trained to have it in higher degree than he can be trained to be taller.'[22]

If we take two other IQ-related claims, the problem becomes clearer. It has been claimed that there is a small positive correlation between IQ and height and, as mentioned earlier, racist psychologists such as J. Philippe Rushton and Richard Lynn have also claimed a positive correlation between head size and IQ. In both cases the correlation tells us nothing. It could be that tall, big-headed people are genetically brighter. It could be that the Almighty chose to dish out brains and inches together. Or it could be that one group is taller and bigger-headed because its members are better fed and suffer fewer diseases, both of which could affect IQ. Or that taller people come from more educated families. Or that shorter people had mothers who smoked and drank alcohol during pregnancy, which reduces body size, head size and IQ. Or, from a different angle, we could ask whether the taller Dutch or Kenyan Masai have higher IQs than shorter Japanese or Chinese. Or we might look at head sizes of our ancestors from tens of thousands of years ago and ask whether the fact that they had brains 10 per cent bigger than today's average means they were smarter than modern humans. I could go on and on. But the point is clear: we can never assume that correlation equals causation.

A related problem with g is that it assumes 'general' intelligence genuinely exists as a distinct entity. It could be that g is simply an average formed by amalgamating two types of information (verbal and arithmetical). A different view was provided by Louis Thurstone, a professor of psychology from the University of Chicago, who believed Spearman failed to identify the 'true' vectors of the mind, which involved independent primary mental abilities. When he took the same data as Spearman, but submitted it to his own methodology, g disappeared altogether. He therefore rejected g, questioning whether it could have a deep meaning if it could shift significantly depending on the test. He also rejected the notion of general intelligence. 'We cannot be interested in a general factor which is only the average of any random collection of tests,' he wrote in 1940.[23] Instead children should be assessed individually for their strengths and weaknesses in each of the seven 'Primary Mental Abilities' (PMAs) he isolated,[24] which would mean 'recognising every person in terms of the mental and physical assets which make him unique as an individual'.[25]

Spearman was undeterred, noting that Thurstone's PMAs were also a product of the tests he had chosen and that he'd achieved no more than providing an alternative maths for the same data. He described g as a physical force that scientists would one day be able to measure. '[O]ne can talk about mind power in much the same manner as about horse power,' he said.[26] Like his American contemporaries, he took the racial averages of the US army tests at face value and praised the 1924 American Immigration Restriction Act for favouring the more intelligent 'German stock' over southern Europeans. He claimed that black people possessed lower doses of 'mind-power' than white, noting that 'the coloured were about two years behind the white,'[27] although he recognised that some from the inferior race group would have more *g* than some in the superior group, an argument used seventy years later by Charles Murray and Richard Herrnstein in their book *The Bell Curve*, to counter charges of racism.

IQ testing came to depend on *g*. Hereditarians using it for racist purposes were particularly protective. In 1981, when Gould's *The Mismeasure of Man* was first published, in which he devoted eighty-four pages to a systematic critique of the use of factor analysis in intelligence testing, IQ fundamentalists could offer nothing more substantial than the reiteration of their viewpoint that *g* was real because general intelligence was real. When Gould repeated his argument in a critique of *The Bell Curve*, Charles Murray defended his fifteen-year silence, saying that the 'factor analysis argument was so irrelevant to the state of knowledge ... that it provoked only perfunctory comment in the technical literature'. He added that he and Herrnstein had learned that *g* 'is associated with brain function at a neurological level. Not bad for a statistical artefact'.[28] Needless to say, no neurological evidence was presented.

Yet, just like the American IQ proselytisers, the man behind *g* had doubts in his later years. In his final book, published in 1950, five years after his death, Spearman acknowledged that the whole business had been a bit of fuss about nothing. He recognised that his factors had no necessary relation to physical reality and that it had all been the folly of youth. He started his *mea culpa* with a quote from Horace: '*Dulce est sespere in loco* [it is pleasant to act foolishly from time to time]', before referring to himself in the third person: 'But for the present purposes he has felt himself constrained to keep within the limits of barest empirical science. These he takes to be at bottom nothing but description and prediction. ... The rest is mostly illumination by way of metaphor and similes.'[29]

CYRIL BURT, IQ AND A-LEVELS

Spearman's successor was an Oxford-educated psychologist, Cyril Burt. At twenty-three years old, Burt had worked on Francis Galton's national survey on the mental and physical characteristics of the British people. He was convinced that differences between social classes were hereditary and that eugenic policies were the future. He was also certain that Spearman was right about g. He went further, using it to find the 'general factor' in other inquiries that intrigued him, from standards of beauty to the paranormal, identifying a 'general paranormal factor' in his studies of parapsychology and extrasensory perception. He added to Spearman's factor analysis by identifying what he called 'group factors' (between g and s) that were secondary to g and trainable rather than innate, but these were extensions rather than revisions of Spearman's theory. Burt shared the view that the factors reflected reality: group factors related to specific areas of the cerebral cortex, he said, while the all-important g represented the 'general degree of systematic complexity in the neuronal architecture'.[30]

Burt's most ardent campaign was to prove this was innate, a quest he started in 1909, when he tested forty-three boys from two Oxford schools. He decided upper-class boys were inherently smarter than lower-middle-class boys, discounting the possibility that their childhood environment might have had an influence. One of his arguments was that the boys' intelligence scores correlated with their parents', but he did not measure parental IQs; instead, he deduced them from their professions and social standing. Burt found himself an ideal laboratory when he worked as a psychologist for the London County Council and was responsible for testing in all London schools. He used factor analysis to establish a close correlation between 'backwardness' and poverty but instead of drawing the obvious conclusion – that poor environments prompt 'backwardness' – he decided that inequality of income was caused by 'inequality in innate intelligence'[31]; that stupid people gravitated towards poverty. 'In well over half the cases', he wrote in 1937, 'the backwardness seems due chiefly to intrinsic mental factors ... that extend beyond all hope of cure.'[32] He insisted that most working-class children were incapable of higher education and that even if they received it, their children would accrue no benefit but instead would inherit 'the original ignorance' (which one would assume most babies are born with).

To emphasise this view of the inbuilt stupidity of the poor, Burt wrote of slum children's 'protruding muzzles' and 'thickened lips' which gave their profiles a 'negroid or almost simian outline'.[33] Given these views on race, it's hardly surprising that, particularly towards the end of his life, he developed close connections with the leading figures of race science. He taught two of the main figures in racist psychology, Hans Eysenck and Raymond Cattell, and retained those connections. He later developed friendships with three other extremists of this calling: Arthur Jensen, William Shockley and Christopher Brand.

After succeeding Spearman as chair of psychology at University College London (UCL), Burt advised government bodies on the introduction of the eleven-plus exam in British state schools. They used his conclusions from studies of separated identical twins, that intelligence was hereditary, to promote tests partly based on IQ and partly on maths and English ability. The top 20 per cent of eleven-year-olds went to grammar schools to prepare for university; the rest to secondary modern or technical schools, to prepare for work. But the results were skewed towards boys (there were more grammar school places for boys) and towards the south (35% of pupils in south-west England secured grammar school places compared to 10% in Nottinghamshire). In line with Burt's prejudices, the main bias was against working-class pupils. Studies of test outcomes show that children on the borderline of passing were more likely to be pushed through if they came from middle-class homes. Working-class children were also not helped by the nature of some questions, such as those about classical composers or household servants.[34] Such criticisms prompted changes in the eleven-plus in the 1960s when they became purer IQ tests. Grammar schools were abolished in 1976 but some survived the cull. The system was again promoted under the Conservative governments of the early twenty-first century, still using IQ tests. Meanwhile, as we shall see in the next chapter, Burt's reputation unravelled when it emerged that the separated identical twin studies he cited did not, in fact, exist.

POST-WAR IQ TESTING

When gurus fall it seldom unsettles the disciples for long. Burt's exposure, Spearman's doubts and the recanting of the US IQ pioneers had no impact on the growth of the IQ industry. It was too big to fail and too useful, not least because the correlation between IQ scores and academic grades made it a valuable tool for educators. Although its usefulness in predicting job performance varied from profession to profession, it was widely used to weed out applicants for many jobs.

From inside the IQ industry the postwar developments might seem significant but they look less so from outside. Tests retain 'general intelligence' as their single number centrepiece, although some internal criticisms, particularly Thurstone's, were integrated, making room for something like primary mental abilities, albeit below almighty g in the hierarchy. Jim Flynn does not accept Thurstone's rejection of g but suggests that general factors don't work if IQ is considered over a period of time, 'when real world cognitive skills assert their functional autonomy and swim freely of g'. He adds: 'If you want to see g, stop the film and extract a snapshot; you will not see it while the film is running. Society does not do factor analysis.'[35]

A major figure after the war was the US-based Romanian psychologist David Wechsler, who defined intelligence as 'the global capacity to act purposefully, to think rationally and to deal effectively with the environment'.[36] Although he retained the notion of general intelligence, he was sceptical about restricting it to a single score. Flynn pointed out that average IQs in most countries were rising year after year; the Wechsler tests were the first to accommodate this by constantly upgrading their questions. Another of Weschler's changes was the abandoning of the quotient dimension (dividing mental age by chronological age); instead he used 100 as an arbitrary mean value and added or subtracted fifteen points for each standard deviation above or below it.

A parallel development started with Burt's protege Raymond Cattell, who, in 1941, revised Spearman's g by dividing it into what he called 'fluid intelligence' (Gf) and 'crystallised intelligence' (Gc). The former was defined as the ability to solve novel problems through reasoning, the latter as knowledge-based ability linked to experience and education. One of his students, John L. Horn, revived this in 1966, adding other broad cognitive abilities. Then, in 1993, the American psychologist John Carroll came up with the 'Three Stratum theory', which involved levels of testing: specialised abilities (such as spelling) at the bottom; 'broad abilities', which he viewed as different 'flavours' of g, in the middle; and general intelligence at the top. These were merged to become the Cattell–Horn–Carroll theory, with g at the top, ten broad abilities in the middle and seventy narrow abilities at the bottom. Contemporary IQ tests give an overall score for g and most also give scores for 'broad abilities' but they often deliver different scores for the same people. One study looking at three of the most popular IQ tests used in the USA showed individual variations of up to twenty-two points among a sample group.[37]

It is worth noting where these IQ theorists came from. Unlike Flynn, who is a critic of racist IQ theory, Cattell's views on race and eugenics were flamboyant. His name features on the masthead of the racist journal Mankind Quarterly and he frequently warned that Western countries were in danger of genetic deterioration. He proposed a scientific religion, 'Beyondism', founded on group selection and eugenics, and decided the future would require colonisation of the solar system or the division of humanity into separate species through genetic engineering.[38] Cattell and Carroll were signatories to a twenty-five-point public statement, written by the psychologist Linda Gottfredson in 1994, to support the positions advocated by the race science Bible, The Bell Curve.

NEUROLOGICAL EVIDENCE AGAINST *G*

One of the arguments against g is that it exists only in the minds of the testers, because intelligence is not amenable to capture by a single number. It is not a thing. It is an abstraction, and when we try to pin down abstractions, they become slippery.

It has become commonplace to talk of different forms of intelligence. We all know smart people but when we think about it, most are smart in some ways but not others. Our colleague the computer geek may never have read a novel; our literary friend may be useless at arithmetic; someone we consider deep and wise may flounder in practical skills; another who is incisively clinical in argument may lack the capacity to discern the emotions of those their caustic eloquence is upsetting; one who is artistically brilliant may be dyslexic. For me, whatever academic pretensions I may have, I know full well they are more than balanced by serious thickness. I have no sense of direction or practical *nous* and I'm useless at following simple instructions. I've always known that I've had these weaknesses and that they would influence my career options. And I've always assumed that biology played a role. After all, no one aims to have a poor sense of direction.

IQ tests measure a certain kind of intelligence; the kind that helps people do well in activities that require abstract logic. Someone with a high IQ score might well excel in a maths test and might also do well in certain other academic disciplines and certain professions. But IQ tests do not measure 'emotional intelligence', or creativity, or innovation, or wit, or wisdom. They do not measure artistic or musical ability, or talent for languages, or intellectual curiosity. They do not measure our way with words, or our ability to engage with existential issues. They do not measure long-term memory, practical ability or business acumen. Yet these are all dimensions of what we regard as intelligence.

The problem relates to semantics. Intelligence is a single word with different meanings. Because these meanings fall under the umbrella of one word, we are inclined to assume they are all one thing. But, really, they are many things. The 'general factor' that gets you both a high IQ score and an A* in maths is not the same as the intelligence needed to create an artistic masterpiece, to read the social dynamics of a dinner party, understand the nature of the emotional pain someone is experiencing or consider the world in existential terms. Some who are good at maths are also good at languages, art, music, social dynamics and existential thought. But one doesn't necessarily follow from another.

In the 1980s the Harvard-based psychologist Howard Gardner devised a theory of multiple intelligences in the Thurstone tradition, based on his view that there were different ways of learning and processing. He identified eight intelligences: linguistic, logical-mathematical, musical, spatial, interpersonal, bodily/kinaesthetic, intrapersonal and naturalistic. 'When we talk about a person being "smart"', he explained:

> we typically mean that he or she is good in school and that they have strong linguistic and logical mathematical skills. But my theory holds that individuals can be strong in other areas as well, ranging from music and spatial abilities to understanding other persons and that these multiple intelligences constitute a better description of a range of human cognitive capacities. Just because a person is smart in one area, we simply can't predict how they will do in other areas of life.[39]

This view hardly delighted those involved in the huge IQ industry, but it has been reinforced by neurological discoveries in a large-scale international cognitive study published in 2012 in the neuroscience journal *Neuron*.[40] More than 100,000 people from around the world were asked to complete twelve mental tests measuring different dimensions of cognitive ability. The researchers narrowed this to a representative sample of 46,000 people and found there were three distinct components to intelligence: short-term memory, reasoning and verbal ability. They also found that no single component (such as *g*) could explain variations in the results. They then analysed the brain circuitry of a sample of the volunteers using MRI scans and found that these three components of intelligence corresponded to distinct patterns of neural activity, using different nerve circuits.

'The results disprove once and for all the idea that a single measure of intelligence such as IQ is enough to capture all of the differences in cognitive ability that we see between people,' said Roger Highfield, one of the scientists involved. 'Instead, several different circuits contribute to intelligence, each with its own unique capacity. A person may well be good in one of these areas but they are just as likely to be bad in the other two.' This had implications for intelligence testing: 'It has always seemed to be odd that we like to call the human brain the most complex known object in the universe,' Highfield said:

> ... yet many of us are still prepared to accept that we can measure brain function by doing a few so-called IQ tests. For a century or more, many people have thought that we can distinguish between people or indeed populations, based on the idea of general intelligence which is often talked about in terms of a single number: IQ. We have shown here that's just wrong.[41]

None of this has direct bearing on whether we see intelligence as mainly biological or mainly environmental. Thurstone and Gardner viewed their 'primary mental abilities' or 'multiple intelligences' as primarily genetic in origin. Even if the entire IQ industry abandoned g, it's likely that race-based claims for each of these multiple intelligences would re-emerge; that someone would say Asians were better at logical-mathematical or Africans better at musical and spatial. However, the significance of this critique is that it allows for assessments of people's individual abilities. g can be useful in predicting certain kinds of academic performance or certain kinds of career success but is less useful when it comes to analysing an individual person's cognitive strengths and weaknesses. IQ leaves out a great deal that comes under the banner of intelligence.

10

WHAT CAN TWINS TELL US ABOUT NATURE AND NURTURE?

Jim Springer and Jim Lewis were identical twins, born to a fifteen-year-old unmarried girl. They stayed together for a month before being adopted by families of similar backgrounds who lived about forty miles apart. Lewis's adoptive mother visited the local court-house to settle the adoption papers and was told by an official: 'They named the other little boy "Jim" too.' Before then, she hadn't known her son was a twin. Eventually, she told her son, as did Springer's adoptive mother. In February 1979, when the twins were thirty-nine, they had an emotional meeting. After the case was reported, a Minneapolis-based psychologist, Thomas Bouchard, who had studied twins for a decade, saw his chance. Two weeks later he met them in his office, and a series of detailed interviews began.

It turned out that both had worked at McDonald's, and at petrol stations, albeit at different times. One had worked as a security guard, the other as a deputy sheriff. They had both taken holidays in the same part of Florida, drove Chevrolets, had dogs called Toy and had married women called Linda and Betty. One had a son called James Allan, the other James Alan. Both preferred maths to spelling, enjoyed carpentry, chewed their nails, smoked Salem and drank the same beer. Both had haemorrhoids, gained 10lb at the same age, started experiencing migraines at eighteen and had similar heart problems and sleep patterns. The story spread around the globe, finding its way into school textbooks and all over the World Wide Web. Some saw this as a case of telepathy between identical twins, others as one of the extraordinary pull of genetics.[1]

THE JIM TWINS, GIGGLE SISTERS AND NAZI-JEWISH TOILET FLUSHERS

Reality was rather less sensational. Over the half-century before 2018, 1,894 separated identical twins have been tested but none of the others have shown anything like the level of overlap of the Jim Twins.[2] There are clear genetic links to heart problems, migraines and tendency to gain weight, and no doubt to sleep patterns, preference for maths and nail-biting too. But Salem cigarettes, dogs called Toy and wives called Linda and Betty? Pure chance, just as there are serendipitous coincidences in most people's lives that might feel meaningful when considered in isolation. Further investigation suggested they were hardly carbon copies. They lived with similar families, of similar backgrounds, in the same neck of the woods, so we would expect parallels in the choices they had made. But they had different stories to tell. They had changed to different kinds of jobs (Lewis a furnace worker; Springer a meter reader). They had very different hairstyles and facial hair. One turned out to be better at writing, the other at speaking. Springer was 'more easy-going' and Lewis 'more uptight'.[3] Springer stayed with Betty; Lewis divorced and married for the third time, to a woman called Sandy.[4] Also, curiously, we were never told whether they had the same IQs.

In other cases highlighted by Bouchard, the differences were more marked. But inevitably, the facts chosen by the news media are the headline-grabbers. The Jim Twins always take pride of place but if back-up is needed, their story tends to be followed by that of the former Nazi twin and his Jewish-raised brother. In Trinidad, Oskar Stohr and Jack Yufe were separated at six months old. Oskar was brought up as a Catholic in Germany and joined the Hitler Youth. Jack remained behind, brought up as a Jew. They were reunited in their forties; they had similar speech patterns, gaits and food tastes, and both flushed the toilet before using it. Then there were the 'Giggle Sisters', Daphne Goodship and Barbara Herbert. Again, one was brought up Jewish and the other Catholic and they were reunited in their forties. Unlike their adoptive families, they were both incessant gigglers and had a fear of heights.

Details pointing in the opposite direction were played down. In one of Bouchard's early cases, one twin was adopted by a fisherman, brought up in a home with few books and was not educated beyond early high school; the other was adopted by a cosmopolitan family, lived all over the world and became an electronics expert for the CIA. His IQ was twenty points above his brother's. In another pair, the IQ gap was twenty-nine points, without any hint of extreme deprivation.[5] It seemed that significant differences in educational levels, social class and exposure to knowledge could prompt significant differences in adult IQ.

Most cases showing gaps between separated identical twins are ignored by the media. One much-covered exception involved two pairs of Colombian identical twins, who were mixed up in a hospital error and brought up separately as pairs of fraternal twins, one pair in a poor rural area, the other in a lower-middle-class urban area. When they met, after discovering the error as young adults, the initial reports focused on similarities. Then Nancy Segal, a professor from California State University, formerly Bouchard's lead researcher, subjected them to interviews, questionnaires and IQ tests. She expected similarity among the identical twins but although their IQ results were not released, she acknowledged they were significantly less alike than anticipated. 'The Columbian pair really made me think hard about the environment,' she said.[6] Elsewhere she added: 'I came away with a real respect for the effect of an extremely different environment.'[7]

GALTON, MENGELE AND THE TRICKY BUSINESS OF COMPARING FRATERNAL AND IDENTICAL TWINS

Twin studies begin with the eugenicist Francis Galton. In the 1870s he wrote to thirty-five pairs of apparently identical twins and twenty-three pairs of apparently fraternal twins ('apparently' because Galton, like his cousin Charles Darwin, was unaware of genetics, so judged on looks alone). He used their anecdotes to draw conclusions: the twins who looked alike had similar personalities, ailments and life interests throughout life, whereas those who looked different became even more different as they aged. Galton claimed that with both sets the 'external influences have been identical; they have never been separated' and therefore the results proved 'nature prevails enormously over nurture.'[8]

Galton's sole method of determining whether the twins were identical was appearance. But, as we now know, some fraternal twins look very similar and some identical twins do not look identical, and his conclusions were drawn entirely from what his group said about themselves in response to his letters. Still, today's hereditarians are inclined to give him a thumbs-up. The science writer Matt Ridley, for one, endorses Galton's results:

> He would have to wait more than a century to see that the study of twins did in the end prove much of what he had suspected. To the extent that they can be tested apart, nature prevails over one kind of (shared) nurture when it comes to defining differences in personality, intelligence and health between people within the same society. Note the caveats.[9]

Galton used his conclusions on twins to promote eugenic beliefs about purifying the population, which, a decade after his death, became the essence of the Nazi credo. It was under the Nazis that the next twin studies took place, through Josef Mengele's notorious research in Auschwitz that involved 1,500 twin pairs. The physician, known as the 'White Angel', the white-coated man who determined who went to the gas chambers, would order any twins among incoming prisoners to step out so that he could experiment on them. He ordered two Jewish doctors to assist him. Twins were treated better than the rest – sometimes given sugar and little gifts – but invariably as a prelude to horrendous experiments. These included injecting their eyes with chemicals to try to change their colour, amputating limbs, removing organs and sterilising women without using any anaesthetic. Most died, or were killed later. In one case, one of his assistants injected chloroform into the hearts of fourteen pairs of Roma twins, after which Mengele dissected their bodies. He sewed a pair of Roma twins together to create conjoined twins; they died of gangrene. He connected one girl's urinary tract to her colon. Sometimes he simply shot them and then dissected them.

These revelations gave twin studies a bad name but they continued, mainly because those wanting to find whether a particular trait had a genetic origin had little alternative. Twin studies have since been used in several countries to test anything from whether vitamin C can prevent colds (it can't) to whether homosexuality has a genetic origin (minor influence for gay men and even smaller for lesbian women).[10]

All these studies, like Galton's, involved comparing the behavioural traits of monozygotic (single egg, identical) twins and dizygotic (two egg, fraternal) twins. This method was also used by a team headed by Robert Plomin, the British psychologist we met in Chapter 9, who acknowledged failure when it came to finding a single gene for intelligence. His subsequent research prompted a wave of media publicity, including a lead story in the *Guardian* entitled 'Genetics accounts for more than half of variation in exam results',[11] which concluded that 'genetics accounts for almost twice as much of the variance of GCSE scores (53 per cent) as does shared environment (30 per cent).'[12] Their method was to use a national sample of 11,116 sixteen-year-old twins, and compare the marks of identical and fraternal twins, on the assumption that both shared the same environment but only the identical twins were genetic carbon copies. Towards the end of their published paper, the team noted potential problems, including 'the equal environments assumption – that environmentally-caused similarity is equal for MZ and DZ twins – and the assumption that results for twins generalize to non-twin populations'.[13]

To illustrate why this method is more than just a minor concern, meriting no more than a sentence at the end of a paper, I'll use two examples of twins in my classes at school. One pair – I'll call them the Thompson twins – were so identical that I never managed to tell them apart. They always seemed to get the same marks, be in the same sports teams and have the same friends. Then there were the Wellingtons, Amy and Mary, whom I knew in Texas. They didn't look anything like each other, and everyone related to them differently. It never seemed a big deal to Amy, but she was regarded as unusually pretty and everyone wanted to be her friend. Without having to try hard she was the life of the party, and was effortlessly sporty. People gravitated towards her and everything came easily, so she didn't put effort into her schoolwork and preferred hanging out, smoking cannabis and having fun. Mary was regarded as average-looking and very 'straight'. She had a small group of friends, steered clear of cannabis, focused on academic work and excelled. I have no clue what their IQs were, but I'd expect a gap between the Wellingtons, whereas it would surprise me if the Thompsons were more than a point apart.

At first blush this seems to confirm Plomin's assumption that the gap between identical twins is narrower because of their shared genes. But it could also be influenced by real environmental differences. Everyone treated the Thompsons alike but the Wellingtons less so; although they were born just a few minutes apart, their life experiences were distinct. Partly because they looked so different, they were treated differently. Amy was doted on by teachers, coaches, schoolmates and family members. Mary did not get the same attention and grew up without any expectation of adoration. They carved out different paths early in life; Mary became more studious, more focused, less casual. She read more, studied harder and excelled in areas her sister didn't bother to contest.

I've known fifteen sets of twins and triplets since childhood. Eight said they were identical and seven fraternal, which is unusual.[14] The identical twins looked alike, were treated alike by their teachers, had the same friends, went to the same parties. They saw themselves as part of a set. It was different for the fraternal twins. There were significant distinctions in how they were treated by classmates, teachers and parents. They found separate niches and were regarded as individual people, leading to different life experiences. Genetic determinists might say: 'Aha! Your fraternals were treated differently because they *were* genetically different.' And they'd be right. But this raises a problem for the hereditarian assumption that identical and fraternal twins share each other's environments in the same way. It doesn't matter *why* they were treated differently. The mere fact of it – whether through looks, abilities or inborn cognitive differences – means we cannot assume the differences in the abilities are primarily genetic, which is precisely what is assumed by people such as Plomin, who insist on devising percentages for the nature-nurture components of IQ and academic performance.

To test this, we might ask if any of the differences in the fraternal twins' experience affect IQ scores. Amy drank more alcohol, smoked cannabis and spent less time on school work than her sister, but had more affirmation and was more self-confident as a result. As we shall see, all these factors could affect the results of IQ or academic tests. For fraternal twins there is also a significantly higher chance of different experiences that can affect outcomes – an inspiring mentor, a life-changing overseas trip, a bump on the head – and these different experiences are likely to be more marked as they grow older.

There's an additional environmental factor that twin studies ignore: the experiential gap starts in the womb because fraternal twins each have their own placenta, and one twin might be bigger than the other, whereas three quarters of identical twins share a placenta, and some share the same amnion (the membrane that protects the embryo). They are also more likely to share viruses and to be exposed to the same accidents. One meta-analytical study that focused on the impact of the foetal environment on IQ concluded it accounted for 20 per cent of the explanation for IQ differences or, in the case of twins reared apart, it explained 20 per cent of the reason for their similar IQs.[15] Although this is a biologically related explanation, it is not a genetic one, yet the design of twin-based research studies ignores the likelihood that the life experiences of fraternal twins, both in the womb and subsequently, are less similar than those of identical twins. These studies start with the false assumption that the twins' shared environment is the same and their conclusions suffer as a result.

CYRIL BURT AND THE IMAGINARY TWINS

The other method of studying twins to determine heritability – finding separated identical twins – started in the UK in the 1950s under

Charles Spearman's errant disciple, Cyril Burt. When Sir Cyril died of cancer in 1971 his reputation in his profession was unmatched, but it soon collapsed. In retrospect, the clues might have been picked up earlier, when he made the false claim that he, not Spearman, was the father of factor analysis in IQ testing. Burt first tried this claim before Spearman's death in 1945 but apologised after being put down by the older man. Once Spearman died, however, Burt campaigned to undermine his mentor's influence, and stopped citing his publications. Also, he was in the habit of writing letters, under various aliases, to his own journal, and shortly before his death he ordered that all his notes and records be burnt. But this was small potatoes compared to what was to come.

Burt made his claims about the heritability of IQ in a series of academic papers reporting studies of separated identical twins that he claimed to have conducted. Unlike in China, where the one-child policy led to many cases of separated identical twins, in the UK these were and are notoriously hard to find, for the simple reason that parents and adoption agencies keep twins together. Yet somehow, Burt found fifty-three sets. Initial queries were raised by the psychologist Leon Kamin, of Princeton, whose team investigated Burt's work, publishing their study in 1974.[16] Kamin noted that Burt had increased his claimed twins' sample from twenty-one in 1955 to fifty-three in 1966, yet the average correlation in their IQs remained unchanged, down to the third decimal place. In the 1950s Burt claimed a correlation in the IQs of separated identical twins of 0.771; in the 1960s a paper published under the name of his supposed assistant, Margaret Howard, also cited a correlation of 0.771. Later, Burt and another supposed assistant, J. Conway, published a paper with the same correlation, 0.771. Kamin's Princeton team deemed this coincidence so unlikely as to fit into the category of impossible.

Some of the details of the separated twins Burt claimed to have found were more suggestive of a fertile imagination than real life. They included a pair of illegitimate identical twins from a well-to-do birth mother; one twin grew up on the Scottish country estate of a prosperous family, the other was brought up by a shepherd, a case that sounds as if it emerged from a reading of Perdita's story in Shakespeare's *A Winter's Tale*.[17] What is even more curious is the close correlation in their reported IQs (118 and 121).[18] As we have seen, when separated identical twins are brought up in different class and educational backgrounds their IQs tend to vary considerably, by twenty-nine IQ points in one case.

Two years after Kamin's exposé, the medical correspondent of the *Sunday Times*, Oliver Gillie, made the direct accusation that Burt had faked his data. He also wrote that Burt's two collaborators, Margaret Howard and J. Conway, either didn't exist or were not in contact with him at the time, even though Burt had included their names as authors or co-authors of some of his papers and they had reviewed his books positively. There was no record of either woman having worked at UCL, where Burt was based, of either having published anything independently, and no one remembered them working with Burt. Towards the end of his life, when asked, Burt said they had both 'emigrated' but was unsure where, which is odd; one would think he would ask where they were going when they said they were leaving the UK. Subsequent research suggested it is likely that both women did exist, although it is doubtful either wrote the papers or reviews attributed to them. And there is no evidence of the existence of a third woman, Miss M. G. O'Connor, whom Burt also named as a collaborator.[19]

Burt's backers leapt to his defence, particularly those who'd used his twin research. They included his former student Hans Eysenck, who wrote to Burt's sister that it was all a plot by 'left-wing environmentalists'.[20] Arthur Jensen and J. Philippe Rushton (who relied on Burt's twin data to support his claims about innate intelligence) claimed that the correlations were unsurprising, while the eugenicist sociobiologist W. D. Hamilton said Burt's opponents were either wrong or had made gross exaggerations. Two other Burt supporters, Ronald Fletcher and Robert Joynson, wrote books backing him.[21] Fletcher provided evidence of Howard and Conway's probable existence, while conceding the absence of evidence for O'Connor.[22]

But by then the killer blow had landed. It came from a surprising source: Burt's close friend, Leslie Hearnshaw, whom Burt's sister commissioned to write an official biography. Hearnshaw started as a Burt enthusiast, but his 1979 book, *Cyril Burt: Psychologist*,[23] shattered Burt's reputation. Hearnshaw concluded all the allegations against Burt were true, that the fraud had begun nearly thirty years before Burt's death, and that all his studies on separated twins were invented. He also accused Burt of fabricating other claims, including those about declining intelligence levels in Britain. Today, it is widely accepted that none of Burt's postwar research can be trusted. The entry in the *Encyclopaedia Britannica* puts it gently: '[A]ll sides agreed that his later research was at least highly flawed and many accepted that he fabricated some data.'[24] Others, less gentle, regard Burt and his career as a salutary example of scientific fraud and establishment gullibility.[25]

Burt's descent into pumping out fake news may have been influenced by the effects of suffering from Meniere's disease, a hearing and balance disorder that can cause heightened 'anxiety and neuroticism',[26] as well as by personal setbacks and his fear that the hereditarian case was being undermined by evidence of the brain-altering impact of early intervention education schemes. These interventions were based on Mark Rosenzweig's research in the early 1960s, which showed that an enriched environment increased the volume of the cerebral cortex of rats.[27] Stephen Jay Gould, however, softened his earlier condemnatory approach, arguing that 'the very enormity and bizarreness of Burt's fakery forces us to view it not as the "rational" programme of a devious person trying to salvage his hereditarian dogma when he knew the game was up (my original suspicion, I confess) but as the actions of a sick and tortured man.'[28]

BOUCHARD'S PIONEER-FUNDED STUDIES OF SEPARATED IDENTICAL TWINS

The next big wave came in 1979, with the Jim Twins case. After this fortuitous success, Thomas Bouchard received generous grants from the overtly racist Pioneer Fund, set up in 1937 to promote eugenics. The fund was inspired by a visit to Nazi Germany made by its anti-Semitic billionaire founder, Wickliffe Preston Draper, who later said his mission in life was to 'prove simply that Negroes were inferior'.[29] Draper funded segregationist causes and opposition to civil rights and backed the compulsory sterilisation programme in North Carolina. He was succeeded by fellow Nazi-backer Harry Laughlin, who campaigned for eugenics and segregation in the American South. The Pioneer Fund's current stated aim is to research or aid research 'into the problems of heredity and eugenics ... and research and study into the problems of human race betterment'.[30] In recent years it has funded people promoting racist psychology with the aim of emphasising IQ variation among races.

Bouchard approached the Pioneer Fund himself. And you don't pick an overtly racist funder naively. Nor was it a one-off. Bouchard joined the usual suspects of racist psychology in signing Linda Gottfredson's race-based 'Mainstream Science on Intelligence'.[31] Presumably he did so because he agreed with its agenda. He also offered an enthusiastic back cover endorsement for Rushton's overtly racist book, *Race, Evolution and Behavior*. Some of Bouchard's cheerleaders, such as Ridley, have played down the Pioneer Fund issue:

> Their motive in supporting Bouchard's research is presumably that they want to believe genes influence behaviour, so they give money to a researcher who seems to be getting results which support such a conclusion. Does that mean that Bouchard and all his many colleagues ... have faked their data to please funders? Seems pretty far-fetched.[32]

Ridley is right, but we have only to look at how the racist agendas of the nineteenth-century skull-fillers and twentieth-century IQ pioneers distorted their use of data to make us wary of research that carries an unacknowledged political load. As we shall see, Bouchard consistently overestimated the genetic contribution and understated the environment.

Bouchard's Minnesota Center for Twin and Adoption Research was building its twins larder when a new book by clinical psychologist Susan Farber was published.[33] She went through several major studies of identical twins brought up separately (ninety-five sets in all) and found that in 90 per cent of the cases the twins were selected *because* they were notably similar. The reason for this bias was that the easiest way of finding separated twins was to advertise; those who responded had already contacted their siblings and believed they were identical because of their similarities. Many had been reunited years before and had substantial contact since, reinforcing their points of similarity. Farber also questioned what 'reared apart' meant, finding that most spent years together, often being adopted by relatives, which meant that they grew up in the same kinds of families. When she examined their reported IQs, she found that the more separately they were brought up, the greater the IQ gap. The largest reported study (other than Burt's fraudulent one) involved forty four pairs of separated identical twins. The data show thirteen pairs lived together for at least their first two years, and five of those for four years or more. Of the remaining thirty-one who were separated before their second birthdays, nine were reunited by the time they were twelve. Most were brought up in similar environments.[34]

The late Leon Kamin, the Princeton psychology professor who first exposed Burt's twin studies fraud, was dismissive of the publicity-seeking Bouchard and the entire twins industry, saying it was 'not serious scientific work'. He went on to argue:

> If you lived in a science fiction world, you could scatter these twins randomly and then go back and find them but they don't get put into homes by storks. They are put into homes by adoption agencies and sometimes the parents. ... [T]here is a grotesquely underestimated degree of previous contact and ... an enormous amount of pressure on the twins to come up with cute stories. The whole thing is a media show-business hype that simply strains credibility.[35]

Bouchard absorbed the criticisms and tried to avoid the potential pitfalls. Because of the publicity he'd courted, separated twins had started contacting his Pioneer-funded team of eighteen researchers. The twins were given honoraria, subjected to a battery of fifty hours of tests, including of IQ, and asked thousands of questions about family, childhood, personal interests, jobs, values and aesthetic judgements. By 1990, Bouchard's 'Minnesota Study of Twins Reared Apart' had analysed data from fifty-six separated identical pairs. He took into account the time his twins had spent together (an average of 5.1 months before separation and 20.3 months following the first reunion) but, contrary to Farber, concluded that there was 'little evidence' that the degree of contact before or after separation 'counted for much'.[36]

DOES THE HERITABILITY OF IQ REALLY INCREASE WITH AGE?

Adult IQ was 70 per cent heritable, Bouchard claimed, explaining that his figure was higher than previous estimates because his studies involved adults and the genetic component of IQ increases with age. This is a common perception among twin researchers; Plomin tells us that the heritability of intelligence increases from 40 per cent in childhood to 65 per cent in young adulthood and to 80 per cent at the age of sixty-five.[37] But the explanation offered for this increasing 'heritability' (a term relating entirely to genetics) is environmental. Plomin, for example, explains that genetic effects 'could be amplified as we increasingly select, modify and create environments correlated with our genetic propensities'.[38]

The IQ theorists Jim Flynn and William Dickens saw it differently, using an example of bright separated identical twins who go to the library, get into top-stream classes and attend university. Flynn asks: 'What will account for their similar adult IQs? Not identical genes alone – the ability of those identical genes to co-opt environments of similar quality will be the missing piece of the puzzle.'[39] Unlike Bouchard and Plomin, he places this 'multiplier effect' in the environmental column, saying that if we use a figure of 45 per cent for the heritable contribution to childhood IQ, it will have to be reduced because the children would have already done some matching of their genes to their environments. 'Let us say that one-fifth of the value is due to matching. Then the direct effect of genes on IQ accounts for only 36 per cent of IQ variance.'[40] And this becomes more marked as the child grows up. The change is therefore due not to increasing heritability but rather to environmental matching; to the way the person finds peers, classes and experiences that match their genetic inheritance.

This 'positive feedback loop' works better in a middle-class *milieu* than a deprived one in which opportunities for intellectual stimulus are limited. Evidence comes from a 2003 study by Eric Turkheimer, professor of psychology at the University of Virginia, who examined data on twins from a study of children, many from working-class families, who had taken IQ tests at the age of seven. For the poorer children, the IQs of identical twins differed just as much as in fraternal twins. Turkheimer concluded that the impact of growing up poor trumps genes: 'If you have a chaotic environment, kids' genetic potential doesn't have a chance to be expressed. Well-off families can provide the mental stimulation needed for genes to build the brain circuitry for intelligence.' He added that the wealthier the family, the more likely it is that the child's genetic potential will be 'maxed out'.[41] But again, there's another way of seeing it. Instead of speaking of genetic potential being *realised* in wealthy homes, you could talk of the well-to-do environment *enhancing* an inherited edge.

These different readings of the same data illustrate the difficulty of attaching percentages to the heritability of behaviour. Some geneticists warn against following this route. One of Britain's leading neuroscientists, Steven Rose, put it bluntly: 'Heritability estimates become a way of applying a useless quantity to a socially constructed phenotype and thus, apparently, scientising it – a clear-cut case of Garbage In, Garbage Out.'[42]

Full DNA analysis has given us more tools with which to measure heritability but it does not necessarily get us much closer to finding precise nature-nurture ratios for behaviour. Even under experimental conditions, in which we can draw on mathematical data about the comparative contributions of genes and environment, it is tricky, because of their interaction. The Stanford biologist Paul Ehrlich notes that this 'cannot be decomposed into nature or nurture because the effect of each depends on the contribution of the other'.[43]

This view is reinforced by a study, published in *Nature Genetics* in 2018, in which eleven scientists looked at data from 54,888 Icelanders.

It found that heritability estimates for traits such as height, which, they say, is really 55.4 per cent, and educational attainment (17%), have been seriously overestimated (Plomin, for instance, puts the heritability of school achievement at 60%[44]) because environmental effects have been included in the genetic category through the mis-reading of twin studies, which placed nurture in the nature column by the underestimation of the contribution of family and other environments. The *Nature Genetics* study used a different method, relatedness disequilibrium regression, of estimating heritability, which, the authors say, avoids the problem of conflating environmental and genetic contributions under the genetic heading, prompting far lower heritability estimates.[45]

TWIN STUDIES AND TOLSTOY'S HAPPY FAMILIES

Another problem with Bouchard's figures is that they are based on 'the assumption of no environmental similarity' in the circumstances of separated twins.[46] This is impossible, so he concedes similarity could account for 3 per cent of the result, even though almost all were brought up in middle-class homes. Only a few were 'reared in real poverty and none were "retarded" [*sic*]', which means 'this heritability estimate should not be extrapolated to the extremes of environmental disadvantage still encountered in society.'[47] He adds that non-middle-class environments might increase the environmental percentage, recognising that his data merely suggests that for adults 'in the current environments of the broad middle class, in industrial societies, two-thirds of the observed variance of IQ can be traced to genetic variation.'[48] He concedes that 'in individual cases, environmental factors have been highly significant,' referring to the case in which the twins' IQs were twenty-nine points apart. Richard Nisbett, a psychology professor and IQ theorist at the University of Michigan, believes twin researchers are wedded to the idea that adopting families are different, when the opposite is true; they're mostly middle class and well motivated to give their children a good start in life. 'Adoptive families, like Tolstoy's happy families, are all alike,' he says.[49]

In Bouchard's later studies, several aspects of common human behaviour (sense of humour, religious affiliation, food preferences, social and political attitudes) were found *not* to be heritable. By the late 1990s, he was toning down his claims: 'There probably are genetic influences on almost all facets of human behaviour,' he said, 'but the emphasis on the idiosyncratic characteristics is misleading. On average, identical twins raised separately are about 50 per cent similar – and that defeats the widespread belief that identical twins are carbon copies. Obviously, they are not. Each is a unique individual in his or her own right.'[50]

Bouchard's earlier figure of 70 per cent heritability for adult IQ is at the high end of the scale. What do other IQ hereditarians say? Plomin has plumped for 50 per cent as a lifetime average[51]; others go for 40 or 45 per cent. Flynn suggests 36 per cent for children, while Kamin provocatively suggested it could be zero.[52] Surveying the twin-related literature in this area, Ridley said that studies have 'converged' on the same conclusion: that IQ is 'approximately 50 per cent additively genetic; 25 per cent influenced by shared environment; and 25 per cent influenced by environmental factors unique to the individual'.[53]

With the possible exception of Kamin – who was being flippant – no one seriously disputes that intelligence, however measured, is heritable. Despite all the problems with twin studies, we can assume that identical twins will usually have more similar IQs than fraternal twins or siblings, and that one of the reasons is that identical twins have identical DNA. But it would seem that the heritability figure of 50 per cent, in a middle-class population, is based on a failure to consider the limitations of twin studies and particularly the inclination to put environmental influences in the genetic column as the twins grow older. It therefore overstates the genetic contribution to individual IQ, which is not the same as denying the role of nature.

Where does this leave us? Ridley talks of the 'sweeping successes'[54] of twin studies. But, at least with respect to IQ, the opposite conclusion seems more apposite. None of the studies managed to eliminate, or sufficiently reduce, the inbuilt problems of using twins to devise a convincing heritability figure. If we lived in a dystopian world in which all identical and fraternal twins were separated and randomly scattered around the globe, some among the American middle classes, rather more among the Chinese working class, many in the slums of Africa, Asia and the Americas, in rural areas, rain forests and deserts, and compared their IQs, what would the heritability figure be: 20 per cent; 10 per cent? Or would the idea of deducing percentages be abandoned altogether?

Heritability percentages relate to the question of how much genetic factors play a role in individual differences in a particular trait in a particular population. This raises the question of how to define populations when assessing the validity of statements such as: 'IQ is highly heritable.' If the population is white middle-class American, that might appear to be the case. Widen it to include all white Americans, and the genetic contribution shrinks. Widen it further to include *all* Americans, and it shrinks some more. Make the population the whole world and the genetic contribution to differences in IQ would be tiny. Look dispassionately at the record of twin studies and it is hard to avoid the conclusion that they arrive at heritability percentages by little more than a thumb-suck. All too often, twin studies tell us less about IQ than they claim and more about those who choose to follow this branch of research.

HOW ADOPTION CHANGES IQ

Another way to assess the genetic impact on IQ is to study adopted children who are *not* twins. Two French psychologists, Christiane Capron and Michel Duyme, perused thousands of records from public and private adoption agencies. Their results showed how genes are expressed through environment. Children from well-off families placed in well-off families had average IQs of 119.6, but children from well-off families placed in poor families had IQs averaging 107.5. Children starting out with poor parents before being adopted by poor families had an average IQ of 92.4 but those from poor families placed in well-to-do homes averaged 103.6.[55] Incidentally, there are several possible environmental reasons why children with poor biological parents had lower IQs; poorer mothers are more likely to smoke, drink, take drugs and experience stress and violence when pregnant, and poorer children are more likely to be neglected in the months or years before adoption.

One hereditarian shibboleth is that only extreme poverty affects IQ, and other environments make little difference. As Ridley put it: '[L]iving on a few thousand dollars a year can severely affect your intelligence for the worse. But living on US$40,000 a year or US$400,000 a year makes little difference.'[56] This was put to the test in a study of severely underprivileged children who were adopted between the age of four and six. Most had been abused and neglected, going from institution to institution or foster home to foster home. Their IQs averaged 77. Nine years on they were tested again; all had improved but some had improved more than others. Those adopted by farmers or labourers scored 85.5, by middle-class families, 92 and well-off families, 98.[57] Contrary to the idea that IQ is fixed by nature, this illustrates that it can be hugely influenced by social circumstance. And, contrary to Ridley's idea that the level of wealth made little difference, there was a six-point gap between the middle class and the rich, and a 12.5-point gap between the stable working class and the rich. This fits with the best-known adoption study, carried out by the American psychologists Skodak and Skeels, in 1949. They tested ninety-eight poorly educated working-class biological mothers whose IQs averaged 85.7. Their children, whose adoptive parents were better educated, had an average IQ of 116.8 at two years old and 108 by the age of 13.[58]

These studies discredit claims that family environment, schools and income make little difference to educational and IQ outcome. Perhaps the most extreme claim is Robert Plomin's, who has been on a perpetual quest to find ballast for his nature over nurture convictions.[59] In *Blueprint*, published in 2018, he argues that the most important environmental factors, such as families and schools, 'account for less than 5 per cent' of the explanation for academic performance. Genetics – 'the blueprint that makes us who we are' – accounts, he says, for 50 per cent. The other 45 per cent he puts down to 'random experiences' whose effects don't last:

'after these environmental bumps we bounce back to our genetic trajectory.'[60]

When we consider the results of studies showing long-term IQ changes as a result of adoption into wealthier, more-educated families, this claim seems ridiculous. Assessing the evidence on adoption studies, including those involving twins, Nisbett said that IQ theorists such as him had consistently underestimated the impact of nurture and of family in particular: 'Unfortunately, for many years I bought the claims of the hereditarians that family environments don't matter much.' Illustrating this point by referring to the impact of cross-class adoptions, he added: 'Raising someone in an upper-middle-class environment versus a lower-class environment is worth 12 to 18 points of IQ – a truly massive effect.'[61]

Overall, the impact of environment on IQ has been understated and the genetic impact overstated, which is not to say it doesn't exist. Even Stephen Jay Gould, a fervent critic of hereditarianism, said it would be hard to find any aspect of life that had no genetic component. 'The hereditarian fallacy is not the simple claim that IQ is to some degree "heritable". I have no doubt that it is, though the degree has clearly been exaggerated ...'[62] Instead, he said, a major fallacy was to conflate heritable with inevitable. 'The claim that IQ is so-many-percent "heritable" does not conflict with the belief that enriched education can increase ... "intelligence". A partially inherited low IQ might be subject to extensive improvement through proper education. And it might not. The mere fact of its heritability permits no conclusion.'[63]

WHY DO POPULATION GROUPS HAVE DIFFERENT IQ AVERAGES?

There is a second, even more serious, fallacy concerning race: the conflation of IQ inheritance *within* a population group with IQ inheritance *between* population groups.[64] Studies of the heritability of IQ have all been 'within-group', examining people brought up in the same population group and social class. But to assume that IQ might be, say, 50 per cent heritable among white middle-class Americans does not mean that *differences* between the IQs of white Americans and black Americans are also 50 per cent heritable, because there is no reason to assume that variation within a group is the same thing as variation between groups. Variation within the group may involve a mix of genetic and environmental factors; the precise mix will be influenced by the size of the group, economic well-being, age and socio-economic range. Differences between groups may be entirely environmental.

I'll paraphrase an illustration suggested by the Harvard biologist Richard Lewontin: let's say we have a bag of mixed seeds taken from genetically diverse varieties of wheat. We take two cupfuls and grow them separately, in the same conditions, except for one detail: one batch receives special nutrients. After a month the plants are measured. Within each batch there is some variation in size, a result of the different genes; within each population, heritability is high. There are height differences between the batches; the batch that got the nutrients is taller, on average. The differences *between* the batches are entirely a result of the nutrients – in other words, an environmental cause.[65] The relevance to the variation in IQs between black and white Americans does not need spelling out.

Let's consider something that is more heritable than IQ: height. If we take the tallest nation in South-East Asia, whose men average 1.74 metres in height, and compare them to the shortest, whose men are thought to average 1.66 metres, we can't assume the difference is genetic simply because of the hereditability of height. The fact that the 'within-group' heritability is high tells us nothing about the possibility that receiving better nutrition and suffering fewer diseases might increase the average height of the shorter nation. (The two nations are South Korea and North Korea, and the average height of the 8cm-taller South Koreans continues to rise because of better nutrition.) Another example would be Japan, where the average height of eleven-year-old boys has increased by 12cm since the mid-twentieth century. 'Heritability' refers to the degree of variation in a trait directly caused by genes in a population but it doesn't relate to the proportion of that trait attributed to genes. If the environmental factors relevant to height change to affect all members of a population (such as improved nutrition in South Korea), the trait's mean value might be altered (taller South Koreans) but its heritability will remain the same. However, degrees of heritability can change if some members of a population experience environmental effects. For example, if some North Koreans are malnourished and others well fed, then variation in that population would be greater.

If we apply this to IQ, we can see why group comparisons are facile. The environmental factors relating to IQ might differ substantially between groups. As with the height of the North and South Koreans, the assumption that the differences were the result of genetics would be wrong. What is the equivalent of better nutrition that could prompt a gap in average IQ scores between populations? IQ differences are not the same as height differences in wheat or people; one factor X is not enough. We need a more comprehensive answer.

11

DO RISING IQS SUGGEST GENETIC OR ENVIRONMENTAL CHANGE?

I once wrote a magazine story about a group of young men hoping to become army officers. To make it to Sandhurst Military Academy (the British Army's initial officer training centre) they spent three days being put through their paces by the Army Officer Selection Board. They were tested on their physical preparedness, willingness to take orders that put them at risk, ability to make decisions under pressure and to devise solutions to problems. And they took an IQ test. 'You can't have a dim-witted officer commanding intelligent soldiers – doesn't work,' the elegantly moustachioed commanding officer, Colonel Peter Ashton-Wickett, drawled when I asked why. Then he back-tracked, drew me closer, and quietly said: 'Actually, we had a fellow whose test revealed a real lawnmower of a brain but we got him through in the end, by God, and it worked rather well because, you see, he'd been exceptionally well-schooled: one of our top public schools.'[1]

It seemed that being 'exceptionally well-schooled' could get you through a lot of problems in life, even a lawnmower IQ. And if that didn't work, the social connections, family contacts and largesse that come from 'our top public schools' helped open doors that might otherwise be closed. In the 1990s, the British comedian Harry Enfield invented a posh but clueless character, Tim Nice-but-Dim, who was forever being ripped off by sharper blades. The point is that Tim Nice-but-Dim, or well-schooled army recruits, can flourish quite happily with their lawnmower brains. That's just the way the system works.

At the other end of the social scale, the impact of a lawnmower brain is dire. This point was stressed to me by a deputy head teacher in a mainly working-class secondary school in east London when we discussed whether children from chaotic underclass backgrounds could be 'rescued' through education: 'Look, my dad was a fireman and I come from a working-class family,' Brian Deal began:

> But it was a stable working-class background. Children I teach from that kind of background have a fair chance if we can get them to apply themselves. But with those from the underclass with damaged backgrounds, well, I won't say it's impossible but very few break through. By the time they get to secondary school, it's usually too late.

I wondered why they went astray:

> Part of it is that they lose hope and confidence and have no concept of themselves succeeding academically, unlike middle-class children who arrive full of hope and aspiration and are exposed to debate and inspiring conversation, which affects them every second of every day. And there are also specific reasons why underclass children fall behind, including their use of language. If you have a limited vocabulary and don't speak like the mainstream, it's going to hold you back.[2]

Educational exposure, language use and intellectual confidence can make significant differences to IQ scores. If you come from the working class and don't develop the cognitive skills that IQ tests measure, your prospects are not rosy. If you come from the upper classes, this might be less of an impediment, although many professions will still be beyond your reach. IQ can make a huge difference to your prospects, particularly if you start lower on the economic ladder. But it is not just that your score will influence your position on that ladder. More significantly, it works the other way around: your position on the ladder will also have a strong influence on your score. Put simply, poverty is IQ-diminishing; wealth is IQ-promoting.

Beyond social class, if you grow up in a society or community that attaches a high premium to abstract knowledge, your IQ is likely to be higher than if you grow up in a community that puts a premium on practical knowledge. As we shall see later in this chapter, nutrition, pollution, your mother's pregnancy experience, your self-confidence when taking a test; all these can also affect IQ scores. Put these factors together and consider how they change from generation to generation and differ from community to community and you arrive at what is now known as the 'Flynn Effect'.

THE FLYNN EFFECT: HOW THE ENVIRONMENT SHAPES IQ

Someone who scored 100 on an IQ test in 1970, when I completed my first IQ test at primary school, would score about 83 today. Or, looked at from the opposite angle, someone with an IQ of 100 today would have scored around 117 in 1970. In the early 1980s, the IQ theorist Jim Flynn (whom we met in earlier chapters), noticed a long-term increase in IQ scores. More specifically, he found that between 1947 and 2002, g scores rose by between 18 and 27.5 per cent, depending on the test (0.3–0.5% per year). As a result, those who set IQ tests had to periodically increase their difficulty to maintain a mean of 100. This observation was dubbed the 'Flynn Effect' by the authors of *The Bell Curve*, and it stuck.

No one could reasonably claim genetic intelligence could improve in fifty years. Flynn accepted both implications: first, that the rise was due to changing environmental factors and second, that IQ tests did not measure the full breadth of intelligence but rather, as he put it in 1987, a form of 'abstract problem-solving ability'.[3] He later added that IQ tests 'do not measure intelligence but rather correlate with a weak causal link to intelligence'.[4]

Evidence for the 'Flynn Effect' was so conclusive that hereditarians could not simply wish it away. If we project back to 1904, when IQ testing began, average scores, measured against current norms, would be about 70. In some countries the rise has been particularly rapid; Dutch IQs in 1982 were twenty points higher than in 1962.[5]

But the most rapid gains were seen in rural Kenyan children, whose tests showed a rise of 26.3 points in fourteen years, which researchers attributed to their parents' literacy, family structure and the children's nutrition and health.[6] Sudanese children experienced a similar rise, and the IQs of Askenazi and Mizrahi Jews, and Chinese Americans, also increased sharply. Richard Lynn and Tatu Vanhanen openly declared their racist genetic determinism, but the studies of children they cited, conducted before German reunification, go against their hereditarianism, showing significant IQ rises in a generation. A 1967 East German study showed an average of 90, compared with a range of 99 to 107 in four West German studies conducted at about the same time – a gap of between nine and seventeen points for children from the same genetic background. A later East German study, conducted in 1984, gave an average IQ of 99, suggesting that changes in the environment were increasing IQs. The conservative writer Ron Unz said these results 'constituted a game-ending own goal against the IQ-determinist side'.[7]

Flynn underscores the implications by going back in time: 'Our ancestors in 1900 were not mentally retarded [sic],' he notes, adding that we differ from them in that we can use abstract logic and hypothetical scenarios; science has helped to free our thinking from being bound by the concrete. 'Since 1950 we have become more ingenious in going beyond previously learned rules to solve problems on the spot.'[8] People in 1900 used practical logic: asked what dogs and rabbits had in common, they would say 'you use dogs to hunt rabbits,' whereas today we'd probably say 'they are both mammals.' We classify the two animals in scientific terms that would have seemed trivial to people at the dawn of the twentieth century. Flynn explains:

[O]ur ancestors found pre-scientific spectacles more comfortable than post-scientific spectacles, that is, pre-scientific spectacles showed them what they considered to be important about the world. If the everyday world is your cognitive home, it is not natural to detach abstractions and logic and the hypothetical from their concrete referents. ... Today we have no difficulty freeing logic from concrete referents and reasoning about purely hypothetical situations.[9]

When Flynn looked at the subtests involved in modern IQ testing, he found the biggest gains came in areas demanding abstract, non-verbal logic, which suggests that today's children are better at solving such problems, a skill encouraged by modern maths teaching and demanded by IQ tests. We also have more leisure time devoted to visual pursuits, we spend more than double the time in education than people did in 1900, the quality of our education is higher and we're more exposed to aptitude testing.

Flynn and his colleague William Dickens further explained rising IQs in terms of the 'social multiplier' effect. Parents seek out ever-higher levels of education for their children, more people are in professional jobs and some leisure activities are more cognitively demanding. Each gain opens fresh possibilities and sets up ever-higher feedback loops. People respond by raising their game, which pushes the average higher still. As Flynn puts it: 'You get a huge escalation of cognitive skills in a single generation.'[10]

One example involved a study of Brazilian children aged between seven and eleven, who were given what is known as the 'Draw-a-Man' children's IQ test. One city-based group was tested in 1930. The researchers compared their results with those of a city-based group tested in 2002. They found the city children had gained seventeen IQ points. They then compared other test results from the 2002 urban group with a 2004 rural group and found a gap of thirty-one points on one of the IQ tests (Raven's) and fifteen on the other (WISC). The reason related to the lifestyle of the rural community: no televisions, no postal service, no banks, no hospital and no electricity in almost all (92%) of the houses.[11] This community had a pre-industrial lifestyle and its members did not look at the world through the 'scientific spectacles' worn by contemporary urban children.

We live in a more cognitively complex world than our parents and they lived in a more cognitively complex world than theirs. The contemporary post-industrial world demands that we classify what we see and hear and read in abstract terms. We work in symbols that have no corporeal reference point, consider hypothetical scenarios and take imaginary leaps. This is our everyday reality, and it's what IQ tests measure. The more we're exposed to abstraction, the better we are at embracing it. But some communities have a more profound exposure to this modern way of thinking than others, which explains why different groups have different average IQs and why the IQs of some are increasing more rapidly than others.

HOW TO BOOST YOUR IQ (OR YOUR CHILDREN'S)

Until the twenty-first century there was a conviction among heredi-tarians that fluid intelligence (*Gf*), the component of IQ testing said to be most '*g*-loaded', because it measures the ability to reason and solve fresh problems without previous knowledge, was fixed for life at an early age and could not be improved. Their only concession was that practising for IQ tests could prompt a rise in *Gf* but they insisted that nothing else could help.

Not everyone agreed. The media theorist Steven B. Johnson made a Flynn-type counter-argument in his book *Everything Bad is Good For You: How Today's Popular Culture is Actually Making Us Smarter*, arguing that the increased complexity of contemporary culture is making people smarter. He cited video games (the mapping of *Grand Theft Auto*, the spatial geometry of *Tetris* and the engineering puz-zles of *Myst*) and the multilayered plots, constantly shifting scenes and variety of characters that make contemporary television dramas more cognitively demanding than those of earlier generations.[12]

This idea is supported by research that showed that adults who practised a computer game involving increasingly demanding tasks of memory showed IQ increases, specifically in 'fluid intelligence'. The Swedish neuroscientist Torkel Klingberg tested fourteen children who had attention deficit disorder. Half were given computerised tasks designed to improve their memory, while the others played less demanding games. After five weeks, the computer group had increased their IQ scores on the Raven's Progressive Matrices IQ test (one of the best tests for measuring fluid intelligence). Three years later, Klingberg replicated the results with a group of fifty children.[13]

Drawing from this research, the Swiss academics Susanne Jaeggi and Martin Buschkuehl designed a game they called 'N-back training', based on working memory and cognitive control. They recruited seventy students, half of whom practised the game for at least fifteen minutes a day for between eight and nineteen weeks. All who played improved their fluid IQ scores and did significantly better than the control group. The study showed that the more the volunteers trained, the more their *Gf* rose: the sub-group that played for the full nineteen weeks showed most improvement, indicating 'that it is possible to improve *Gf* without practising the [IQ] testing tasks themselves'.[14] Buschkuehl spelled out the implications: 'Mental exercise ... can enhance important abilities and is most likely [to be] the most efficient way to improve specific cognitive processes.'[15]

These conclusions have been replicated in other forums with similar results. In each case, IQ scores increased, and the gains lasted several months. 'Do we think they're now smarter for the rest of their lives by just four weeks of training?' Jaeggi asked. 'We probably don't ... We think of it like physical training: if you go running for a month, you increase your fitness.'[16] However, this prognosis might be overcautious, particularly for children. Because children's neural pathways are still developing, it is possible this kind of training could prompt longer-term gains. Certainly, there's strong neurological evidence to show that by 'exercising' the brain, various webs of synapses can be strengthened, or parts of the brain enlarged, even among adults. Studies of the brains of the drivers of London's black cabs show that they have an enlarged hippocampus, the part of the brain used for navigating three-dimensional space.[17]

Many other claims to raise IQ relate to music. Glenn Schellenberg, professor of psychology at the University of Toronto, examined 144 six-year-olds. Half (72) were given music lessons for a year (half playing keyboards, the other half singing), a quarter had drama lessons and a quarter had no extra lessons. The IQs of the music groups rose by 7 per cent compared to 4.3 per cent in those in the drama or no-lessons control groups. All the children had started grade school between the two tests, which explains the 4.3 per cent rise.[18] He then tested adults on a similar basis, studying two groups of undergraduate students (106 in total); just under half had received at least eight years of private music lessons; the other half had had no lessons. The IQs of the musically trained group were significantly higher, even when sex, parental education levels, family income and first language were taken into account.[19] Another study found that university students who spent ten minutes listening to Mozart shortly before being tested saw short-term gains of between eight and nine points on the spatial subtest of one of the major IQ batteries, the Stanford–Binet Intelligence Scale.[20]

As we'll see later in this chapter, there's also plenty of research showing that the expectation of doing well may help you to do well and that the expectation of doing poorly will hinder you. In other words, if you start the test believing you'll excel, your results are likely to be better than if you start the test believing that you're likely to fail. This can also apply if you believe that the population group you belong to is good or bad at the test or task in hand.

There are also health-related ways of increasing IQ. It has been claimed that breast-feeding for at least nine months can raise your child's IQ by seven points, although it's not clear whether it's the breast milk or the necessary close contact between mother and child that causes the effect. It might just be that the type of mother who breast-feeds is smarter than the type who doesn't.[21] It has also been claimed that a diet that avoids saturated fats and includes fish, olive oil, citrus fruit and vegetables is ideal for brain development, because it improves the blood supply to the brain.[22] A meta-analysis of Chinese studies on more than twelve thousand children whose mothers were given iodine supplements during pregnancy, and of children given iodine supplements, showed increases of between 4.8 and 12.45 IQ points over the no-iodine control groups.[23]

Various claims have been made for the cognitive benefits of exercise, including a University of Montreal study that found that pregnant women who exercised for at least twenty minutes three times a week boosted their babies' brain development in a way that could benefit them for the rest of their lives. The women joined the study in their first trimester and were randomly assigned to a sedentary or an exercising group. A few days after birth, the babies' brain activity was measured while they slept. The newborns from the physically active group of mothers were found to have a more mature cerebral activation, 'suggesting that their brains developed more rapidly', said one of the researchers, Elise Labonte-LeMoyne.[24]

Even more significant is the quality of the relationship between parents and young children. One of the best ways to improve IQ if you are from a poor family is to get adopted as a baby; adoption from a poor family to a well-off family tends to prompt long-term IQ gains of twelve to eighteen points, according to a number of studies.[25] Also, choose parents who give you plenty of attention, particularly by reading to you. A British study of 14,853 children born in 2000 (the 'Millennium Cohort Study') found that children who are read to every day at the age of three are more likely to succeed in a wide range of primary school subjects at the age of five.[26]

But the surest way of increasing your score is to practise IQ tests. The more familiar you become with how they work, the more you will improve. No one disputes this. It follows from the correlation between performance in IQ tests and performance in academic work, particularly maths, that if you spend plenty of time doing calculus and geometry – and perhaps if you spend plenty of time on other forms of logic and lateral thinking – your IQ performance will improve. This is part of Dickens' and Flynn's social multiplier effect.

Which cat is more likely to get most of the IQ-enhancing cream? Step forward yummy mummy. She always remembers to take her iodine pills, maintains her gym habit during pregnancy, breast-feeds her baby and reads bedtime stories to her child. She ensures her child has a balanced diet, plays educationally enriched video games, takes music lessons, goes to a good school, practises plenty of aptitude tests, does lots of maths and enters the test room full of self-belief. And yes, all these happy things are significantly more likely to accrue to the child of the well-to-do and not so much the children of the ne'er-do-well.

HOW TO STUNT YOUR IQ, OR YOUR CHILDREN'S

The damage can start in the womb. Numerous studies show that smoking tobacco[27] and cannabis[28] and drinking alcohol during pregnancy are associated with a reduction in the children's IQs. The more you smoke, drink or take certain drugs during pregnancy, the more serious the effect, although the reasons might not relate entirely to the direct impact of those chemicals on the foetus; they could also involve their impact on parenting. Whatever the balance, the outcome is clear. A study of 4,167 mothers and children, conducted by scientists from the Universities of Oxford and Bristol, found that the consumption of more than three units of alcohol a week during pregnancy could affect the baby's brain, leading to an eight-point reduction in IQ. In larger doses, drinking alcohol could cause more severe learning difficulties.[29] And it is poorer people, including poorer pregnant mothers, who are more likely to smoke, drink or take drugs to excess.[30]

The parts of the brain relating to cognition are connected to those governing stress and emotional reactions. There is evidence that early emotional trauma in childhood, including chaos and chronic neglect, can undermine a child's problem-solving ability.[31] Where neglect is severe, the results can be profound, with long-term changes in the way parts of the child's brain are wired. One example comes from studies of Romanian orphans, who were cut off from all but the most perfunctory human contact during their early years. The brains of those most isolated and abused for the longest time showed unusual developmental patterns. In some it was reported that these differences could be detected in scans of the orbitofrontal cortex (the part of the brain concerned with how people manage their emotions and relate to each other, experience pleasure and appreciate beauty). In addition, some had unusually high levels of the stress hormone, cortisol, in their bodies, and these high levels persisted in subsequent tests, long after the children were adopted.[32]

Even when the parental attention deficit is in the 'normal' range, the outcome can be dire. The Millennium Cohort Study also found that those who failed to reach key development stages at nine months were more likely to fall behind in terms of cognitive development, and to have behavioural problems, by the age of five. Another study, carried out by the Sutton Trust, found that in their acquisition of vocabulary, children from the poorest homes were more than one year behind children from well-off homes when they started primary school.[33]

At all ages, it appears, poorer children suffer academically, and in IQ, by having a less educationally stimulating home and social environment. For most children, their performance in maths, science and in IQ tests falls during the long summer school holidays, but this decline is more marked in poorer students. A thirteen-year-long study of 790 children from twenty schools in Baltimore found that the achievement gap between working-class and middle-class students was explained mainly by the poorer students' cumulative loss of ground during school holidays.[34] The reason seems to be that the richer children's parents helped them exercise their brains during the holidays, whereas children from poorer homes, often with single parents, were left to their own devices; one of the reasons why a two-parent family is more likely to provide a more IQ-stimulating environment than a single-parent family, according to research by Flynn and others.[35]

There is also evidence that specific child-raising behaviour can diminish IQ. A New Hampshire University study tested nine hundred children at one and four years old. It found that being 'spanked' at least once a fortnight had a direct negative impact on IQ, reducing it by an average of three points.[36] Many studies show that working-class parents are more likely to hit their children and to do so more often; research by the Joseph Rowntree Foundation prompted the conclusion that poorer parents, under greater stress, 'are less likely to be able to provide optimal home circumstances and more likely to use coercive and harsh methods of discipline'.[37]

Pollution also affects IQ. A 2018 Chinese study of twenty thousand people found that high levels of nitrogen dioxide and sulphur dioxide in the air prompted a significant decline in mathematical and verbal test scores; the equivalent of a year's education. Previous studies had found that pollution affected students' performance, but this study found the cognitive impact was even more severe on people aged sixty-five or older. And if they lived near busy roads, they had an increased risk of dementia. Pollution levels also seem to be linked to higher levels of mental illness among children. Researchers found that the more and longer people are exposed to polluted air, the greater the damage to their intelligence. Pollution also harms men more than women, particularly less-educated men.[38] And working-class people are more likely to live near busy roads where pollution levels are high.

I could go on to offer further examples of environmental factors that reduce IQ. But my point should be clear: poverty not only can have this impact, it's also likely to have it. It's not so much that low IQ leads to poverty; it's more that poverty leads to lower IQ.

But even high IQ is no guarantee of a leg-up. A study by researchers from Oxford University and a Swedish research institute found that of the brightest children born in the UK in 1970, those from the poorest homes had only a 40 per cent chance of gaining A-level qualifications (the Year 13 exams taken by about 70 per cent of eighteen-year-olds, usually in just three subjects, making it the equivalent of the first year of university in most countries), half the chance of children from well-off homes. 'Even the very brightest children are hampered if they come from a disadvantaged background,'[39] the lead researcher, Dr Erzsébet Bukodi, commented. A wider study conducted by the Institute of Education, UCL, found that in the UK by the age of fifteen bright pupils from poor backgrounds lagged two years behind those from well-to-do backgrounds in reading, while in Germany, Finland and Iceland, the gap was one year.[40]

Tim Nice-but-Dim might have an IQ of 80 but he's not going to end up living on state benefits. Social mobility in most countries in the advanced industrial world diminished in the late twentieth century, but in the few countries where the income gap narrowed, it appears that the IQ gap followed. For example, in social-democratic Norway, which kept national IQ data for children between 1952 and 2002, average IQs rose across the board, but the gains were larger in the lower half of the IQ and income range.[41]

There has been much debate about whether special intervention programmes for poor children, such as Head Start in the USA and Sure Start in the UK, support long-term gains in IQ and educational performance. Some studies suggest that once the programme ends, the original environmental disadvantages take over and the gains fade. However, if those disadvantaged children continue to receive high-quality education, their odds of completing secondary school and going on to university rise by around 33 per cent, according to the IQ theorists Eric Turkheimer, Kathryn Paige Harden and Richard Nisbett.[42] It also appears that if intense educational intervention is made early enough, the gains persist. In one American study, black infants from low-income families, whose mothers had an average IQ of 85, received eight hours a day of exercises to improve their cognitive, linguistic, perceptual and social development. By the age of twelve, 87 per cent of the children exposed to this intervention were in the normal range (85 to 115) compared with 56 per cent of a control group.[43] Other studies suggest that very early intervention programmes raise the IQs of underprivileged children by an average of 4 or 5 per cent and that these gains do not evaporate as they reach their teen years.[44]

The point that poverty diminishes IQ should be so obvious as to be not worth making. But hereditarians prefer the opposite argument, rooted in a conveniently smug analysis of wealth and poverty. Your economic standing reflects your IQ, so rich people are rich because they're smart and their children do well because they inherit smart genes; poor people are poor because they're thick and their children do badly because they inherit thick genes. There is no point putting taxpayers' money into improving the educational lot of the poor; you'll simply be pissing into the genetic wind.

WHY WHITE WORKING-CLASS BOYS STRUGGLE ACADEMICALLY

Every year, the UK government publishes a breakdown of performance in GCSE exams (taken at sixteen) by ethnicity. Every year, the results are the same: white working-class boys languish at the bottom of the ladder. Data for 2016 show that thirty-one per cent of white boys who received free school meals (a proxy measure for working class) obtained a grade C or better (A* is the highest grade, G the lowest).[45] This result was down from 34 per cent three years before.[46] Then came working-class boys from Caribbean background (33% grade C or above). Above them were working-class boys from mixed-race backgrounds (39%). Then came working-class boys from African backgrounds (49%), followed by Asian working-class boys (51%). Right at the top are girls from Chinese backgrounds – 83 per cent from the poorest families achieve grade C or above. Incidentally, working-class girls outperform boys in every subject, in every broad ethnic group. White working-class girls were 8 per cent ahead of boys in GCSE maths and English. Girls also outperform boys at A-level, in university applications for prestigious courses such as medicine, dentistry and law, and in going to university.

There has been much publicity about the dire performance of white working-class children, particularly boys. UK governments periodically make statements about this being a priority, and effected some policy changes, such as reducing the contribution of assessed coursework in favour of final exams. Yet unlike black working-class boys, whose performance has improved (in 2016, 44% of black working-class boys achieved at least a C, up from 40% in 2013), that of white working-class boys has fallen. Despite everything, Britain's white working-class boys stubbornly remain at the bottom of the pile.

When class and sex are removed from the picture, overall, 63 per cent of white GCSE students achieve grade C or above, which puts them on the same percentile as African and mixed-race students (though still behind Asian students).[47] A similar pattern is seen at A-level; 13 per cent of all students achieved the top grade (A or A*) in three subjects in 2016. Grades at this level would potentially make students eligible for entry to the best of the elite universities. White British students of both sexes and all classes were slightly below this level, at 11 per cent, as were mixed-race students, but were significantly below Chinese, Indian and Irish A-level students.[48] More girls than boys achieved these elite grades, although the gap is narrowing.[49]

Data from early education suggest the problem starts when children are very young. A national survey of under-fives in Britain, which included assessments of language use, problem-solving, numeracy and creativity, found that only 36 per cent of poor white boys reached the expected targets, compared to 56 per cent of poor white girls. Again, the results were better among working-class children from other ethnic groups: 42 per cent for Pakistani boys, 44 per cent for black Caribbean boys, 48 per cent for black African boys and 50 per cent for Indian boys.[50]

It persists as children move through primary school. Among six and seven-year-olds receiving free school meals who achieved the required level in mathematics in 2017, the pattern was similar, but this time with Africans on top. White British children averaged 58%, black Caribbean 60, mixed race 63, Asian 67 and African 69.[51] In reading, the averages were white British 58 per cent, mixed race 65, black Caribbean 67, Asian 68 and African 72.[52]

Why is this happening? Innate racial intelligence? There is a close correlation between IQ and academic performance, particularly in maths. If, as race science advocates claim, group-based IQ differences are partly genetic in origin, it seems African (mainly Nigerian) children, Asian, mixed-race and Caribbean children must be significantly more innately intelligent than white British children. Yes? Oddly enough, the race science advocates are silent on this. If they do comment, it is to point to environmental reasons for this particular achievement gap.

It goes without saying that the answer has nothing to do with population genetics. It seems to lie in the learning culture among white British working-class children, especially boys. Some social theorists relate white male working-class educational failure to a masculine ethos that emerged when men expected to have a job for life and 'book learning' was considered effete, even effeminate. Mining and factory jobs have long gone but what has replaced them is often despair, rather than recalibration. These attitudes have lingered and seem more ubiquitous among white working-class men than among other groups. The British example illustrates the huge role culture can play in intellectual performance. It can affect whole population groups, even from the majority culture; we should never assume that population genetics is part of the picture.

THE NARROWING AMERICAN BLACK-WHITE IQ GAP

Let's cross the ocean and consider African Americans. In the next chapter, I'll focus on the specific claims made by Murray and Herrnstein in *The Bell Curve* but for now I'll restrict myself to the claims about IQ made by Linda Gottredson and others in 'Mainstream Science on Intelligence'.[53] This document was published by a group of race science psychologists to back *The Bell Curve*'s claims, and serves as the founding document of contemporary race science. It starts by claiming that general intelligence is real, that it can be measured accurately by IQ tests and that 'genetics plays a bigger role than the environment in creating IQ differences.' It adds:

- American white IQs average 100; black IQs average 85
- There is 'no persuasive evidence' that IQs for different racial groups are converging
- 'Racial-ethnic' differences in IQ are the same when children start and leave school: 'black 17-year-olds perform ... more like white 13-year-olds'
- 'Racial-ethnic' differences are 'smaller but still substantial for individuals from the same socioeconomic backgrounds'

First, let's begin with the 85/100 claim. Even if the fifteen-point gap between white and black Americans was correct as a 1994 national average, it did not apply as a consistent average for children from grades 1 to 12. IQ testing from this era shows that at ten months old, there was only a one-point difference between average black and white IQs, explained by the impact of poverty on IQ in the womb and in the early months of a child's life. This rose to 4.6 per cent by four years old, and 16 per cent by age twenty-four.[54] Life experience was increasing the gap as these children grew older.

Second, the 'Flynn Effect' shows that average black American IQs have been rising faster than white American IQs. The black average of 85 in 1994 was the equivalent of the white American average in the early 1950s, or perhaps even the mid-1960s, depending on the test used. Since then, there is persuasive evidence that the black-white IQ gap has narrowed. Dickens and Flynn showed that black Americans gained 5.5 IQ points on white Americans between 1972 and 2002, putting the 2002 gap at 9.5 points.[55]

Fellow IQ theorist Richard Nisbett, from the University of Michigan, drew on a welter of data to test this claim. He used scores from aptitude testing between 1965 and 1994, which were highly correlated with IQ, showing that in mathematical ability, reading and vocabulary the gap had closed by almost 6 per cent. He reinforced this with comparable data from elementary, junior and high schools. He estimated that by 2005 the real IQ gap was about ten points, rather than fifteen (fractionally more than the Flynn-Dickens estimate). 'The best evidence we have indicates that the value is out of date and that the black-white IQ gap has lessened considerably in recent decades,' he wrote.[56] This conclusion is reinforced by steady improvements in the academic results of black Americans, suggesting the gap has narrowed further over the past fifteen years. The white-black reading gap for American seventeen-year-olds fell by 9 points between 1975 and 2012 and the maths gap by 4.5 points. As Nisbett put it in 2017: 'IQ is highly correlated with these measures of academic achievement, so it is almost surely the case that the black-white IQ gap has been very substantially reduced'.[57]

Some in the race science camp like to point out that black scores in the SAT (a national test of aptitude and learning) have not increased significantly over the last two decades but this ignores a key demographic fact: the African-American population rose by 4 per cent between 1996 and 2015 but the number of SAT-takers increased by 50 per cent in this period, compared to a 17 per cent white increase. Nisbett explained it like this: 'If the average black IQ is increasing but the black adolescents from the lower portion of the IQ distribution are increasingly likely to take the test, this will result in a static mean score.'[58]

The point that there are IQ differences between black and white Americans from the same social class does not tell us what the 'Mainstream Science' signatories assume it does, because they seldom grow up in the same environment. They may go to the same school or college class but are brought up in different subcultures, live in different areas and have parents with different education levels. Black families earning $100,000 a year typically live in more disadvantaged neighbourhoods than white families earning $30,000.[59] Among African Americans there's also a higher proportion of single-mother families, and children of single mothers may start school with a lower IQ. Flynn controversially says this suggests that 'a worse marriage market within the black community is one of the missing factors.'[60] Flynn continues:

> If other environmental handicaps take over at each stage of life history – a teenage subculture that is atypical linguistically and more prone to gang membership, an adult world in which many go to prison and most have jobs that make few cognitive demands – then an environmentally induced IQ gap between the two groups will persist into old age.[61]

Middle-class African Americans are not immune from these 'environmental handicaps', even if they affect poorer black Americans more profoundly.

Charles Murray claimed in a debate with Flynn that 'by the 1970s you had gotten most of the juice out of the environment that you were going to get.'[62] It's an astonishing statement when we consider the range of environmental factors that still affect black American IQs, starting with the widening income gap. In 1979, black men earned 80 per cent of white men's income levels. By 2016, that had slipped to 70 per cent ($18 per hour vs $25 per hour). For black American women the fall was from 95 per cent to 82 per cent ($16 vs $20).[63] In 2016, the average black man's annual income was $29,376, compared to $40,632 for white men.[64] Considered in terms of household net worth, the gap was a gulf. The median net worth of white families in 2016 ($171,000) was nearly ten times that of black families

($17,600); for Hispanic families it was $20,700. Nearly one in five black families had zero or negative net worth, twice the rate of white families.[65] In 2016, 8.8 per cent of non-Hispanic whites were said to be living in poverty, compared to 22 per cent of blacks, 19.4 per cent of Hispanics and 10.1 per cent of Asians.[66] In 2018 the black American unemployment rate was almost double that of whites.[67]

Two American economists conducted an experiment that drew out one reason for the employment, income and wealth gap. They created five thousand CVs, half with typically black American names (Jamal, Tyrone, Lakisha and Latoya) and half with typically white American names (Emily, Alison, Greg and Brendan). They divided the CVs into high and low quality and sent them out. Those with white-sounding names were 50 per cent more likely to be invited to interview. And while high-quality 'white' CVs were preferred to low-quality 'white' CVs, the quality of 'black' CVs made no difference.[68] To the employers, white applicants seemed to come in two categories, black applicants in just one.

These very different experiences of living in America also play into relations with the state. In 2016, black citizens were 5.87 times more likely to be in prison than white citizens.[69] Black and white young people take cannabis and other drugs at about the same rate, but black people are arrested for drugs offences far more frequently; in some states, ten times more.[70]

It might seem glib to conflate the reasons for these gaps under the heading of 'legacy of slavery'. But if you take the long view that is what it amounts to: a population ripped from its geographical, historic, linguistic, religious and cultural roots, to live and die under brutally inhuman conditions for hundreds of years, followed by a century of vicious discrimination entrenched in that legacy. Even today, 150 years after the abolition of slavery in the USA, and more than fifty after the repeal of the 'Jim Crow' laws, African Americans experience higher levels of racial suspicion than other groups and have found it harder to integrate into the cultural mainstream. Only a generation ago, a survey found that 53 per cent of American non-blacks believed blacks were less intelligent.[71]

This relates to another element of prejudice: how prejudice is absorbed by those on the receiving end. I'll touch on just one dimension; how a perception of intellectual inadequacy affects those taking tests. For comparative purposes, I'll start with two of many sex-based studies that led to similar outcomes.[72] In a test of mental rotation ability (traditionally seen as a 'male' skill), one group was told it was linked to abilities in aviation, engineering, navigation and 'undersea approach and evasion'. Men came out on top. The other group, given the same test, was told its purpose was to predict ability in dress design, interior decoration, needlepoint and knitting. Male performance collapsed and female performance soared.[73] In another study, more than a hundred American university students, at the same level of grade achievement, were divided into two groups and given an advanced calculus test. The first lot were told it was designed to measure innate maths ability; the second lot was also told that no difference had ever been found between the sexes in the test. In the first group, the male and female students scored the same. In the second group, which received the additional information, the women significantly outperformed the men. The researchers concluded that when women were assured that they weren't inferior, it 'unleashed their mathematics potential'.[74]

In both these tests, when the female students were released from what the psychologist Claude Steele called 'stereotype threat', their performance improved, but when stereotypes were reinforced their minds shifted from expecting success to avoiding failure, which seemed to clog up their working memory. Steele noted that those burdened with negative stereotypes 'know that they are especially likely to be seen as having limited ability. Groups not stereotyped in this way don't experience this extra intimidation. And it is a serious intimidation, implying as it does that they may not belong in walks of life where the tested abilities are important.'[75] Of course, women are not the only group facing 'stereotype threat'. Steele gave one group of his undergraduate students a standardised test, informing them it was designed to measure intellectual ability. White students outperformed black students. But when he gave the same test to another group of students, portraying it as a routine exercise without mentioning intellectual ability, black and white scores evened out.[76]

These examples suggest how racist perceptions can be absorbed by those who experience them and how this can have a direct impact on results. This might well be part of the explanation of why the gap between white and black IQs seems to grow as children grow older. The longer black children have to absorb racist perceptions about their intellectual capacity, the more their confidence diminishes when faced with intellectual challenges such as IQ tests.

If we wanted to remove the post-slavery environment from the equation, we might consider testing black and white American children who were adopted in places that did not have America's history of black slavery. In 1961, a German psychologist, Klaus Eyferth, examined the IQs of 181 adopted German children who had been fathered by black and white GIs during the post-1945 occupation (98 mixed race and 83 white). Remarkably, given that Germany was hardly devoid of racism, there was no statistically significant difference between the average IQs of white and mixed-race adoptees. Among the girls, those of mixed race had an average IQ three points higher.[77] When we remove the legacy of black slavery, the most significant environmental reason for the IQ gap disappears. Flynn has argued that this study suggests that the black-white IQ gap in the United States has nothing to do with genetics.[78]

If differences between IQs of population groups were partly hereditary, and upbringing made little difference, mixed-race children would have about the same average IQ, regardless of which parent was white or black. But if the gap was environmentally prompted, IQ differences among mixed-raced children might depend on which parent was black and which white. This assumes that in most families the mother plays a bigger nurturing role, and the socialisation of children by white mothers might be more conducive to IQ development, because of education and wealth levels. Based on these environmental assumptions, we could anticipate that children with black fathers and white mothers would have higher IQs than those with black mothers and white fathers. And this is precisely what a study along these lines found: the difference was nine IQ points.[79]

Another test used to support claims that the white-black IQ gap is innate is the 'digit span' test, in which subjects repeat a series of numbers either forwards or backwards. Jensen and others claimed that the white edge increases for repeating backwards because reversing numbers is more mentally demanding. Herrnstein and Murray asked: 'How can lack of motivation (or test willingness or any other explanation of that type) explain the difference in performance on the two parts of the same test?' The answer, they suggest, must lie in genetics.[80] But the fact that the gap is bigger on the more difficult test in no way implies that the gap itself has a genetic origin. If we assumed that the white-black gap on the forward digit span test was prompted by environmental factors, we could draw the opposite conclusion. We'd anticipate a larger gap in the more demanding test for the same environmentally related reasons. Another study showed that Chinese children do better than English children in the digit span test.[81] But the reason related to how many digits the children could say in Mandarin and in English. It takes 20 per cent less time to say the numbers in Mandarin, which is why English children remembered only 80 per cent as many numbers as Chinese children. This was confirmed when children who were fluent in both languages were tested: they still took 20 per cent longer for the English section. As Flynn notes: 'This issues a warning to those who make cross-cultural comparisons without making a functional analysis of what is going on.'[82]

Jensen, Herrnstein and Murray make similar points about reaction time, which they say is faster among white people than black people, while 'movement' time is faster among black people. They claim this correlates with g in IQ tests because '[s]marter people process faster than less smart people,'[83] a view hereditarians have clung to since the time of Francis Galton. But, as Flynn has shown, this component of IQ tests has seen larger rises than IQ scores more generally (the equivalent of 4.75 IQ points between 1989 and 2001). Flynn suggests that the speeded-up tempo of events we view using visual media has had an impact on individuals' speed at processing information. Since not all population groups are equally exposed to visual media, this is one reason why Flynn warns against cross-cultural comparisons about reaction time.

Nisbett, who surveyed the results of all these studies, reached a definitive conclusion: 'the genetic contribution to the black-white IQ gap is nil.'[84] Flynn avoids absolute verdicts unless the evidence allows them but notes that the history of cultural deprivation and discrimination faced by African Americans has not given them the same opportunities to don 'scientific spectacles' and that the narrowing gap between the different average scores is therefore 'probably environmental in origin'[85]; nothing to do with genetics.[86] Reich has not reflected on the specific question of the American black-white IQ gap, for the simple reason that there is no relevant genetic evidence, but warns that from past scientific errors and what he calls the 'racist' leaps of logic 'with no scientific evidence' made by the likes of Nicholas Wade that 'we should take a cautionary lesson not to trust our gut instincts or the stereotyped expectations we find around us.'[87]

Marrying the American, German and British test results together, the first showing a narrowing black-white IQ gap, the second showing no differences between the IQs of white and mixed-race children, and the third showing a widening educational achievement gap between white working-class children at the bottom and those from African and Asian backgrounds at the top, it seems there is persuasive evidence to suggest that the reason for these differences falls entirely in the cultural/environmental space.

12

THE BELL CURVE: WHAT'S IT ALL ABOUT AND WHY IS IT BACK?

Before 2017, Sam Harris had a reputation as an intrepid atheist campaigner and author, a penetrating podcast interviewer and an all-round man of ideas. He was thought to be more on the liberal side of the American political divide, if anywhere, and on some issues that is still where he would place himself. But something happened that year that turned Harris into a vociferous advocate of race science. It was all about one man.

Charles Murray, one of the authors of *The Bell Curve*, was making a speech at Middlebury College, Vermont. He was shouted down, and his chaperone was assaulted. Murray's star had faded from his heyday in the mid-1990s, when he had given expert evidence on welfare reform to a Senate committee. Since then, he had been branded a scientific racist and faced boycotts and protests. Although he earned a whack from various think tanks and had pressed on with books and papers on his pet subjects of libertarian welfare reform, race and IQ (including a very odd one on Jewish intelligence that we'll peruse in the next chapter), his public reach was curtailed. *The Bell Curve* continued to influence right-wing politicians – the Republican congressman and Speaker of the House, Paul Ryan, cited Murray as an expert on poverty – but his high profile seemed largely confined to alt-right forums such as Stefan Molyneux's YouTube channel.

The Vermont incident changed all that. Murray became an alt-right hero and martyr to the cause of free speech. It goes without saying that the alt-right and free speech are not the same, or even similar, things. Most advocates of free speech hate racism but some found their way to the alt-right's Web-based forums through Murray, whose public career was given a boost; Molyneux's half an hour massage of Murray, originally made in 2015, received a flood of new viewers, passing the 300,000 mark. Suddenly everyone wanted Murray.

One of them was Harris. He read Murray and Herrnstein's book and swallowed it whole. Having once declined to participate in a symposium alongside Murray, he invited the seventy-five-year-old political scientist on to his popular 'Waking up' podcast, calling the episode 'Forbidden Knowledge'. He portrayed Murray as a valiant truth-seeker and his critics as cowardly, dishonest, hypocritical witch-burners. In his introduction he parroted some of *The Bell Curve*'s key propositions on race and intelligence: that IQ tests genuinely measured intelligence, that 'average IQ differs across races and ethnic groups', that 'genes appear to be 50 to 80 per cent of the story' and that 'there seems to be very little we can do environmentally to increase a person's intelligence, even in childhood.' These, Harris insisted, were facts. There was 'almost nothing in psychological science for which there is more evidence', he said, adding that there was 'virtually no scientific controversy' over Murray's claims.[1] Not only was this spectacularly wrong – they are not facts and there is much controversy over each of these points – considering the weight, detail and authority of the many scientific critiques of Murray's work and these claims in particular, it was extraordinary. It seemed that while Harris had absorbed *The Bell Curve*, he hadn't read much further.

For 138 minutes Harris gave the race science libertarian a more-than-respectful belly-rub that attracted 500,000 viewers. Murray used the opportunity to repeat his claim that genetics played a role in racial IQ averages before launching a vituperative attack on his many academic critics, claiming they 'lied without any apparent shadow of guilt because, I guess, in their own minds, they thought they were doing the Lord's work'. Harris listened, affirmed but did not challenge. Only once, tentatively, did he pose what might be classed as a penetrating question when he noted that *The Bell*

Curve was much-loved by white supremacists and asked about the purpose of exploring race-based differences in intelligence. Murray didn't miss a beat, using the moment to plug his political views. Its use, he said, came in countering policies such as affirmative action in education and employment, which he found 'morally repugnant' because it was based on the false premise that 'everybody is equal above the neck ... whether it's men or women or whether it's ethnicities. When you have that embedded into law, you have a variety of bad things happen.'[2]

This interview transformed Harris's reputation from being the 'fourth horseman of the atheist apocalypse'[3] into what the uncharitable might call a useful idiot of the alt-right. Clearly, at a personal level he is no racist but in defending and advocating this viewpoint he put himself in the same boat as Murray, even finding himself blacklisted by the Southern Poverty Law Center as a 'propagator of hate', alongside Murray. He didn't help by getting into a year-long skirmish with Ezra Klein, the then editor-in-chief of *Vox*, who had criticised the 'soft-soap' conduct of Murray's interview. The spat began with an email exchange that Harris unilaterally published, although he later admitted 'it backfired on me.'

One of the reasons it backfired was that Harris, the advocate of free speech, was refusing to debate with Klein, having initially invited him on to the podcast. Eventually, Harris changed his mind. Their debate covered much relating to their motivations, and those of various other players, but not so much to *The Bell Curve*'s claims. However, Harris did reiterate the point that ethnic differences in IQ are partly genetic and that at an individual level IQ is virtually immutable. As he put it: 'The problem is, yes, it's hard to change your IQ. We don't know of an environmental intervention that reliably changes people's IQ. Murray is right about that.'[4] Klein disputed this, pointing out that adoption can raise IQ by twelve to eighteen points, and he explained IQ differences between black and white Americans in terms of the history of slavery and racism, and current disparities in wealth and experience. And that is, in essence, the core of the dispute over race and IQ that has waxed and waned for fifty years.

ARTHUR JENSEN AND THE RETURN OF RACE SCIENCE

The discovery of the full horror of the Holocaust, which was in part inspired by eugenic theory, forced race science into retreat. It was kept alive – barely – by the openly racist journal *Mankind Quarterly*, whose board had once included Josef Mengele's mentor, and by the white supremacist Pioneer Fund, whose stated *raison d'être* is 'racial betterment'. This changed in 1969, when the *Harvard Educational Review* published a paper by the educational psychologist Arthur Jensen, entitled 'How much can we boost IQ and scholastic achievement?',[5] which sparked the furious and concerted debate on race and intelligence that led to the publication of *The Bell Curve* twenty-five years later.

Jensen's case was not much different from that of the racist IQ advocates of the first quarter of the twentieth century but because such views had been kept quiet, his intervention looked like something new. He claimed IQ was 80 per cent heritable and suggested differences between black and white IQs were genetic in origin. He argued that remedial schemes, such as the Head Start programme, had failed to boost the IQs of black Americans and that further interventions of this kind were also likely to fail, because black Americans had lower genetic intelligence than white or Asian Americans. In fact, he began the article with a statement of political intent: 'Compensatory education has been tried and it apparently has failed.'[6]

Jensen's paper met an overwhelmingly negative reaction – 'an international firestorm', as the *New York Times* called it forty-three years later, in an obituary.[7] It prompted twenty-nine rebuttals and critiques in academic journals and a wave of student protest, much of it at the University of California, Berkeley, where he was based. Students burned him in effigy, and he received death threats and had to be accompanied by bodyguards. His publishers refused to permit reprints or to allow Jensen to respond to letters of criticism. But the paper had what many considered to be its intended impact; it allowed other racist psychologists to crawl out of the woodwork, while Jensen continued to publish papers with increasingly overt racist content. He dipped his toes in political waters, questioning the point of programmes giving support to poor black people, and was heavily backed by the Pioneer Fund (to the tune of at least US$1.1 million by 1994).[8] But, unlike his collaborators, he steered clear of extreme right groups or writing for *Mankind Quarterly*.

Among the new wave was the UK-based German psychologist, Hans Eysenck. In 1971, his book *Race, Intelligence and Education* argued that the selection of slaves for the plantations made black Americans a less intelligent African sample, and implied that Italian, Spanish, Greek and Portuguese immigrants to the USA were not up to scratch in the IQ stakes. He said his research pointed to the 'overwhelming importance of genetic factors in producing the great variety of intellectual difference which we observe in our culture and much of the difference observed between certain racial groups'.[9] Eysenck, who studied under Cyril Burt, had enjoyed a lofty reputation until his zeal in backing extreme rightist causes embarrassed his colleagues. He contributed several pieces to neo-Nazi publications such as *National-Zeitung* and *Nation und Europa*, including an anti-Semitic one about Freud, and wrote racist prefaces for books by the French fascist Pierre Krebs and British fascist Roger Pearson. An interview with Eysenck was published in the British National Front's mouthpiece, *Beacon*, and he was a member of the advisory council for *Mankind Quarterly*. Eysenck's journal, *Personality and Individual Differences*, offered a launch pad for fellow travellers such as Richard Lynn and J. Philippe Rushton. He played the role of Jensen's wing-man but the American did the heavy lifting, pumping out books and papers on his beliefs in racially based intelligence differences.

A flood of academic critique focused on the details of their underlying theory. Stephen Jay Gould criticised Jensen for not understanding heritability; Jensen used it as a measure of difference between populations, rather than within a population, as intended.[10] Richard Nisbett summed up the research methods of Jensen and his collaborator, J. Philippe Rushton: [they] 'ride roughshod over the evidence ...'.[11] Yet, despite the flaws and caveats, Jensen was taken seriously in the esoteric realm of IQ psychology. Even beyond it, his reputation was less flaky than that of his most ardent supporters.

One supporter was the physicist William Shockley (inventor of the transistor and joint winner of the Nobel Prize for physics in 1956), who later turned his mind to intelligence and population growth. Shockley, who featured as a source of inspiration for the Ku Klux Klan in Spike Lee's 2018 film *BlacKkKlansman*, said higher rates of reproduction among unintelligent people – blacks in particular – were leading to a decline in civilisation. (To offset this calamity, he generously donated some of his sperm to a Nobel Prize-winner sperm bank.) Among his odd outbursts on race was: 'Nature has colour-coded groups of individuals so that statistically reliable predictions of their adaptability to intellectually rewarding and effective lives can easily be made and profitably used by the pragmatic man in the street.'[12]

The oddest supporter of the lot was Rushton, a British-born, South African-raised Canadian, who went from collaborating with Jensen to producing some weird outpourings of his own. He showed none of Jensen's caution, writing for the white supremacist magazine *American Renaissance* and addressing its conferences. He praised the 'scholarly' work of the segregationist Henry Garrett, and in 2009 was a keynote speaker at a conference on 'Preserving Western Civilization', held in Baltimore. In his speech, he argued that Islam was a genetic problem; not just a belief system but a reflection of an innately aggressive Muslim personality that involved a simple, closed mind not amenable to reason.[13]

Rushton indulged his obsession with racial difference in sexual behaviour at the University of Western Ontario. He surveyed his male students, asking about penis length, sex partners and ejaculation distance (not something most men would measure, presumably). His dubious methods – such as bullying first years into participating – caused the university to ban him from using students as research subjects. Instead, he trotted off to a Toronto shopping mall, where he recruited fifty white, fifty black and fifty Asian men to answer the same questions. He was reprimanded again for a 'serious breach of scholarly procedure', partly because he paid his recruits.[14] His secondary research was also idiosyncratic; he used *Penthouse* as a source of data, and repeatedly dipped into an 1896 anthro-porn book, *Untrodden Fields of Anthropology*, which includes advice for sexual tourists such as 'black women smell like crocodiles'.[15] Rushton's theory was that Mongoloids, the smartest humans, were socially disciplined, emotionally inexpressive, relatively sexually inactive and had smaller penises and testicles but bigger brains. Negroids, the least smart, were emotionally expressive, socially undisciplined, hypersexual and had the biggest penises and testicles but the smallest brains. Caucasians were in the middle.[16]

This related to another of Rushton's ideas, borrowed from E. O. Wilson: r/K selection theory. Organisms with an 'r' strategy reach sexual maturity quickly, have short gestation and produce many offspring who are left to survive on their own. Those pursuing a 'K' strategy reach sexual maturity slowly, have longer gestation and invest heavily in the survival of a few offspring. Sociobiologists such as Wilson, who believe all male and female behavioural differences are evolved, characterised male mammals as r-type and female as K-type. The theory was popular in the 1970s but was critiqued for its faulty premises and lack of predictive power. Rushton embraced it, and applied it to humans, arguing that some branches of humanity were more evolved. K-people, such as Mongoloids, became more intelligent as a result of living in cold climates, reached sexual maturity later, had fewer children and invested more heavily in them. R-people, such as Negroids, reached sexual maturity early, started reproducing early and did not invest heavily in their offspring.[17] Higher crime rates and lower marriage rates were a result of Negroids being r-people. Rushton looked for physiological evidence: he felt the high correlation between molar-tooth eruption and brain size (the later the tooth eruption, the bigger the brain) was significant. The problem was that there were several Mongoloid groups whose molars erupted earlier than those of Africans.[18]

Astonishingly, E. O. Wilson gave Rushton's research a sympathetic pat on the back, noting that 'Phil' was an 'honest and capable researcher'. Wilson commended Rushton for his 'solid evolutionary reasoning' but issued a curious warning: 'If he had seen some apparent geographic variation for non-human species ... no one would have batted an eye,' he began, before getting to the nub of the matter. '[W]hen it comes to [human] racial differences, especially in this country, special safeguards and conventions need to be developed.'[19] It seemed Wilson, who had also faced accusations of racism, backed Rushton's conclusions but cautioned against spelling them out quite so brazenly. Richard Herrnstein and Charles Murray, who quoted Rushton liberally in *The Bell Curve*, also defended him, claiming in a footnote that his views, which coincided with their own, were 'not that of a crackpot or a bigot as many of his critics are given to charging'.[20] Others were less sympathetic, including those using r/K theory. The biologist Joseph Graves, who applied r/K to fruit flies, said the theory was of little use in explaining human evolution and that Rushton didn't apply it correctly and showed scant understanding of evolution.[21] The evolutionary psychologist David Barash, who had some sympathy with the application of the r/K theory to humans, said Rushton merely tried to fit the theory to his racist agenda. 'Bad science and virulent racial prejudice drip like pus from nearly every page of this despicable book,' he wrote in a review of *Race, Evolution and Behavior*.[22]

Rushton was able to continue to pump out papers until his death in 2012 at the age of sixty-eight, partly because of generous grants from the Pioneer Fund. By 1994, he had received at least $770,738;[23] in 2000 he received a grant of $473,835, 73 per cent of the fund's total grants for that year. In 2002 he became president of the fund, a post he held for ten years.

Overall, in the quarter-century between Jensen's 1969 paper and the publication of *The Bell Curve* in 1994, race science had captured considerable ground in the esoteric *milieu* of IQ studies but made scant progress among the public. Jensen had softened the American psychological soil but Eysenck, Rushton, Lynn and Shockley were too far to the extreme right, and simply too flaky, to survive in the mainstream for very long. In the scientific community, excoriating critiques of the biology and methodology of race science from big hitters such as Gould and Lewontin seemed to have settled the issue. *The Bell Curve* changed all that.

THE BELL CURVE DIGESTED

The title of this bestseller (which refers to the bell-shaped curve of the distribution of IQ in a population) was certainly snappy but the arguments set out in its 845 pages were nothing new. They were essentially those of Jensen, Eysenck, Lynn and Rushton, with a bit of right-wing libertarian politics thrown in. The difference lay in presentation: the impression of objectivity, the tight chapter summaries, the graphs and tables, the chirpy and authoritative tone.

Herrnstein and Murray argue that America is being stratified according to intelligence, drawing high-IQ people to the top and pulling low-IQ people to the bottom. Because IQ is primarily genetic, they suggest, high-IQ couples have high-IQ babies (and *vice versa*), with this new elite 'taking on some characteristics of a caste'.[24] They worry that America will morph into Latin America, with high-IQ whites and Asians enclosed in fenced enclaves, protected from the 'menace of the slums' by armed guards. Meanwhile, the low-IQ masses, living in high-tech versions of Native American reservations, will be kept going by welfare handouts. This represents 'something new under the sun'.[25]

The appendix-filled pages of part two of the book argue that IQ determines how we live, not the other way around; a point Herrnstein and Murray claim is original, although it is not. They say they are 'clearing away some of the mystery that has surrounded the nation's most serious problems'[26] using data from the National Longitudinal Study of Youth conducted in the 1980s by the US Department of Labor. The authors say these test scores are a better indicator than parental class status of where children will end up on the economic ladder, as well as whether they'll go to jail or be divorced. The correlation between poverty and low IQ is seen as proof.

Part three concentrates on race and IQ. The authors immediately state their premise: 'Given cognitive differences among ethnic and racial groups, the cognitive elite cannot represent all groups equally.' They add that a 'substantial difference in cognitive ability' between blacks and whites is reflected in public and private life.[27] They review selected literature on race and IQ, drawing from Lynn (24 times), Jensen (23), Rushton, Gottfredson and *Mankind Quarterly*. They say Asians have IQs slightly above those of white people, with black people fifteen points below white, and dismiss the idea that the reasons for this might be entirely environmental. In a circular argument, they refute the view that the lower economic status of black Americans is the reason for the lower average IQ by repeating their own view that low IQ causes poverty. They acknowledge the black-white IQ gap is shrinking but doubt it could ever close. Drawing from Jensen, they insist there must be a genetic dimension, although they are 'resolutely agnostic' on the precise nature-nurture mix.[28] But this doesn't matter because 'realised intelligence, no matter whether realised through genes or the environment, is not very malleable.'[29]

Next, they present their case from a different angle, claiming that black people with the same IQ as white people earn as much, or more, than they do: 'Racial and ethnic differences in this country are seen in a new light when cognitive ability is added to the picture.'[30] In fact, because of affirmative action, blacks 'overachieve' on the job market. But blacks with the same scores as whites are still more likely to be unmarried parents, unemployed, and on welfare. (This claim prompted a *New York Times* reviewer to declare that the book 'is just a genteel way of calling somebody a Nigger'.[31]) They go on to discuss their belief that Americans' IQs are falling because poor, stupid people are having more children than rich, bright ones (which they call *dysgenesis*). This contradicts the 'Flynn Effect'; their way around this conundrum is to make the obvious point that the 'Flynn Effect' does not reflect a real rise in intelligence but rather the growing sophistication of those taking IQ tests, which is at odds with their faith in the immutable *g*. They speculate on a genetic explanation (bright Baby Boomers had more babies) and toy with some unexplained environmental factor but in the end, stick to their guns: dysgenesis is thriving because low IQ people are breeding more quickly than high IQ people. They add another argument, used by Yerkes in 1917: less-intelligent immigrants are making their homes in the USA, lowering IQ averages. Flynn notes that the average American IQ has risen by around 9 per cent per generation; Herrnstein and Murray say the average IQ has fallen by one or two points per generation.

In part four, they indulge Murray's right-wing libertarianism, posing the question: can people become smarter if given the right kind of help? They answer with a big fat 'no' but contradict themselves by noting that improving infant nutrition and going to school can raise IQs. They insist compensatory education makes little difference but again contradict themselves, acknowledging a Venezuelan study in which extra tuition did indeed raise children's IQs. They also note that coaching can improve test results and reference adoption studies in which children who moved from deprived to well-to-do backgrounds added twelve points to their IQ scores. (In other words, their big fat 'no' would seem more like quite a big 'yes'.) Perhaps it would be possible to raise the IQs of those at the bottom, they concede, but it is 'tough to alter the environment for the development of general intellectual ability'.[32] They go with their instincts, opposing affirmative action because it unfairly promotes people with lower IQs, giving black people better positions than their intelligence merits. Instead, they say, it would be more productive to spend money on gifted children.

After warning of the social consequences of the growing divide between the high-IQ elite and the expanding low-IQ underclass, they end with some rather quirky policy proposals, including one that the state should abandon attempts to create equality of outcomes and instead recognise genetic inequality, allowing people to find their 'valued places in society' according to their innate intelligence. 'It is time for America once again to try living with inequality,' they say, as part of their attack on affirmative action. They plump for a small-is-beautiful solution, arguing there is too much state centralisation. Simplification – less red tape for small businesses, punishments that fit crimes, marriage as a requirement for parental rights, careful screening of immigrants, cheap birth control and replacing welfare with cash supplements – should be 'a top priority in reforming policy',[33] so that everyone can find their IQ-determined place.

THE CRITIQUES FLOOD IN

The Bell Curve was an instant bestseller, boosted by the controversy generated by its race science section. Reviews in the mass media focused more on its conclusions than its premises, perhaps because the reviewers felt ill-equipped to critique the use of data. But once the academic reviews started coming in, the premises were picked to pieces.

Stephen Jay Gould argued that the book's entire edifice would collapse if any of these four premises were false: (i) intelligence must be reducible to a single number, (ii) it must be capable of ranking people in linear order, (iii) it must be mainly genetically based and (iv) it must be effectively immutable.[34] He showed why all these premises were wrong, noting that the book 'contains no new arguments and presents no compelling data to support its anachronistic social Darwinism'.[35] Gould noted that even if one accepted the raw data, they 'permit no conclusion that truly equal opportunity might not raise the black average to equal or surpass the white mean'.[36] He accused Herrnstein and Murray of 'spinning' their cause, by exaggerating the genetic case and playing down the strong evidence that IQ is malleable. He also refuted their view that IQ, as a number, can measure a real quality in the brain. 'How strange', he wrote, 'that we would let a single false number divide us when evolution has united all people in the recency of our common ancestry...'[37]

Another wave of criticism looked at the authors' reliance on the work of a tiny coterie of race psychologists.[38] Leon Kamin, professor of psychology at Princeton, reviewed IQ data relating to Southern Africa, borrowed from Richard Lynn (whom Herrnstein and Murray called 'a leading scholar of racial and ethnic differences', saying they 'benefited especially' from his advice). Kamin showed that some data did not even involve IQ tests, that Lynn had drawn unwarranted conclusions from other data and had ignored tests in which black people outperformed white people. He concluded that the calibre of data in *The Bell Curve* was 'pathetic' and that its authors 'frequently fail to distinguish between correlation and causation and therefore draw many inappropriate conclusions'.[39] Likewise, the biologist

Joseph Graves placed *The Bell Curve* in the tradition of unscientific racism, arguing that its claims were not supported by its data, that calculation errors had reinforced its bias and that the authors had ignored data that contradicted their core beliefs.[40]

Other critics noted that the authors 'corrected' data to achieve the results they favoured, allowing them to overstate their case that IQ is better than socio-economic standing as a predictor of poverty.[41] Some focused on Herrnstein and Murray's reliance on the Armed Services Vocational Aptitude Battery tests for their suggestion that IQ was a better predictor of social outcome than family background, when this battery of tests does not measure IQ.[42] A team from the University of California found the Armed Services tests were a 'poor measure of innate intelligence and instead reflected the social environment that shaped people's academic performance, largely their schooling'.[43] Backing this view, the economist Sanders Korenman and the sociologist Christopher Winship cited data that indicate family background was, in fact, 'more important than IQ in determining socioeconomic success in adulthood'.[44]

Several critiques drew on data from studies that contradicted the authors' faith in the predictive power of IQ, including one subtest that turned out to be the best predictor of future earnings, left out by Herrnstein and Murray because it did not back their conclusions.[45] Robert Hauser, a sociologist from the University of Wisconsin, and his colleague Min-Hsiung Huang analysed the results of 12,500 adults in a verbal ability test, taken from the General Social Survey, to show that the *Bell Curve*'s assertion of a predictive relation between IQ and the professions was not borne out. All that the results showed was that selected highly educated occupation groups had grown rapidly since 1940.[46] A correlation between aptitude examinations and economic status was hardly surprising when so many employers subjected applicants to IQ tests.[47]

In his review of Flynn's book *What Is Intelligence?*, Murray said it was a 'gold mine of pointers to interesting work, much of which is new to me'; a surprising comment, given that Flynn had been developing his argument for more than a quarter of a century. But perhaps it means no more than the fact that Murray was less *au fait* with intelligence theory than his IQ-obsessed partner, Herrnstein. Murray added that all those wrestling with questions about intelligence 'are in Flynn's debt'.[48] This is generous, given that Flynn explicitly attacked *The Bell Curve*'s theories, offering a critique of their notion of race while noting that their meritocracy thesis is 'incoherent'.[49]

What is clear is that if you accept Flynn's theory, something has to give in Herrnstein's and Murray's. If IQs have been steadily rising, and this is not a temporary aberration, then 'real intelligence' is unlikely to be falling. And if intelligence isn't falling, despite the poor having more babies than the rich, then the *Bell Curve* must either have overstated the genetic component of IQ, or its authors' assumption that the poor are poor because of their lower intelligence must be wrong. Or both. Which brings us back to Gould's point that if any of the work's key premises are shown to be invalid, its entire case collapses.

THE *BELL CURVE*'S COUP

If you're under the age of forty-five or were not part of the American cultural *zeitgeist* of the mid-1990s, you might be perplexed by the publishing phenomenon that was *The Bell Curve*. Read it cover to cover, particularly the chapters dealing with race, poverty and IQ, and if you've read anything else on this subject, you can't avoid concluding that, despite its claims to novelty, there is nothing new under this particular sun. In essence, it combines the anti-welfare political perspective of one of its co-authors with the racist IQ perspective of the other. That is all.

Murray, who was fifty-one at the time, had a track record of working for right-wing think tanks. One of his previous books, *Losing Ground*,[50] was much admired by the Reagan administration. The sixty-four-year-old Herrnstein, a controversial psychology professor from Harvard, had a track record in race-based IQ studies going back to the early 1970s, although he had not achieved the same infamy as Jensen, Rushton and Lynn, but it was these men's data he embraced.

He showed no qualms about drawing uncritically from psychologists who were heavily financed by the Pioneer Fund and had contributed to *Mankind Quarterly*. In fact, Herrnstein and Murray cited seventeen authors who had written for this expressly racist journal, ten of whom served on its advisory board. They also directly cited five articles from *Mankind Quarterly*, including one by Nathaniel Weyl, an erstwhile communist who had turned to the far right and, on the basis of their IQ scores, had decided that white Rhodesians were an intellectually 'elite element within the English-speaking world in terms of psychometric intelligence'.[51]

Herrnstein died shortly before *The Bell Curve* was published; Murray, with no background in IQ studies, was left to mount the defence. 'Here was a case of stumbling onto a subject that had all the allure of the forbidden,' he admitted, in the tone of one who had chanced upon a secret stash of porn. 'Some of the things we read to do this work we literally hide when we're on planes and trains. We're furtively peering at this stuff.'[52] Perhaps, for Murray, the furtive thrills were real; for Herrnstein, most of it was old hat. What this pair skilfully did was to package 'this stuff' as something fresh and daring.

Now and then they referred to evidence that contradicted their case, burying most of it in footnotes and appendixes but doing enough to ensure their tone appeared more measured than that of Rushton and his ilk. For instance, at the beginning of Chapter 13 they make this point about IQ: 'It seems highly likely to us that both genes and the environment have something to do with racial differences.' If you had read nothing else on this subject, this might seem a reasonable acknowledgement of both sides of the story. In fact, it is the opposite. No one, not even Lynn, denies that environment has something to do with racial differences. But most mainstream academics in the field agree there is no evidence of a genetic role in this gap.

The American Psychological Association set up a task force to investigate the book's claims, publishing a document entitled 'Intelligence: knowns and unknowns'. On the American white-black IQ gap, this text noted that the evidence 'fails to support the genetic hypothesis'. It added that the gap may be diminishing, and the reasons for the remaining gap may relate to caste and culture.

'There is certainly no such support for a genetic interpretation,' it concluded.[53]

The Bell Curve was an audacious bid to break that consensus. It was a meticulously planned guerrilla campaign, from its carefully edited presentation, complete with short chapter summaries and impressive-seeming graphs and tables, to its publicity, the fanfare of its publication and Murray's energetic television appearances, articles and interviews.

Linda Gottfredson, who was developing her own track record for race-based IQ studies, got busy with 'Mainstream Science on Intelligence', which she sent to 131 hereditarian academic psychologists: 52 signed and 31 ignored it. Of the 48 who declined to sign, 11 went public with their view that it did not represent a mainstream perspective. The text's publication in the *Wall Street Journal* gave *The Bell Curve* a significant boost; two months after its publication 400,000 copies had been printed.

Gould said that behind the 'catchy title' and 'brilliant publicity campaign', the book's argument was no different to those he had critiqued in the first edition of *The Mismeasure of Man* fourteen years before.[54] He said the main reason for its success was the political *milieu* of its suggestions to slash social services and cut taxes for the rich. A biologically based argument that the poor were poor because they were unintelligent, and that money diverted to improve their lot was money wasted, had strong appeal to a Republican constituency. An additional reason for Gould's criticism related to the enthusiasm for finding single-gene solutions to all sorts of issues, including intelligence; an enthusiasm, Gould said, that was founded on an ignorance of genetics.[55]

ANDREW SULLIVAN PLAYS THE RACE CARD

One detail was missing in Gould's assessment of the reasons for *The Bell Curve*'s success: the extraordinary role played by one man. Andrew Sullivan has taken on several journalistic roles over the years: editor, columnist, super-blogger. In each, he has gone out of his way to promote *The Bell Curve*. Whatever else he's achieved, he'll be remembered as the man who put himself at the service of race science. Without Sullivan's help, the book would not have been the publishing sensation it was. And his subsequent efforts to promote it – even twenty-five years later – have helped maintain its presence.

Sullivan grew up in a Catholic family in Surrey, UK. As a result of an IQ test, he won a place at a top grammar school, going on to Oxford University, where he became president of the Oxford Union and a confirmed Tory. He later shared a flat with a future Foreign Secretary, William Hague, and a future Minister of State, Alan Duncan. He interned at a Thatcherite think tank, before moving to the USA, coming out as gay and HIV-positive, studying for a PhD at Harvard and launching his journalistic career at *New Republic* magazine. At twenty-six, he was appointed its editor, a position he held for five years. After that, he wrote for the *New York Times Magazine* (until he was fired four years later) and was a columnist for the *Sunday Times*. In 2000 he launched his blog, *The Daily Dish*, hosted by Time.com from 2006 and later by TheAtlantic.com and Daily Beast, which he claimed had eight million page views a month. Sullivan, who currently contributes to *New York* magazine, has written or edited seven books and regularly appears on American and British television. To say that he is one of the most influential journalists in the world would be an understatement. I've long admired the quality of his prose and his willingness to change his mind. In 2000, he was a cheerleader for George W. Bush and the invasion of Iraq but recognised the failings of the adventure and came out in support of John Kerry and Barack Obama. Yet he remains resolute in his backing for *The Bell Curve*: 'I had no interest in this subject until I saw the data in Murray's and Herrnstein's book,' he said. 'I was, frankly, astounded by it. ... And yet, it turned out it was undisputed. Merely the interpretation of it was open to real and important debate.'[56]

Sullivan agreed to publish an extract of *The Bell Curve*'s most contentious chapter in the *New Republic*, making it the cover story, under the banner 'Race and IQ', together with an on-message editorial. He did this, he has since boasted, 'before anyone else dared touch it'.[57] That edition was the biggest seller in *New Republic*'s history. His staff threatened to resign, so he published nineteen rebuttals before resigning himself, admitting he was a 'lousy manager of people'.[58] The nineteen writers criticised the book's premises, methodology and conclusions. Some went for the jugular, declaring it an out-and-out racist venture; Alan Wolfe began by asking: 'What's the difference between thinking that the black male next to me is dumb and thinking there's a 25 per cent chance he's dumb?'[59] Sullivan's decision to publish brought racist psychology out from its academic *purdah*, restoring it to the realms of respectable debate. You might have expected that once the peer reviews, almost all deeply negative, came in, and the book's academic credibility was shredded, Sullivan would have retreated, but after a spell of silence he bounced back, not only defending his decision but also endorsing *The Bell Curve*'s race-based argument, without reservation.

Sullivan said in 2005 that his decision to publish that extract was 'one of my proudest moments in journalism', adding that 'the book held up and still holds up as one of the most insightful and careful of the last decade' and he was 'proud of those with the courage to speak truth to power, as Murray and Herrnstein so painstakingly did'. He endorsed the book's race-based premises, telling his readers that 'human inequality and the subtle and complex differences between various manifestations of being human – gay, straight, male, female, black, Asian – is a subject worth exploring, period,' adding that liberalism 'should not be threatened by empirical research into human difference and varied inequality'.[60]

In 2011 he dipped in again: 'No one is arguing "that black people are dumber than white,"' he said, 'just that the distribution of IQ is slightly different among different racial populations and these differences also hold true for all broad racial groups.'[61] Actually Herrnstein, and those he quoted most often, such as Lynn, Rushton, Gottfredson and Jensen, have argued that black people are dumber than white. And Sullivan's 'different distribution' argument is regularly invoked by race science enthusiasts: it is not racist to say white and Asian IQs are higher than black IQs because you get people at all IQ levels in these groups. This is another way of saying 'I can't be a racist even though I think that most white people are smarter than most black people because I accept that at an individual level there are smart black people and dumb white people.' To put it differently, when meeting a black person for the first time the perception would be: 'It is possible that you're more intelligent than average but likely that you are, in fact, less intelligent.' Not racist?

Sullivan complained that 'political correctness and racial squeamishness have hindered the study of intelligence,'[62] mentioning in particular Jensen's standing among 'many traditional intelligence researchers' while claiming that Jensen's studies on race and IQ had been 'strangled by PC egalitarianism'.[63] It is worth recalling that Jensen's view was not just that black Americans' IQs were fifteen points below white Americans' but also that the main reason for this gap was genetic. In other words, black people were inherently dumber than white people. Sullivan, like Jensen, was quick to add the usual rider that his own views on racial distribution of IQ didn't *only* apply to black people.[64]

Returning to the fray in 2013, Sullivan admitted he should 'know better than to bring this up again' before bringing it up again. He announced he'd come to doubt the existence of *g* and to recognise that IQ should not be conflated with intelligence. This was a breakthrough and a clear departure from *The Bell Curve*'s view. But just when it looked like he might be siding with the angels, Sullivan slid back, with this curious analogy: 'We remain the same species, just as a poodle and a beagle are of the same species. But poodles, in general, are smarter than beagles and beagles have a much better sense of smell.'[65] Sullivan was savvy enough to acknowledge that breeding and natural selection were different but then ignored this huge distinction by applying his dog analogy to humans, noting that humans evolved for skin colour because of exposure to different environments. He prefers to talk of poodles and beagles, but we don't need to guess who are the beagles and who the poodles.

Sullivan's next big epistle on race and IQ was a 2018 comment piece for *New York* magazine. He began by caricaturing the 'blank slate' and 'Utopian' ideas that environment and culture play a major role in human affairs. He then leapt from endorsing the geneticist David Reich's uncontested view that there are genetic differences between human populations to claiming, without evidence, that there will therefore be group 'differences in intelligence tests',[66] which Reich neither said nor implied. And from there to once again defending his decision to publish the excerpt from *The Bell Curve*: 'My own brilliant conclusion: group differences in IQ are indeed explicable through both environmental and genetic factors and we don't yet know what the balance is.'[67] 'Balance' implies it is a bit of both, which he suggests, appropriately, is the same as Murray's view but, inappropriately, also Reich's and Flynn's. (Reich never mentions IQ in his book; Flynn believes the US white-black IQ gap is 'probably environmental'.) Sullivan then gives vent to his conservative ideals, saying that an assumption that genetics plays no role in racial IQ differences could lead us to 'over-shoot and over-promise in social policy' and may lead to affirmative action, which he says is 'racial discrimination'.[68]

The Bell Curve argues that IQ is primarily genetically based, that black IQs are lower than white, that low IQ is the reason why there are more poor blacks than whites and that the way to deal with this is to endorse inequality, abolish affirmative action, cut red tape for business and reduce the role of the state. This is the cause that Andrew Sullivan has chosen to endorse, over and over and over again, for a quarter of a century.

THE BELL CURVE IN THE UK

Twenty-five years after its publication, *The Bell Curve*'s key pillars have fallen, so why bother with it? The answer is that it continues to exert influence, as seen in Charles Murray's resurrection. In the UK, such sentiments tend to be more muted. It may no longer be a tome to display on the coffee table but right-wingers are still drawn to its premises. Unlike Andrew Sullivan, most are savvy enough to keep its racist component safely tucked away but the rest is allowed to roam free.

Dominic Cummings, senior advisor to Michael Gove during his time as Secretary of State for Education, and later campaign director of the 'Vote Leave' Brexit drive, wrote a 250-page tome attacking the government's Sure Start programme, which aimed to improve poor children's prospects; the programme was subsequently abandoned. Using arguments that sound as if they had been extracted live from *The Bell Curve*, Cummings said there was little scientific evidence that this kind of thing worked and that children's performance in school was based more on IQ and genetics than teaching, giving a figure of 70 per cent, which is at the extreme end of the genetic-determinist spectrum. 'There is strong resistance across the political spectrum to accepting scientific evidence on genetics,' he wrote. 'Most of those that now dominate discussions on issues such as social mobility entirely ignore genetics and therefore their arguments are at best misleading and often worthless.'[69] Given the fervour of his view that school performance is mainly genetic, it would be intriguing to hear his perspective on why poor British African, Caribbean, mixed-race and Asian children consistently outperform poor white children. From his perspective, we might expect a sad declaration that white children are lacking in IQ genes but, oddly enough, this is not an issue that has exercised his typing fingers.

This was followed by a speech by Boris Johnson, then Mayor of London. He mocked the '16 per cent of society' with IQs below 85 and implied the poor had lower IQs, suggesting the state should put more effort into the 2 per cent with IQs over 130, another *Bell Curve* proposition. Johnson concluded: 'The harder you shake the pack, the easier it will be for some cornflakes to get to the top'[70] – cornflakes such as the Eton-schooled Johnson, presumably.

Then there's Rod Liddle, a former BBC producer, controversialist and columnist, who defended remarks he made about race and IQ, writing under a pseudonym on the website of Millwall FC. 'It's true that 97 per cent of intelligence tests put whites 7 per cent ahead of black Africans and that we're behind Asians and particularly east Asians,' he was quoted as saying. He added the usual note of qualification: 'there's a greater division in races than between races' and then qualified the entire sentiment, saying, 'you can't trust any of them because they're culturally determined.' To justify why he nevertheless chose to make the race-and-IQ point, he added: 'I'm merely being accurate.'[71]

The pseudonymous Liddle aside, open venting of scientific racism remains the love that dare not speak its name, at least in the house-trained corner of the British right, which is why the action is mostly subterranean. In 2018, the *London Student* newspaper exposed the clandestine 'London Conference on Intelligence', bankrolled by the Pioneer Fund and held for three years at University College London, without the university's knowledge.[72] Its organiser, James Thompson, an 'honorary' psychology lecturer and prolific far-right tweeter on race and gender, declared the conference's focus was on eugenics and how IQ was inherited between different races,[73] with topics including whether 'racial admixture' reduced population quality. There were also three papers (all by men) which argued that women were innately stupider than men.[74] Richard Lynn and Gerhard Meisenberg, editor-in-chief of *Mankind Quarterly*, were among the conference's notable speakers. Another was the right-wing gadfly Toby Young, who supports eugenics, arguing that poor people with low IQs should be helped to 'choose which embryos were allowed to develop, based on intelligence'.[75] His proposal relies on the *Bell Curve*'s assertion that poor people are poor because they're stupid and that 'IQ genes' will eventually be identifiable. He described the elaborate measures of the conference's organisers to prevent leaks but said these were reasonable 'considering the reaction that any references to between-group differences in IQ generally provoke'.[76] Young said he attended to 'gather material' for a speech he was to make at an event in Canada, hosted by the International Society for Intelligence Research, on 'the history of controversies provoked by intelligence researchers'. Soon after, a public petition protesting against Young's tweets on eugenics, gays and women gathered 220,000 signatures, forcing him to resign from both his post on the Fulbright Commission and a new position leading the government's Office for Students.[77]

A postdoctoral researcher, Noah Carl, was another whose presence at the conference caused him problems. He spoke there at the 2015 and 2016 gatherings. Carl, a fervent supporter of race science, whose work is published on the website *OpenPsych*, which promotes race science, was awarded a research fellowship at St Edmund's College, Cambridge. An open letter protesting against this appointment, which was quickly signed by more than two thousand academics, accused Carl of, among other things, focusing on 'ethically suspect and methodologically flawed' research involving race and genetics and noted his participation in the London Conference on Intelligence.[78] This nudged Toby Young into action; he launched a pro-Carl campaign, which included a comment piece in *The Spectator*,[79] an editorial in *Quillette* and a counter-petition, and the college withdrew the fellowship.

And just one more: in 2020 Dominic Cummings employed another London Conference on Intelligence attendee, Andrew Sabisky, as a Downing Street advisor. Sabisky had previously blogged on Cummings's website in favour of eugenics, while making the claim, based on Bell Curve-like data, that black Americans were innately less intelligent than white Americans, incorrectly saying their average IQ scores were 15 points lower – remarks Downing Street refused to condemn even after Sabisky's forced resignation two days later, as a result of negative publicity.

I use these examples to illustrate that although *Bell Curve*-flavoured ideas have less of a public airing in the UK than in the USA they are, nevertheless, bubbling away and looking for outlets. These sometimes come in muted form, as with Boris Johnson and Dominic Cummings, sometimes in provocative form as with Rod Liddle and Andrew Sabisky, and occasionally in clandestine form, as with the London Conference on Intelligence, in which the racist content was so overt and extreme that it could only be sold from under the counter.

13

ARE JEWS SMARTER THAN EVERYONE ELSE?

Jordan Peterson is not above saying extraordinarily wacky things in his role as a YouTube celebrity. Ask about feminism, political correctness, Hitler, the Holocaust, Jungian archetypes, witches, enforced monogamy, birth control, involuntary celibates, red meat or lobsters and he'll take you into territory both weird and disturbing.[1] But on race, he treads more carefully, preferring to shake his head in despair, skirt around the edges, cherry-pick his answers and move on. Interviewed by the alt-right podcasting star Stefan Molyneux, Peterson said that 'the IQ literature reveals that which no one would want to be the case.'[2] In another YouTube interview, he sighs heavily and says: 'This is something you can't say anything about without immediately being killed so I am hesitant to broach the topic.'[3]

But Peterson can't resist, so broach it he does, even if his answers are carefully calibrated. He begins with an extremist perspective on IQ, basically all-nature-no-nurture. He equates IQ with intelligence, and declares it to be biological, permanent and to carry profound practical consequences. It is also 'irremediable', meaning that cognitive training 'does not produce an increment in actual IQ and general intelligence – doesn't happen!'[4] (He likes to end his sentences with shouty bits.) Peterson says the army 'has experimented with IQ tests since 1919' – actually, 1917 – 'and in the last twenty years a law was passed that it was illegal to induct anyone into the armed forces that had an IQ of less than 83.'[5] There is no such law, and the US military uses the Armed Forces Qualification aptitude test, not IQ tests. Peterson adds, with a look of sad resignation, that his sub-83 figure represents 10 per cent of the population and reflects the 'cognitive stratification of society that was laid out in *The Bell Curve* in the 1990s'.[6]

I've included these little corrections to show that Peterson doesn't always get his facts right. But it's more than that: none of his outpourings on IQ give any hint of any awareness of the overwhelming evidence that contradicts his assertions; of how adoption into a middle-class home can substantially increase IQ; of how separated identical twins can have huge IQ gaps when brought up in backgrounds of different class; of how practising IQ tests, and even playing some computer games, can increase your IQ score considerably. Peterson ignores the 'Flynn Effect' of generational rises in IQ and how this affects different populations differently. One can only conclude that his reading on IQ is very limited and very dated.

Peterson has a way of sounding convincing to those, mostly young American white men, who seek conviction, certainty and intellectual self-confidence. There's a fair amount of aggression in his delivery and he's not beyond comical bravado. He once tweeted a reply to a critic: 'You sanctimonious prick. If you were in my room at the moment, I'd slap you happily.'[7] When he declaims, he uses emphatic hand movements and angry assertions, based on an absolute conviction that he's right about everything, and an impatient dismissing of contrary ideas. This makes his talk of IQ appear authoritative to anyone who likes that kind of thing and has not delved into the subject.

Yet despite his self-righteous swagger, Peterson treads on tip-toes when extending his general points about IQ to 'ethnic' groups. He likes to start by drawing from his rich seam of sex-related analogy, telling his listeners the reason most engineers are men is because 'men are more interested in things and women are more interested in people'; about 15 per cent more, he insists. I investigated this claim in my book on genes and gender;[8] the evidence against it is persuasive but for Peterson, counter-arguments, such as the fact that a hundred years ago almost all doctors were men or that the proportion of female engineers is rising, are not worthy of consideration.

His point is that greater male interest in things, on average, means that at the top end of the scale, where people are hyper-interested in things and become engineers, most will be men because you get 'walloping differences at the tails and the tails are important because we draw exceptional people from the exceptions'.[9]

This opens the way for him to bring out the one ethnic example he feels safe to use, that of Ashkenazi (European) Jews. Whenever he's asked about race and IQ, Peterson reaches for the Ashkenazim. The reason this argument is safe, and therefore used so often, is that Jews are widely believed to be smart, so asserting this is not likely to upset too many people; it's not like saying that white people are smarter than black people. But its convenience for those who believe in race science is that it works well as a cat's paw for a wider acceptance of innate racial differences in intelligence. Accept that Jews are innately smarter than the rest, and you're accepting the founding principle of race science, and its inevitable corollary that other groups are innately less smart.

Peterson asks why Jews are so over-represented in 'positions of competence' and declares only two answers are feasible: a Jewish conspiracy (which he dismisses) or superior Jewish IQ. 'Now it's not like we have more geniuses than we know what to do with', he says:

> ... and if Jews happen to be producing more of them, which they are by the way, then that's a pretty good thing for the rest of us, so let's not confuse competence with power and authority even though that's a favourite trick of the radical leftists who always fail to make that distinction.[10]

He claims Ashkenazi Jews have a '15-point IQ advantage over the rest of the Caucasian population' (most studies suggest a 7 or 8 per cent gap). This is 'sufficient to account for their over-representation in positions of authority and influence and productivity' because those at the top percentiles all come from the group with the small edge.[11]

In Peterson's two-tone world, the only compensation for those 'ethnic groups' with lower IQs is that it doesn't imply they aren't nice people, because intelligence and human value aren't the same thing:

> You can be pretty damn horrific as a genius son-of-a-bitch. There doesn't seem to be any relationship whatsoever between intelligence and virtue so if it does turn out that nature and the fates do not align with our egalitarian presuppositions, which is highly probable, we shouldn't therefore make the mistake of assuming that if group A or person A is lower on one of these attributes than group B or person B, that is somehow reflective of their intrinsic value as human beings.

He leaves us to guess who group A and group B might be.

A final point on Peterson's logic: even if we accept his group-based hereditarian premises and dismiss the case that Jewish learning cultures explain the IQ gap, the fact of scoring higher on IQ tests on average would not imply that those at the top would be Jewish. The 33 per cent greater genetic variability among sub-Saharan Africans could negate that; the group with more genetic variability could have the highest IQs even if its average was lower. Peterson does not seem to be aware of this genetic variability point, but more seriously, he seems unaware of any of the environmental arguments relating to the higher IQs of Ashkenazi Jews, the subject of the rest of this chapter.

JEWISH IMMIGRANTS: THE 'FEEBLE-MINDED' 83 PER CENT

Probing the roots of Jewish intellectual success, and asking why it persists, may be a worthwhile line of inquiry. The same could be said for inquiries into the intellectual success of Germans, Asians and others. But the answers won't be found in genetics, IQ scores or brain measurement. To illustrate this point, it's worth going back in time to see how Jews scored in the tests of earlier eras.

Jews didn't do well in the nineteenth century when the skull-filling George Morton was at work on brain measurement (something still embraced by Richard Lynn). He claimed Jewish brains were smaller than Teuton and Anglo-Saxon brains, reflecting their intelligence levels. The same is true of early IQ tests, including those given to immigrants at Ellis Island, before the First World War. If we take those tests at face value and retain Henry Goddard's faith that IQ measures innate intelligence, we might conclude that Jewish immigrants were dullards. Goddard tested thirty-five immigrants (via a Yiddish translator) in 1913 and found that 83 per cent were 'morons'. Jews came second to last on his 'feeble-minded' scale, just above the Russians. The nation of Tolstoy, Chekhov, Dostoevsky, Stravinsky and Nabokov kept the people who produced Maimonides, Kafka, Einstein, Mahler and Roth off the bottom rung.[12]

Next came the First World War army tests. Charles Yerkes and Carl Brigham found that Jewish recruits scored low on IQ tests. Brigham explained that Jews were distinctly inferior to Nordics in terms of average innate intelligence but were also a more variable bunch. The fact that the Jewish average IQ was low didn't mean there were no geniuses, although Brigham suspected people were rather too keen to recognise exceptional Jews. 'The able Jew is popularly recognised not only because of his ability but because he is able and a Jew,' Brigham suggested, before asserting: 'Our figures ... would rather tend to disprove the popular belief that the Jew is highly intelligent.'[13]

The conservative economist Thomas Sowell subjected *The Bell Curve* to a withering critique for underplaying the low IQ scores of Jews and other immigrants early in the twentieth century. 'Strangely', he wrote:

Herrnstein and Murray refer to 'folklore' that 'Jews and other immigrant groups were thought to be below average in intelligence.' It was neither folklore nor anything as subjective as thoughts. It was based on hard data, as hard as any data in *The Bell Curve*. These groups repeatedly tested below average on the mental tests of the World War l era, both in the army and in civilian life. For Jews, it is clear that later tests showed radically different results – during an era when there was very little intermarriage to change the genetic makeup of American Jews.[14]

After the war, results did indeed change. A paper published in 1928, referring to American IQ studies, notes that 'four of them find that Jews are not more intelligent than the non-Jewish population and four of them find that they are.'[15] By the latter half of the twentieth century Jews were routinely scoring above average on IQ tests. Were they becoming genetically more intelligent? Or was something else involved? Perhaps the answer could be gleaned from studies of two generations of Mizrahi (Middle Eastern) Jews, conducted in Israel in 1972. The first generation had a mean IQ of 92.8 but the second generation, more acclimatised to Israeli life, had a mean IQ of 101.3.[16]

Despite overwhelming evidence that tests of Jewish IQ in early-twentieth-century America and late-twentieth-century Israel showed lower than average IQ, the myth of perpetually high Jewish IQ persists. Steven Pinker said the Jewish IQ advantage was 'long standing and has been around for as long as there have been IQ scores of Jewish populations'.[17]

THE PARALLEL RISE OF ASIAN AMERICAN IQ

Ashkenazi Jews are not the only American population group whose IQs rose more quickly than average. The IQ theorist Jim Flynn focused on Chinese Americans born in 1948. In their senior high school year (aged seventeen or eighteen) their non-verbal IQs showed a below-average score of 97. Yet they matched their white peers on high school grades and surpassed them thereafter. Similar figures emerge from university entrants. In 1966, American-born Chinese-American entrants to UC Berkeley had IQs averaging seven points below those of white Americans, yet their university grades matched their white contemporaries'. By 1980, when they were thirty-two years old, 55 per cent of the Chinese members of the class of '66 were in managerial, professional or technical occupations, compared to 34 per cent of white members, and their average income was 20 per cent higher.[18] Flynn said the reason lay in cultural differences relating to learning. Their parents surrounded them with more cognitively demanding environments than was typical for white

American children, creating a passion for educational achievement. 'If an Irish lad qualifies for an elite university and his fiancée wants him to stay at home, he may do so,' Flynn noted. 'A Chinese youth is likely to get a new fiancée.'[19]

The children of the Chinese-American class of '66 grew up in wealthier environments, living through preschool years that were more cognitively demanding. The result was that their IQ rose faster than those of white Americans. Among the class of 1990, Chinese Americans had average IQs of 108.6 at the age of six and 103 at eighteen, three points above the white average.[20] White American IQs were also rising, due to the 'Flynn Effect'; but Asian IQs were rising more quickly. However, high achievement preceded high IQ, rather than the reverse. The fact that the adult IQ average of Chinese Americans is three points above that of white Americans is, says Flynn, 'a good measure of the cognitive advantage conferred by their distinctive sub-culture'.[21] And their economic presence in the most prestigious professions is magnified. Asian Americans comprise fewer than 6 per cent of the US population but made up 21 per cent of the numbers in American medical schools in 2017,[22] three and a half times the percentage of twenty-five years earlier.[23]

There is an obvious logic: bring up each generation of a population in a more educationally stimulating environment, in which learning is highly valued, they are challenged intellectually, exposed to various forms of abstract logic from their early years and parents have the resources and time to focus on their children, and IQs will rise faster than those of a population in which these things are less valued. Yet, as we have seen, since 2000, a group of ardent hereditarians has come to the opposite conclusion regarding Ashkenazi Jews: their rise in IQ is entirely the result of genetics.

RICHARD LYNN ON ASHKENAZI BRAINS

The notion of Jewish intellectual superiority tended to be kept in-house until recently, partly because claiming superior smarts might not go down so well outside but also because it might open

Pandora's box. My own father, who was Jewish, believed Jews were blessed with special smarts, citing favourite examples of intellectuals to prove his point, from his mother, who'd been the youngest-ever science graduate of the University of Cape Town, to his political heroes Benjamin Disraeli and Golda Meir. But he never entered this territory beyond the confines of the family. Apart from questions of manners, there's the problem of where it might lead. If Jews evolved superior brain-power, what other traits evolved? And if one group is brighter, it follows that others would be dimmer.

After I left home, at seventeen, I thought some more about this question, and the counter-view made more sense. It was well put by one of my favourite Jewish intellectuals, the historian Tony Judt, in response to claims that Jews were more intelligent than others: 'My own, statistically naive impression these days is that intelligence, ignorance and bigotry are distributed among Jews in proportions comparable to their presence in society as a whole.'[24] Outside households such as mine, the question of the reasons behind the disproportionate intellectual success of Jews was left to lie, until Murray, Lynn, Pinker, Peterson and Henry Harpending got in on the act.

Lynn was first, noting the number of Ashkenazi Jews who had been awarded Nobel Prizes. 'Jews must have had a high IQ to have achieved this astonishing over-representation,' he said. The average American Jewish IQ was 7.5 points above the white average, he added, meaning there would be, relative to population, four times as many Jews with IQs over 130. This edge, he said, was mainly genetic in origin.[25] In a later paper he wrote that Jewish high achievement was unrelated to 'work ethic'; it was all down to IQ.[26]

However, he emphasised this was restricted to Ashkenazi Jews and did not apply to Sephardic (Spanish) or Mizrahi Jews. He said European and American Jews had IQs averaging 108 but the Israeli Ashkenazi average was 103, because of migration by non-Jews from the Soviet Union, while his estimate for Israeli Mizrahi Jews was 91. He pulled this together in his book *The Chosen People*,[27] in which he argued that Mizrahi Jews were less intelligent than Ashkenazi Jews because of a differentiated gene pool, playing down the fact that the Mizrahi had lived for centuries in cultures with lower education levels, more authoritarian families and less exposure to cognitive stimulation and abstract logic. Lynn's work on Jewish IQ remained on the intellectual fringes; on the rare occasions it received mainstream reviews, such as in the *Times Higher Education Supplement*, it was slated.[28]

COCHRAN, HARDY AND HARPENDING'S 'NATURAL HISTORY' OF THE JEWS

Three anthropologists from the University of Utah, Gregory Cochran, Jason Hardy and Henry Harpending, received a far rosier reception in 2005 when they published a paper, 'Natural History of Ashkenazi Intelligence' in the *Journal of Biosocial Science* (which once went by the name of the *Eugenics Review*). This was trumpeted in the *New York Times*, *The Economist*, *New York* magazine and *National Geographic News* and received a huge boost when Steven Pinker, the doyen of evolutionary psychology, endorsed it.

We first met Harpending and Cochran in Chapter 2, and their claims that genetic changes nudged a select group of Africans in the direction of Eurasia, where they continued evolving, with cultural innovation prompting fresh selection pressures. Those with favourable genes got moving and continued to evolve intellectually; those without stayed behind. They argued that these adaptations varied across populations, so that among Africans and Amerindians there were fewer genetic changes and among Australian Aboriginals, none.[29]

Harpending is described by the geneticist David Reich as having 'a track record of speculating without evidence on the causes of behavioural differences among populations'.[30] In 2009, Harpending gave a talk on 'Preserving Western Civilization' in which he claimed that people with sub-Saharan ancestry lacked the inclination for hard work. 'I've never seen anyone with a hobby in Africa,' he said, explaining this was because Africans had not gone through the kind of natural selection for hard work that some Eurasians had.[31] He also rallied to the defence of the evolutionary psychologist Satoshi Kanazawa, from the London School of Economics, who got into trouble for writing that black women were innately ugly.[32]

The Utah threesome start with Lynn's point that Ashkenazi Jews have the highest average IQs. In common with Lynn, they claim no similar elevation of intelligence among Jews in classical times, nor among today's Sephardic and Mizrahi Jews. They say higher Ashkenazi IQ kicked in 1,200 years ago and that they now have average IQs of 'one standard deviation' (fifteen points) above the north-western European average (presumably the source of Peterson's figure).[33] Additional evidence that this is innate comes from the number who have received Nobel Prizes, Turing awards and won the world chess championships, although we're told their visuo-spatial abilities are lower than average.[34] The picture is of a brainy, geeky people, which would be news to the scores of outstanding Jewish professional boxers who won world titles in the nineteenth and early twentieth centuries.

After citing a 'who's who' of race-based IQ theorists (Lynn, Jensen, Herrnstein, Gottfredson, Eysenck, Murray), they insist IQ is 'a biological rather than a social variable', proved, they say, by twin studies. Just as mental illness is a 'biological phenomenon', so is IQ. They claim a 'neurobiological' basis, related to brain volume and density and 'brain glucose utilisation rate'.[35] Children's IQs may be influenced by class or home environment but these factors 'are essentially gone in adulthood'.[36] They explain the 'Flynn Effect' by saying that it may reflect 'real improvements in biological well-being' such as improved nutrition, vaccination and antibiotics, and that it also reflects greater familiarity with testing.[37]

Cochran and his colleagues admit little is known about early Ashkenazim but speculate that 'unusual selective pressures' favoured intelligence.[38] They explain that the Jewish requirement for literacy gave them the entrée into more cognitively demanding work, such as money lenders, estate managers, merchants and tax farmers.[39] The high-IQ rich had more surviving children, raising IQ over the generations. Low-IQ Jews would be less likely to find spouses and would drift out of the community. In five hundred years, they guess,

IQ would increase by sixteen points. In contrast, they say, Jews of the Islamic world tended to have 'dirty' jobs that were not cognitively demanding, meaning there was no IQ increase.

This theory is based on the idea that the Ashkenazim was an isolated population with 'very low inward gene flow' because during the period when they became super-bright (800 to 1650 CE) marriage outside the faith was discouraged.[40] A higher gene flow would 'limit the natural increase in locally favourable mutations'[41] but the flow from non-Jewish populations averaged 'less than 0.5 per cent per generation'.[42] However, this imposed costs, reducing genetic fitness for 'more typical environments' such as farming (tell that to the *kibbutzniks*). More importantly, it was a factor in the introduction of Ashkenazi genetic diseases, such as Tay–Sachs, Gaucher's and Niemann-Pick, a range of conditions known as DNA repair diseases, and torsion dystonia, congenital hyperplasia, and others.[43]

There's a twist here, and the authors devote half their paper to it. They claim most of these diseases took root 1,200 to 1,300 years ago when the Ashkenazim were making their mark in the niche of finance, and that the pattern of these diseases suggests 'selective forces at work', relating to intelligence.[44] In other words, Ashkenazi diseases were a by-product of selection for IQ. They say some of these mutations 'look like IQ boosters'[45] and confidently predict that those carrying genes for Gaucher's, Tay–Sachs and Niemann-Pick 'will have higher IQ than control groups, probably in the order of five points'[46]. They acknowledge they lack direct evidence but point to indirect evidence from Gaucher's patients in brainy professions and high IQ scores among some of those suffering from other Ashkenazi diseases. In addition to the 'unusual selective pressures' of their brainy, baby-boosting careers, there's an added dimension: intelligence genes are responsible for horrible diseases. It's a bit like sub-Saharan Africans and malaria: one copy of the recessive allele and people are protected against malaria; two copies and they have sickle cell anaemia. For European Jews, inherit two copies of a recessive allele and they end up with one of the Ashkenazi diseases; one copy, and they have superior intelligence. This, they say, had an impact on the average intelligence of the entire Ashkenazi population.

WHAT'S WRONG WITH THE ASHKENAZI THESIS?

After the first excited wave of media attention, the paper received critical comment from scientists working in this area. Harry Ostrer, who led New York University's human-genetics programme, said: 'It's bad science – not because it's provocative but because it's bad genetics and bad epidemiology.'[47] More recently, Reich has described Harpending's statements as 'racist' because of the way they make speculative jumps to race-based conclusions 'with no scientific evidence'.[48]

Earlier in this book I dealt with their claims about the genetic component of IQ suggested by adoption studies but a key pillar of their argument relates to a different fallacy, that of Ashkenazi genetic isolation. Recent studies have shown that European Jews are less genetically discrete than once assumed. This applies particularly to mitochondrial DNA, which descends through the maternal line. Genetic analysis by nineteen scientists, published in 2013, found that European women, not women from the Middle East, were the main female founders of the Ashkenazi population.[49] The four main sources of mitochondrial DNA all emerged from Europe: 81 per cent was European, 8 per cent from the Levant, 1 per cent from further east in Asia and 10 per cent was ambiguous.[50] The prime source was in Italy, via the Roman Jewish community, which included converts made through mass conversions to Judaism in the Empire, where an estimated six million citizens practised the religion, but the 2013 study also indicates later conversions in Western and Central Europe. The Ashkenazim 'carry a substantial fraction of maternal lineages from their "host" communities'.[51]

Most genetic studies suggest the male lineage is mainly Middle Eastern. Some studies put the European contribution at as little as one-eighth and others up to 50 per cent.[52] This is consistent with what we know of proselytising in the days when Judaism was a patrilineal religion. At least until the time of the historian Josephus, in the first century CE, Judaism was passed on by the father. Christianity and Islam proved to be more aggressive proselytisers, after which Judaism became a matrilineal religion; a surer way of ensuring Jewish identity – you can always be sure of the mother; never the father. But even after this there were periodic conversions, including of the royalty and aristocracy of the Khazars (a Turkic people from the Caucasus) from the eighth to the eleventh centuries. The 2013 DNA study found no evidence of Khazar DNA among modern Ashkenazim, a view disputed by the Israeli geneticist Eran Elhaik, who said mass conversion of the Khazars was the only way to explain the growth of the Ashkenazim from 25,000 in the Middle Ages to more than 8.5 million by the start of the twentieth century. He criticised the study for not analysing the DNA of Jews from the Caucasus and compared the genetic signatures found in Jewish populations using modern Armenians and Georgians as stand-ins for the Khazarians, finding that they were genetically related. Elhaik's results have also been disputed, including by critics who note that Armenians originated in the Middle East, but were welcomed by the Israeli historian Shlomo Sand, who argued that the Ashkenazim descended mainly from converts and that the descendants of the original Jews included the current Palestinians.[53] Either way, what is clear is that the Ashkenazim were not a genetically isolated population.

Cochran, Hardy and Harpending can't avoid acknowledging that IQ scores are rising but they bypass the fact that average IQs are rising at different rates in different populations and ignore the reasons Flynn offers for this. Instead, they say it's of no relevance, because it applies to all groups, and they venture no further than nutrition, antibiotics and familiarity with testing as the reasons. Flynn, in contrast, shows that the prime reason for the rise is exposure to a scientific way of viewing the world. This varies hugely from one population to the next, which is why IQs are rising faster in, for example, Kenya than in Western Europe.

What of the view that money-lending prompted selection for higher intelligence? It might be true that the rich had more surviving children, but they didn't form enough of the population to make much difference. One study of Jews in Breslau, Poland in the fourteenth century estimated the wealthy at 7 per cent of the population.[54] More significantly, were the rich really more intelligent? Cochran and his colleagues paint a picture of a meritocracy but a critique by Brian Ferguson, professor of anthropology at Rutgers University, Newark, shows that Ashkenazi society was highly stratified and wealth was inherited.[55] In each community, specific families dominated for centuries. Class stratification grew over time, as rich married rich.

This stratification also applied to money-lending; a few families did most of the lending and the big lenders were tax collectors, merchants and clergy, who also controlled financial courts, set tax rates, negotiated with political authorities, decided who should live where and made civic appointments.[56] It was not smart genes that made you a big money lender or trader but, rather, the luck of birth. If you didn't have capital, you wouldn't be able to lend. Nor did money-lending require braininess. Loan contracts from the Middle Ages are simple documents that anyone with a background in the business could have drawn up. And money-lending wasn't a constant in Ashkenazi history; it rose and fell at different times in different parts of Europe. There also were other forms of trade and management. Jews were butchers and bakers and candlestick makers, craftsmen and peddlers, while the entrepreneurs employed bookkeepers, clerks and debt-collectors. No doubt some were smarter than their better-heeled bosses. The argument that Ashkenazi society fast-tracked high-IQ people, who then had more children, doesn't hold.

Finally, there's the claim about Ashkenazi diseases as a by-product of selection for intelligence. The Utah authors say they arose through natural selection, but geneticists agree that they do not exhibit the mutation patterns, allele frequencies and geographic distribution of naturally selected genetic diseases. Instead, they spread through genetic drift.[57] As Reich shows, the authors' entire case is contradicted by evidence that these diseases emerged through 'random bad luck – the fact that during the medieval population bottleneck that affected Ashkenazi Jews, the small number of individuals who had many descendants happened to carry these mutations'.[58] The history of the diseases is also consistent with the bottleneck theory; when the Ashkenazi moved around Europe, small groups with commercial assets moved first. Only later did communities develop and persecution, plague and famine reduce particular Ashkenazi populations to isolated islands of settlement. In the period considered by Cochran and his colleagues, populations rose and fell. When they fell to a few hundred, this led to higher rates of intermarriage and therefore a higher chance of particular mutations taking root in the confined gene pool, and subsequently spreading.

But if, for argument's sake, we accept these diseases took root through natural selection, what of their view that this involved selection for higher IQ by reducing inhibitions on neural growth, thus allowing more neural connections and boosting IQ? Reviewing the literature, Ferguson says this suggestion is 'predicated on a simplistic view of neurological development', that most inherited conditions highlighted by the authors do not have 'even a suggested pathway to higher intelligence' and their claimed connection to higher IQ is 'very inconsistent with current research on the genetics of IQ'[59]; in some cases it has been decisively discredited[60] while in others there is no indication of higher IQ.[61] The Utah authors confidently predicted that future tests of those carrying the best-known Ashkenazi disease, Tay–Sachs, would show they had enhanced IQs but as Ferguson puts it, this is 'the dog that did not bark'. He adds that this dog's non-bark 'is particularly loud'.[62] In the 130 years since Tay–Sachs was first researched, no studies have shown a connection with intelligence.

In fact, only a small minority of Ashkenazim carry the alleles for which the authors made IQ-related claims: Gaucher's 1:15, CAH 1:27, Tay–Sachs 1:30; Nieman-Pick 1:90; idiopathic torsion dystonia between 1:2,000 and 1:6,000. None are Ashkenazi-specific diseases: Tay–Sachs is common among Cajuns and French Canadians, non-classical adrenal hyperplasia among Italians and Hispanics, Gaucher's one in a hundred of the non-Ashkenazi population. Yet there are no claims about higher intelligence for any of these other groups.

We can see that none of authors' contentions about Ashkenazi IQ survive inspection. Not their genetic diseases hypothesis, not their money-lender IQs idea, not their isolation idea and not their IQ theory. But this has not stopped hereditarians reacting as if the case had been proved beyond doubt. Andrew Sullivan, for one, endorsed this view, noting that the IQ differential between Ashkenazi and Sephardic Jews was 'also striking in the data'.[63]

STEVEN PINKER ON A 'TRICKY AREA'

The Times once dubbed Pinker the 'stud muffin of science'. Half of that description resonates (actually, he's not *really* a scientist). He's sixty-six but has a more youthful appeal. The sparkle in his blue eyes and his delight in ideas makes for an attractive combination. The shoulder-length curls hint at vanity but also suggest a residue of the sixties, when he grew up in a Jewish-Canadian family in Montreal. Most of the words spouted by this Harvard psychology professor fit the image. Even if, like me, you think his Swiss army knife view of the mind, his Dawkins-prompted view of human nature and his Chomsky-inspired take on language acquisition is all hocus pocus, it comes across as hocus pocus in its most acceptable form. For one thing, he writes with panache.

Despite the fervour of his attacks on what he terms radical feminism and Marxism (often neither), he's by no means a conventional figure of the right. It is only in one area that the thrice-married Pinker is, well, rather more glinty-eyed than we might expect: women. Let's just say he goes a good deal further than *vive la différence*. Like Jordan Peterson, he's a virulent proponent of a faux-scientific version of *Men Are from Mars, Women Are from Venus* and his fervour can take him over the edge. He once joked that the 'technical term' for those stressing nurture rather than nature on gender differences was 'childless'. Then again, when the childless Steven Pinker goes over the edge, he usually does so in a way that sounds so *mensch*-like that the barbs are easy to dismiss as mere over-exuberance.

One thing to admire is his willingness to debate with scientists. He has locked horns with Stephen Jay Gould[64] and Steven Rose[65] and even the most blinkered Pinker fan would be hard-pressed to claim he came out on top. But his most interesting debate was with Elizabeth Spelke, a cognitive psychologist from Harvard. He argued men were the more variable sex – 'more prodigies, more idiots' – and had evolved different intellectual abilities.[66] Piece by piece, Spelke showed that what he had assumed was innate was anything but. She drew an analogy between what Pinker was saying about women and what was once said about race. 'Let's consider who the nineteenth-century mathematicians and scientists were', she began:

> They were overwhelmingly male, just as they are today, but also overwhelmingly European, not Asian. You won't see a Chinese face or an Indian face in nineteenth-century science. It would have been tempting to apply this same pattern of statistical reasoning and say there must be something about European genes that give rise to greater mathematical talent than Asian genes do. If we go back still further and play this debate in the Renaissance, I think we would be tempted to conclude that Catholic genes make for better science than Jewish genes, because all those Renaissance scientists were Catholic. If you look at those cases, you see what's wrong with this argument.

Spelke concluded in a way that should have made the stud muffin of psychology take note. 'What's wrong with the argument', she said:

> ... is not that biology is irrelevant. If Galileo had been switched at birth with some baby from the Pisan ghetto, the baby raised by Galileo's parents would not likely have ended up teaching us that the language of physics is mathematics. I think that Galileo's genes had something to do with his achievement but so did Galileo's cultural and social environment: his nurturing. Genius requires huge amounts of both. If, in that baby switch, Galileo had found himself growing up in the Pisan ghetto, I bet he wouldn't have ended up being the example in this discussion today either. So yes, there are reasons for this statistical bias. But I think we want to step back and ask, why is it that almost all Nobel Prize winners are men today? The answer to that question may be the same reason why all the great scientists in Florence were Christian.[67]

Pinker ignored these wise words, coming out in measured support of the Ashkenazi IQ thesis in a speech entitled 'Jews, Genes and

Intelligence', given at an event hosted by the Institute for Jewish Research in New York.[68] One audience member described it:

> [T]o my ear what Pinker presented was a spirited endorsement of the Cochran-Hardy-Harpending paper ... Pinker emphasized the reasonableness of the authors' hypotheses, the generally better quality of the genetic evidence over the environmental, the non-rational basis of much of the opposition and the paper's strong foundation in the current state of knowledge.[69]

A *Seed* magazine report noted: 'People will hear what they want to hear. And many in attendance were there to hear that Jews are naturally smarter than everyone else.'[70] Pinker followed this with a piece in the *New Republic*, repeating the points he'd made in his speech and describing the Utah paper as 'thorough and well-argued'.[71]

After spelling out his Jewish credentials, Pinker parrots the Utah authors' points on Jewish Nobel Prize winners and on IQ, agreeing it is 'highly heritable' and measures general intelligence. He concedes Flynn's point that differences between groups are not necessarily genetic (in fact, Flynn says they are probably not genetic) but suggests Cochran, Hardy and Harpending were right that genes are involved. 'The Ashkenazi advantage has been found in many decades, countries and levels of wealth,' he writes, ignoring evidence of early-twentieth-century low IQ scores among Ashkenazim, 'and the IQ literature shows no well-understood environmental factors capable of producing an advantage of that,' he adds, echoing the authors' position that families 'have no lasting effect on intelligence'.[72]

Pinker endorses their view that Ashkenazim tended to 'marry their own', allowing intelligence-related genes to proliferate. He also backs the idea that Ashkenazim were concentrated in occupations requiring high intelligence, which led to greater economic success, and richer Jews who had more surviving children. On Ashkenazi diseases, he admits the evidence is 'iffy' but notes that 'a gene might raise a child's IQ but also predispose him to a genetic disease.' He declares that the Ashkenazi paper 'meets the standards of a good scientific theory' but gives himself a get out of gaol card by acknowledging it 'could turn out to be mistaken'.[73] Towards the end of the paper, Pinker takes his nascent race science beyond the Ashkenazim by claiming that 'personality traits are measurable, heritable within a group and slightly different, on average, between groups,' noting that we will soon develop 'the power to uncover genetic and evolutionary roots of group differences in psychological traits'.[74]

Pinker added to this the following year, when he formulated *Edge*'s Annual Question: 'What is your dangerous idea ... dangerous not because it is assumed to be false but because it might be true?' His own answer was: 'Groups of people may differ genetically in their average talents and temperaments.' In his essay on this topic, he cited the Utah paper and one on race differences written by Charles Murray, before complaining that '[l]arge swathes of the intellectual landscape have been re-engineered to try to rule these hypotheses out *a priori*.'[75] He invents a straw man version of those who think differently: '[P]rogress in neuroscience and genomics has made ... shibboleths (such as the non-existence of intelligence and the non-existence of race) untenable.'[76] I'm not aware of anyone who says intelligence doesn't exist but it's hardly surprising that the all-nature, no-nurture Pinker holds a more fixed idea about it. However, his statement also suggests an expansive idea of race that includes innate differences in intelligence and personality, which seems out of sync with his earlier emphasis, set out in his book *The Blank Slate*. 'My own view, incidentally', he said there, 'is that in the case of the most discussed racial difference – the black-white IQ gap in the US – the current evidence does not call for a genetic explanation.'[77]

Pinker has subsequently dug in on his view that different populations might have different innate intellectual abilities, reiterating his Ashkenazi idea and tweeting in favour of other race science promoters such as Murray[78] and Linda Gottfredson[79] who had faced 'no platform' pressure. Gottfredson has a thirty-year history of race science, including being the author of the founding document of the modern version of this calling.[80] Pinker described her as an 'expert on ... intelligence', referencing a story on the right-wing website *Quillette* to back her.[81]

CHARLES MURRAY ON THE GENIUS OF THE 'CHOSEN PEOPLE'

Charles Murray also mounted the Jewish IQ bandwagon. Writing in the right-wing American magazine *Commentary*, he extolled Jewish achievement in the arts and sciences, saying its source was superior intelligence, illustrated by the fact that Jews have exceptional verbal and reasoning skills, with a mean IQ of 110.[82] He claims IQ is primarily biological, and would therefore not be boosted by an educationally rich Jewish home life. Instead, he considers two reasons why Jews evolved for higher IQ: persecution (only the smartest survived) and marrying for brains ('scholars and their children were socially desirable spouses'). He plumps for the latter, combining it with the Utah authors' argument about occupational selection, economic success and reproductive success being closely bound.[83]

However, he also deviates, arguing that high Jewish intelligence is not confined to Ashkenazim, pointing to examples of exceptional Jewish achievement in Spain and the Muslim world.[84] He claims that selection for high IQ happened far earlier than 800 CE, pointing to the introduction of compulsory education in 64 CE, which encouraged a move away from farming and into occupations involving commercial transactions, prompting selection for these IQ-rich activities. Warming to this theme, he says the requirement of being able to read the Torah had evolutionary implications, because 'to be a good Jew meant that a man had to be smart'. Those who struggled moved away from the religion, with many farmers abandoning Judaism, leaving the brighter traders and money men; a process exacerbated by persecution that reduced the world Jewish population by two-thirds.[85]

Murray also proposes an alternative thesis, going back to 586 BCE, when Nebuchadnezzar's Babylonians captured soldiers, craftsmen and artisans and left the rest behind. By the time the exiles returned, sixty years later, many of that rump would have been absorbed into other religions. The smart exiles therefore formed the bulk of the reconstituted Jewish community, he speculates. Or we could go even further back: throughout Jewish history fathers taught their children the complicated Jewish law, which required exceptional verbal skills.

The less intelligent would have left Judaism because of their failure to grasp it. Murray ends by asking why all this learning kicked off; why only one tribe at the time of Moses evolved elevated intelligence. The answer he proposes, acknowledging it is 'uniquely parsimonious and happily irrefutable', is: 'The Jews are God's chosen people.'[86]

None of this survives scrutiny. The information that underpins Murray's Babylonian thesis is drawn solely from the Old Testament. This is not a book on Biblical history but when Murray treats the Bible as an historical document, it's hard to take him seriously. Yet even if, despite writing hundreds of years after the event, the authors of Kings and Ezra reflected a mythologised version of history rather than pure myth, their words can only aid a genetic version of racial intelligence if you start from that position. Who is to say that the captured soldiers, craftsmen and artisans were innately smarter than the poor, the women and the children left behind? Murray is transporting the views he promoted in *The Bell Curve* (the poor are poor because they're stupid) 2,600 years back, to a time when social class was set by birth.

Other items of speculation are nothing but guesswork. Murray contradicts himself, swearing by innate IQ but then rejecting what it's telling him. If you come from Murray's position, you have to accept that superior intelligence applies only to Ashkenazi Jews and that Oriental Jews are below average. If you live by IQ faith, you die by it too. But Murray doesn't like this conclusion, so he ignores its implications and presses on with his 'all Jews are smart' thesis, taking it all the way back to the mythical Moses. If that is true, then there is something wrong with IQ testing. But I guess we knew that anyway.

NICHOLAS WADE AND THE CAPITALIST-ADAPTED JEWS

Perhaps the most objectionable version of this faux-science came from Nicholas Wade, whose book *A Troublesome Inheritance*[87] ascribes just about every racial stereotype to genetics. In 2018, I wrote a long feature in the *Guardian* on the revival of scientific racism, in which I noted that Wade's book repeated three race-science shibboleths: that the notion of 'race' corresponds to profound biological differences among humans, that human brains evolved differently from race to race and that this is supported by different racial averages in IQ scores.[88] Wade objected to this last point, saying he never took sides in the race and IQ debate. Yet that's precisely what he argues when discussing Jews, noting that higher than average IQ 'helps explain why the Jewish population, despite its small size, has produced so many Nobel Prize winners and others of intellectual distinction'.[89]

Like the Utah anthropologists whose paper he embraces, Wade understates the genetic flow into the Ashkenazi population, while echoing Cochran, Hardy and Harpending's money-lending idea and applauding their Jewish diseases thesis without reservation. His only hesitation is that, like Murray, he thinks Jewish braininess is not restricted to the Ashkenazim. His additional explanation relates to the decline of the Jewish population from 5.5 million in 65 CE to 1.2 million in 650 CE, mainly as a result of conversion to Christianity by those who 'lacked the ability or commitment to become literate'; a similar argument to Murray's.

These, and subsequent adaptations, meant the Jewish population includes 'proportionately more individuals of higher cognitive capacity than most others'.[90] He draws direct analogies between the biological adaptations of the Tibetans for living at high altitude and of the 'Eskimos' for living in the Arctic, and Jewish evolution for business and braininess. As he puts it: 'The adaptation of Jews to capitalism is another such evolutionary process.'[91] Which invites the question: where have we heard that one before?

WHY SO MANY JEWS ARE SMART

What factors in Ashkenazi culture might have boosted IQ if we dispute the genetics case? Jordan Peterson can't think of any. Nor can Steven Pinker, who concedes that an environmental explanation 'can't be ruled out' but then dismisses the obvious candidates such as home environment and the impact of enhanced educational expectations: '[I]f wishes were horses, beggars would ride. Mere expectations cannot produce a brilliant mind.'[92] Let's help Professors Pinker and Peterson along.

The place to start is to acknowledge the extraordinary record of Jews in science, maths, literature, psychology and politics and then to ask why Jews have had such consistent intellectual success – and not just Ashkenazim: the great economist David Ricardo and my father's Victorian hero, Benjamin Disraeli, were Sephardic Jews, as were Maimonides, Spinoza and Bohr. One reason today's Ashkenazim have higher than average IQs is that they tend to be wealthier. Most belong to the middle classes, which wasn't the case when soldiers were tested in the First World War. Unless, like Murray, you believe wealth is caused by higher IQ, you must accept it as a well-established independent variable that boosts test results. But this doesn't explain the Jewish intellectual record over centuries and across continents.

We know the Talmudic educational tradition, motivated by the need to learn scripture, started before Christianity, and long before the emergence of the Ashkenazim. The equivalent of what we'd call primary education for boys was close to universal by the first century CE, and many received something like secondary education. This began when most Jews were farmers, not money-lenders, so there was no financial advantage. Centuries later, when urban centres developed, literate and numerate Jews moved into commerce and out of farming and Talmudic academies provided a leg-up for boys. Those who made it were higher status but even Cochran and his colleagues acknowledge that the 'marry a rabbi' argument for Jewish IQ doesn't wash; there were too few rabbis to have a genetic impact. However, the cultural impact was immense, helping build admiration for intellectual achievement and for ideas. Residues of the Talmudic tradition have rippled through Jewish communities over the centuries, influencing even poorer homes. To repeat the question: why are there so many smart Jews? Ferguson's answer is that many Jews 'inherit traditions that foster scholarship and abstract thought to an extent that few other cultures can match'.[93]

To return to my own family, my Ashkenazi Jewish father grew up in a home in which learning was valued and debating was an everyday affair. He was taught by his mother, a science graduate, to be proud of his brain, to embrace ideas, to be intellectually ambitious and to love a good argument, and he passed this on to his children. My mother, who came from Danish and British stock, was not brought up in an intellectually driven family and was less inclined to read, argue and theorise. She wanted to go to university, but her parents said this was for boys. They paid for her sports-mad brother to study abroad, while she attended a training college for nursery teachers. In some ways my mother was sharper and more perceptive than my father but I doubt she'd have done as well in an IQ test. My experience illustrates that a cultural tradition favouring a love of learning, of sharing and debating ideas, and placing a value on intellectual achievement, is more likely to produce intellectuals than one in which books, learning and debating are less valued. Contrary to what Pinker, Peterson, Murray and the rest claim, early learning, exposure to abstract logic and intellectual confidence make a significant difference in IQ tests.

THE ALT-RIGHT, ANTI-SEMITISM AND THE JEWISH INTELLIGENCE THESIS

Spend any time patrolling alt-right websites and social media feeds and it won't take long to strike a thick vein of anti-Semitism running alongside the wider racism. How does this fit with claims that Ashkenazi Jews, or all Jews, have innately superior intelligence? At first blush their views seem to be pro-Semitic. Are they?

Anti-Semitism seldom pictures Jews as stupid. Instead, they are portrayed as clever, cowardly, manipulative, cabalistic plotters. Today's alt-right conspiracy theories revive the ancient trope of a rich Jewish bogeyman pulling the strings and providing the funds (the currently favoured personification is George Soros). The idea of Jews having high IQs hardly contradicts this. Peterson's point that 'you can be pretty damn horrific as a genius son-of-a-bitch' is apposite. Wade went further, claiming that Jews were naturally selected for capitalism, while the Utah authors said that the Jewish brain evolved to deal with money.

The University of Virginia psychology professor Eric Turkheimer extends this logic to demonstrate the problems in leaping to unsupported conclusions, such as that racial IQ gaps have a genetic base. He illustrates the implications with a different assertion: that Jews are more materialistic; not very far from the idea that Jews are adapted to capitalism. Turkheimer, who's Jewish, suggests it's not unreasonable to believe that materialism may be partly heritable, that Jews differ ancestrally from non-Jews and that we may even possibly find different average levels of materialism. 'If you were persuaded ... that the black-white IQ gap is partially genetic but uncomfortable with the idea that the same kind of thinking might apply to the personality traits of Jews, I have one question: Why?' The reason most people don't believe this anti-Semitic trope, he says, is because 'the horrific recent history of false hypotheses about innate Jewish behaviour helps us see how scientifically empty and morally bankrupt such ideas really are.'[94]

This returns us to Pinker, who gave a thumbs up to the Utah authors' Ashkenazi intelligence thesis and yet worried aloud whether this really was good for Jews, expressing his concern that 'someday someone could test whether there was selection for personality traits that are conducive to success in money-lending and mercantilism' and also how such findings 'would be interpreted in, say, Cairo, Tehran and Kuala Lumpur'.[95]

If we follow Pinker and Turkheimer's logic, the 'smart Jews' idea is not only available as a cat's paw for race science more generally; it also lends itself to undisguised forms of anti-Semitism.

14

DO SOME POPULATIONS HAVE DIFFERENT BRAINS FROM OTHERS?

The Soviet psychologist Alexander Luria (1902–77) is known as the father of neuropsychological assessment, because of his work with people suffering from brain injuries during the Second World War. His other source of professional fame comes from his founding role in what is known as 'cultural-historical psychology', which involved probing the unity of mind, brain and culture in real life situations. In one of his books, Luria describes interviews conducted with peasants in remote parts of his country and what they revealed about concrete and abstract knowledge. He had the following conversation with peasant men:

Q: All bears are white where there is always snow; in Novaya Zemlya there is always snow; what colour are the bears there?

A: I have seen only black bears and I do not talk about what I have not seen.

Q: But what do my words imply?

A: If a person has not been there he cannot say anything on the basis of words. If a man is sixty or eighty and he had seen a white bear there and told me about it, he could be believed.[1]

Another example:

Q: There are no camels in Germany; the city of B is in Germany; are there camels there or not?

A: I don't know. I have never seen German villages. If B is a large city, there should be camels there.

Q: But what if there aren't any in all of Germany?

A: If B is a village, there is probably no room for camels.[2]

And one more:

Q: What do a chicken and a dog have in common?

A: They are not alike. A chicken has two legs, a dog has four. A chicken has wings but a dog doesn't. A dog has big ears and a chicken's are small.

Q: Is there one word you could use for them both?

A: No, of course not.

Q: Would the word 'animal' fit?

A: Yes.[3]

In the first two examples, the man's point is that only experience can tell us facts. This would be detrimental in an IQ test. In the third, the peasant uses observation about what a chicken and dog look like but the abstract category, animals, does not come without prompting. This too would hold him back in an IQ test. As Jim Flynn points out: '[P]eople attached to the concrete will not find these categories natural at all. First, they will be far more reluctant to classify. Second, when they do classify, they will have a strong preference for concrete similarities [over] abstract categories.' In modern IQ tests 'the preference for answers that classify the world ... is extraordinary'.[4] Such tests may be appropriate in a post-industrial world, because this kind of abstraction permeates the minds of those immersed in the scientific ethos with its 'detachment of logic and the hypothetical from concrete referents' but not for peasants whose minds are permeated by concrete, pre-industrial categories.[5]

In Chapter 2 I used the example of the hunter-gatherer Piraha community in the Amazon. They show no artistic interest, have no fixed terms for colour, no counting system and live entirely in the present.[6] The children are perfectly capable of learning the things children in the rest of the world absorb, including number concepts, but not the adults. No IQ test has been designed that could accommodate the Piraha system of knowledge and learning. Which brings me to a story about another community with hunter-gatherer roots.

RICHARD LYNN AND THE 'BUSHMEN' OF THE KALAHARI

Piet Rooi adjusts his Union Jack cap, pushes his glasses up his nose and smiles at the inanity of my question about why he used the hoodia plant. 'Why?' the seventy-three-year-old asks. 'Because it helps us survive.' He raises his hand to forestall further interrogation on the obvious:

> I eat the Xhoba to stave off hunger and thirst and then I no longer feel hungry or thirsty. I eat it when I am feeling weak and then I feel strong and virile. I eat it when I have a bad stomach or flu and then I feel better ...

It is quite a trek to find the cactus-like plant that might one day make billions for foreign shareholders and, eventually, perhaps a million or two for South Africa's ten thousand San people. A ninety-minute flight to Upington, where the local newspaper headline was 'Cobra eats whole puff adder', complete with three pictures of this snake-on-snake snack; a two-hour drive passing herds of skittish springbok, mating meerkats, scurrying jackals, house-sized birds' nests and an exotic range of road kill; a bumpy four-wheel-drive expedition and a walk through the desert brush watching out for puff adders (those not already eaten by cobras).

In mid-spring, even at 10am, the temperature is 42°. It hasn't rained for ten months in this remote part of the Kalahari near the Botswana border, and the dry red sand is barren. But Piet has no trouble finding his favourite clump of hoodia. 'We're in drought now, so I'll *braai* [grill] it because it's too bitter to eat raw,' he explains, carefully picking two cucumber-sized branches, 'but when it rains it turns brighter green and has a nice, sour taste and I can chew it. I've been eating it since I was nine and I'm still eating today.'[7]

Piet uses a match to scrape the thorns from one of the branches and hands it to his forty-four year old neighbour, Susanna Witbooi, who will crush it to powder to treat her sister's asthma. I ask what she thinks of the foreign interest in the hoodia's hunger-busting properties; Pfizer, the company that created Viagra, is spending millions of dollars with the aim of using it to create a new weight-loss drug. Putting an arm around the shoulders of her five-year-old daughter, she shakes her head and laughs:

> All the San people use the Xhoba – it's just part of our life. In the old days the men often went three days in mid-summer without food or water when they were hunting and they never felt hungry or thirsty, and now it's going to make life better for me and for children like Kayla.[8]

The San have the world's oldest continuous culture and 'bloodline', according to one study of genetic diversity.[9] It is possible that the people of the Blombos Cave, who were fire-hardening tools and using blended paints 100,000 years ago were their ancestors, although the oldest direct evidence comes from a cave in KwaZulu-Natal, where a set of tools identical to ones used by late-nineteenth-century San was dated at 44,000 years old.[10] The San were once routinely portrayed as primitive – the fantasist Laurens van der Post romanticised them as 'children of Nature' – but it seems likely they 'invented' the bow and arrow and used poisoned arrows to kill their prey. Their colourful rock art, featuring people, animals, god-like creatures and half-human hybrids and later, the arrival of Europeans in ships, suggests a vibrant, imaginative culture. In other respects, aspects of their traditional way of life seem more contemporary than much of what passes as modernity. San women had a high status, were sometimes clan leaders, were frequently involved in hunting and always involved in making group decisions.[11]

This is how the San lived for thousands of years. Game, and edible fruits and plants were plentiful and for much of their history there was no reason for fundamental change, although more than 2,000 years ago some began herding and domesticating animals, branching off to become the Khoekhoe people. Until the arrival of Bantu-speaking tribes after about 200 CE, and the Dutch in the seventeenth century, the San and Khoekhoe were the only people in the region. But by the nineteenth century, genocide, slavery and disease had wiped most of them out, confining the rest to remote parts of what is now the Northern Cape, Botswana and Namibia. Under apartheid, most South African San were classified as coloured (mixed race). Without land for hunting, they became 'farm boys', like Piet, 'house girls', like Susanna, or trackers for the apartheid military. Along the way, San culture was buried and some of the ancient San languages died. San cave paintings have been dated at nearly 30,000 years old, but the last cave painter died early in the twentieth century, ending a practice that, like their ritual dancing, was integral to their ancestor-based belief system.

Those of the 100,000 San people remaining in Southern Africa who want to keep to their hunter-gatherer lifestyle are in a minority. Others have embraced modernity or been burned by it: there are high rates of alcoholism, domestic violence and HIV infection. But their numbers also include professionals and political activists, fighting for San rights. A dwindling number in Botswana and Namibia do their best to maintain the old way of life against huge odds, including the Botswanan government's embarrassment in having Stone Age communities in their midst. This has led the government to try to bully the San into modernity, removing them from their land and cutting off water supplies. In South Africa, the old way of life seemed on the verge of extinction until the country's first democratic elections in 1994. Nelson Mandela – whose colouring and features strongly suggested Khoesan genes – was particularly proud that his country was home to the world's oldest indigenous culture. His government moved quickly to settle two major land claims, which returned tracts of white farmland to San communities, including the Khomani of the Kalahari, where the hoodia thrives. San representatives got busy winning land claims, forging links with other San clans and forming committees to represent their interests.[12] The San Council asserted their rights as the original source of knowledge about the hoodia, eventually leading to a deal with a British drug research company,

Phytopharm, to develop the drug, and with Pfizer to market it in exchange for royalties,[13] although the San have yet to see any benefit because the active ingredient has not yet been legally developed.

Meanwhile, Piet is relieved to be back. 'For most of my life I had to work on white people's farms but now we own our own land,' he says. 'I was born here and I'll die here.'[14] For Susanna, this new life offers more: fresh hope for her children. 'Things have improved,' she says:

> I worked in white people's kitchens since I was 14 but now I'm home and doing well making crafts. But children like Kayla are the future and they need more, so I feel grateful these companies are going to be making pills out of the Xhoba. I'm hoping the money will bring us teachers and computers and work projects. Things will be much better in future.[15]

I use this vignette not to romanticise the San, nor to place them on a pedestal of historical victimhood. Rather, I want to paint a picture of a traditional community struggling to find a place in the modern world. Some among the San (mainly in Botswana) are clinging on to their ancient mode of existence; others, such as Piet and Susanna, have held on to remnants of this lifestyle. But most have abandoned it altogether, including those taking the path Susanna wants for her daughter: embracing education and technology. In this sense, the experience of the San parallels that of traditional communities in other parts of Africa, the Amazon and Australia, and San activists have made common cause with these groups.

What has all this to do with IQ? Were it not for the reigning doyen of racist IQ theory, Richard Lynn, an evolutionary psychologist who is Professor Emeritus at the University of Ulster, despite his age (90) and infamy, the answer would be 'nothing'. Lynn happily calls himself a 'racialist', a 'racist' and a 'scientific racist'[16] and yet is the source most cited by the authors of *The Bell Curve*, whose work was praised by the likes of Andrew Sullivan for its rigour. Lynn's great contribution to racist IQ theory is in devising a race-based world map of what he claims is intelligence, with Mongoloids and Ashkenazi Jews at the top,

Caucasians slightly below them, well below that 'Negroid' Africans, then Australian Aboriginals and right at the bottom are what he calls Bushmen, Hottentots and Pygmies. This is what he has to say about San, whom he claims have an average IQ of 54:

> An IQ of 54 is at the low end of the range of mild mental retardation in economically developed nations. ... An IQ of 54 represents the mental age of the average European eight-year-old child and the average European eight-year-old ... would have no difficulty in learning and performing the activities of gathering foods and hunting carried out by the San Bushmen. ...[17]

To rub it in, he adds:

> There is a range of intelligence among the Bushmen and most of them will have IQs in the range of 35 to 75. An IQ of 35 represents approximately the mental age of the average European five-and-a-half-year-old ... The average five-and-a-half-year-old European child is verbally fluent and is capable of doing unskilled jobs and the same should be true for even the least intelligent Bushmen. Furthermore, apes with mental abilities about the same as those of human four-year-olds survive quite well as gatherers and occasional hunters and so also did early hominids with IQs of around 40 and brain sizes much smaller than those of modern Bushmen. For these reasons there is nothing puzzling about contemporary Bushmen with average IQs of about 54 and a range of IQs mainly between 35 and 75 being able to survive as hunter-gatherers and doing the unskilled and semi-skilled farm work that a number of them took up in the closing decades of the twentieth century.[18]

I presume Lynn has never met a San person, but my experience suggests the notion that their average intelligence is that of a European eight-year-old is absurd. And the idea that a European child could survive alone in the Kalahari is laughable; the kind of statement that could only be made by someone who'd never set foot in a desert.

The San I met seemed like everyone else: some brighter and more articulate, some more ambitious for themselves and their children, some more beaten down. They were all fluent in at least two languages, some in four or more. None were on 'the low end of the range of mild mental retardation'.

Lynn's slanted history of 'Bushmen and Hottentots' reflects his prejudice. He tells how 'Negroids' encroached on Bushmen and Hottentot lands, killed most of the Bushmen and drove the survivors into the Kalahari Desert.[19] In fact, the ancestors of the Kalahari San had been there for thousands of years. And the Khoe and San living in the Cape were wiped out not by 'Negroids' but by white colonists, who treated them like vermin. The last recorded permit for a Bushman hunt was granted in 1936.

What of Lynn's data on San IQ? There's an obvious problem when representatives of the white overlords travel to the Kalahari to test black subjects. The man who conducted the later tests, Helmut Reuning, was a German, from the South African colony of South-West Africa (now Namibia), who fought for the Third Reich and wrote for the racist journal *Mankind Quarterly*. Yet even he doubted the value of his San tests and described difficulties in administering them. But apart from their source, there's also the issue of using tests devised to measure the cognitive potential of Americans, brought up in a world of scientific abstractions, on illiterate, innumerate people who live a subsistence lifestyle, and have a fundamentally different perception of the world. The peasants interviewed by Luria had at least some access to modern life, including education. The Kalahari San tested by Reuning in the 1960s and early 1970s, and those tested by South Africans in the 1930s, had no such access. The average IQ score of 54 is therefore valueless and says nothing about their intelligence.

KAMIN EXPOSES LYNN'S SOUTH AFRICAN DATA

It was not only with the San that Lynn's research wobbled. Leon Kamin, a former chair of psychology at Princeton, and later at Northeastern University in Boston, is best known for his discoveries on associative learning in animals but he was a scourge on racist psychologists over the years. He had a knack for burrowing into the depths of hereditarian research, going back to their sources and their methodology, and exposing them. Kamin was the first to spot the fraudulent nature of Cyril Burt's twin-based IQ studies, and he did it again with Herrnstein and Murray's *The Bell Curve*, picking their basic research to pieces and, in particular, its reliance on Lynn's data on black and 'coloured' South African IQs.

Lynn used data from five studies conducted in South Africa under apartheid – his main African source – to draw the conclusion that the average African IQ is 75, twenty-five points below the white American average and ten points below the black American average. He speculates that the reason for this gap is at least 50 per cent genetic, a view embraced by *The Bell Curve*'s authors. He draws an analogy between black American and 'coloured' South African IQ, because they are both an admixture of white and black genes. He bases this on a 1989 publication by Kenneth Owen, which he describes as 'the best single source of Negroid intelligence'.

Owen's studies compared black, white, Indian and later, coloured pupils, using the Junior Aptitude Tests, which, as Kamin shows,[20] could not be converted into IQ scores; a conclusion confirmed by the man who devised them.[21] According to Lynn, the mean 'Negroid' IQ in this study was 69 but as Kamin points out, 'Owen did not in fact assign IQs to any of the groups he tested ... The IQ figure was concocted by Lynn.'[22] Owen did no more than report test scores and said there was no reason to suppose that the lower scores related to genetics, noting that the 'knowledge of English of the majority of black testees was so poor that certain tests proved to be virtually unusable'. He also noted the tests assumed Zulu pupils were familiar with microscopes, electrical appliances and the 'western type of ladies' accessories'.[23] Owen added that his results 'certainly cannot' be taken as a suggestion of intelligence among blacks in Africa as a whole.[24] Yet *The Bell Curve* used this to claim that black American and coloured South African students had similar IQs.

Kamin reveals further distortions repeated by Herrnstein and Murray. Lynn cites a paper by D. H. Crawford-Nutt on tests given to

Zambian mine-workers, in which they were said to have had an IQ average of 75, but ignores the main part of the paper, which reported that 228 Soweto high school students scored an average of 45 correct responses on an IQ test, whereas a peer group of white students had a mean score of 44.[25] There can be only one reason why Lynn, Herrnstein and Murray ignored this test: because it contradicted the claim of a genetic basis for the differences between black and white IQ scores. If black Soweto students were scoring higher than more privileged white students of the same age, what did that say about race and IQ? Kamin points to several other omissions and distortions in Lynn's African IQ data and concludes that they 'constitute a truly venomous racism, combined with a scandalous disregard for scientific objectivity'.[26]

FURTHER TROUBLES WITH LYNN

Lynn is an evolutionary psychologist and like most of his calling, a genetic determinist. As we shall see in the next chapter, evolutionary psychologists believe behaviour is genetically programmed, in the main. They're particularly obsessed with male and female behaviour. Lynn writes extensively in this area, claiming men are innately smarter than women. But his main passion is his idea that black people are innately stupider than white people and that some Asians are slightly brighter than most whites, except Ashkenazi Jews. The prime reason for differences in IQ scores between different races, he says, is genetics, and this has had huge historical implications. '[T]he Caucasoids and the Mongoloids are the only two races that have made any significant contribution to civilization,' he writes.[27]

The basis of Lynn's race-based theory, set out over the past half century, has not changed. The fullest exposition is found in his 2006 book, *Race Differences in Intelligence: An Evolutionary Analysis*, which draws from 620 published studies. Using these data, he breaks down average regional IQs: East Asians (105), Europeans (99), Inuits (91), South-East Asians and Amerindians (87), Pacific Islanders (85), Middle Easterns, including North Africans and South Asians (84),

East and West Africans (67), Australian Aboriginals (62), 'Bushmen and Pygmies' (54).[28] He further subdivides these categories, placing Ashkenazi Jews above other Jews and everyone else.[29] Mexican IQs are similarly broken down (white Mexicans, 98, mestizos 94, Indian Mexicans 83),[30] Italians range from dumb Sicilians (89) to smart northerners (103),[31] while mixed-race people have IQs between the averages of the two races of their genetic origins.[32]

Lynn points to the evolution of animal breeds after they become geographically isolated and says that 'in the case of humans these different varieties are called races'. He talks of different 'races' of chimpanzee and gorilla while drawing a direct race-based analogy with the differences between red and grey squirrels, lions, tigers and jaguars[33] and even different breeds of dog.[34] It goes without saying that no evolutionary biologist would make this direct analogy; they would point out that genetic differences between human population groups are miniscule compared to those between animal breeds and that almost the full range of genetic variation can be found in each human population. Lynn classifies races according to physical characteristics, blood groups and diseases and rejects the idea that just because all races have interbred, this invalidates the concept: 'Among dogs, clines and hybrids are called mongrels but the existence of mongrels does not mean there are not pure breeds.'[35] Countering views such as that of the American Anthropological Association, that 'race is not a scientifically valid biological category,' he uses examples such as that 'genes for the epicanthic eye-fold are present only in East Asians, Arctic peoples and in some American Indians.'[36] In fact, they are also found among the San and some Europeans. The American Anthropological Association notes that these are superficial genetic differences that cross the boundaries of race and have no bearing on intelligence.

Lynn believes IQ is an accurate measure of intelligence, that general intelligence exists and therefore that the g of IQ theory reflects a biological reality. In common with Murray and Herrnstein, he believes that IQ accurately predicts socio-economic status, crime rates, moral values and a range of other human potentials and drawbacks. He has consistently come out in support of eugenics, including embryo selection and, in principle, a licence scheme for would-be parents involving the compulsory sterilisation of boys or girls, who could apply for a licence to have it reversed when they reach adulthood.[37] He also believes human intelligence continued to evolve after humans migrated to Asia and Europe, although he gets the detail wrong, claiming that *Homo sapiens* first appeared 150,000 years ago when it was actually more than double that. Like Jensen, Rushton and the rest, he believes the cold climates of Europe were more cognitively demanding, prompting selection for higher intelligence because 'less intelligent individuals and tribes would have died out, leaving as survivors the more intelligent.'[38] Alternatively, new alleles for high intelligence could have appeared as mutations in some races – probably large populations under the stress of low temperatures – and not others.[39]

Most of *Race Differences in Intelligence* is taken up with his racial breakdown in IQ scores, nation-by-nation, tribe-by-tribe. He also throws in scores from tests of reaction times, short-term memory and perceptual speed which, as previously demonstrated, are subject to the same range of environmental influences as IQ scores. Lynn's other obsession is brain size, which he believes reflects intelligence: bigger head, bigger brain, smarter person. I have already touched on this argument but, in general, brain size, head size and body size are causally correlated, which is one reason why men have bigger brains than women. There is some variation between population groups in the ratio between brain and body size. The San have larger brains, relative to body size, than the Burmese, Sinhalese and Tamil peoples. Several Native American tribes have bigger brains than the average for the white population, but Lynn claims that the Native American average IQ is thirteen points below the white average. His data suggest the biggest brains belong to the Inuits and other Arctic people, who are nine points below his white IQ average.[40] The 10 per cent larger brains of early *Homo sapiens* from Africa and Europe provide further counter-evidence, as do those of the Neanderthals.[41]

No doubt there's some correlation between brain size and IQ. Brain size can be affected by a number of factors, including if the mother smokes or drinks heavily during pregnancy. The child of a smoking and drinking mother might therefore have a smaller head and might also be denied the kind of IQ-stimulating nurturing another child would enjoy. The cause of the correlation between head/brain size and IQ would be environmental, not head size *per se*. Ian Tattersall, one of the world's leading experts in the brain sizes of our ancestors, put it like this: 'Nobody has yet contrived to show that among living people there is any correlation at all between brain size and levels of general achievement.'[42]

Another problem is the way Lynn interprets the 'Flynn Effect'. He misses one of its central points by making uniform adjustments to his scores. But, as I have described, IQ scores are rising faster in, for example, Kenya than in most other parts of the world, so an acultural application is simply wrong.[43] Lynn acknowledges the reality of steadily increasing IQs but says it can mainly be explained by better nutrition.[44] But there's robust evidence that nutrition-prompted height gains, at least in the West, do not tally with IQ gains, although this does not imply that the IQs of children malnourished for a sustained period will not be affected.[45]

I've already pointed to problems with Lynn's Southern Africa data and he has received scathing critique in other areas. One review of a book he wrote with Tatu Vanhanen, father of a former prime minister of Finland,[46] pointed to their use of dubious data, including that he provided IQ scores for Equatorial Guinea based on those of Spanish children in a home for developmentally disabled.[47] Lynn wrote an article claiming that one of the reasons for Japan's postwar economic success was that Japanese people had a ten-point IQ advantage over white Americans.[48] Two psychologists wrote a rebuttal showing that Lynn's Japanese IQ sample was drawn from well-to-do middle-class parents (which Lynn did not disclose), rather than the population as a whole, as was his American sample.[49] (Incidentally, *The Bell Curve* included Lynn's Japanese data, only mentioning the critique of his 'early work' in a footnote.[50]) Nicholas Mackintosh, an IQ theorist from the University of Cambridge, picked out other examples of Lynn's misuse of IQ data, including of studies by Mackintosh himself. '[T]hey do not increase my confidence in Lynn's scholarship,' he said, adding that it was the 'sort of book that gives IQ testing a bad name'. He went on to say: 'Lynn's preconceptions are so plain and so pungently expressed that many readers will be suspicious from the outset.'[51]

KANAZAWA, IQ AND THE HEALTH OF NATIONS

With this kind of reception, you might think that anyone who valued their academic credentials would avoid Richard Lynn. That's certainly the case beyond the realm of IQ psychology. But in that world, particularly in its energetic hereditarian corner, Lynn is seen as the last of the intrepid iconoclasts who helped push their calling into the mainstream. I have described how *The Bell Curve*'s authors relied heavily on his data, and they weren't alone. His most vocal protégé is the Tokyo-born, London-based Satoshi Kanazawa, an evolutionary psychologist at the London School of Economics (the LSE), who uses Lynn's data and has co-written papers with him, including one proposing evolutionary reasons for Ashkenazi intelligence[52] and another on sex differences in IQ.[53]

Kanazawa first came to public attention in 2006, when one of his papers, published in the *British Journal of Health Psychology*, used Lynn's IQ data to claim that poor health in Africa was caused by low intelligence. He started by asserting that evolutionary psychology theory shows that 'general intelligence evolved as a domain-specific adaptation to solve evolutionarily novel problems.' In other words, the intelligence of different populations varies because their ancestors encountered different environmental problems. Brains can't easily cope when these environments change, although 'more intelligent individuals are better able to recognize and deal with such dangers and live longer.' Inequality, poverty, water quality, lack of healthcare, low education levels and economic underdevelopment have little to do with poor health: it's mainly down to low IQ. Therefore 'average IQ has a very large and significant effect on population health.' Kanazawa focuses mainly on sub-Saharan Africa which, he claims, had little scope for the evolution of intelligence and is therefore less healthy. For example, in Ethiopia, average IQs were only 63 and people only expected to live until their mid-forties but people in wealthier countries 'live longer and stay healthier, not because they are wealthier and more egalitarian but because they are more intelligent'.[54]

This nakedly racist paper created little controversy. Other than the *Observer*, which published a substantial and critical story,[55] most UK newspapers covered his claims in short news reports, and academic criticism was confined to psychology journals. One critique, by the American evolutionary psychologist Kevin Denny, picked on Kanazawa's claim that intelligence was associated with geographical distance from the point of human origin, as a result of further adaptations. 'A particular problem is that in calculating distances between countries it implicitly assumes that the earth is flat,' Denny notes. 'This makes all the estimates biased and unreliable.'[56] A more trenchant critique, by George Ellison, a biostatistician from the University of Leeds, subjected the paper to a thirty-six-page mauling before homing in on the familiar territory of causation: 'Kanazawa mistook statistical associations for evidence of causality and falsely concluded that populations of sub-Saharan Africa are less healthy because they are unintelligent and not because they are poor.'[57]

Asked to comment, the LSE blandly noted that Kanazawa's research was based on empirical data, had been published in a peer-reviewed journal and that they took no 'institutional view' on his work.[58] However, the LSE adopted a different stance on Kanazawa's next big splash. In 2011, he wrote in *Psychology Today* that black women had evolved to be innately less attractive (and black men more attractive), while Asian women were the most attractive of the lot. His results were drawn from an American survey of white, black, Asian and Native American men and women, who were given photographs and asked to rate their subjects on attractiveness.[59] Considering the reasons for these results, what came to Kanazawa's mind was not socialised images of beauty, or perceptions of a dominant culture, but hormones. 'The only thing I can think of that might explain the lower average level of physical attractiveness among black women is testosterone,' he was quoted as saying. 'Africans on average have higher testosterone than other races ... Women with higher levels of testosterone have more masculine features and are therefore less physically attractive.'[60]

This article prompted social media petitions calling for Kanazawa to be dismissed. *Psychology Today* withdrew the article from its website, admitted that it had not been reviewed before publication and apologised to readers. The LSE leapt into action in defence of the attractiveness of black women and launched an internal investigation, which found that Kanazawa had brought the institution into disrepute. Kanazawa duly apologised to the director of the LSE, saying he deeply regretted the 'unintended consequences' of his article and accepting that some of his arguments 'may have been flawed and not supported by the available evidence'. The LSE responded with a slap on the wrist, prohibiting him from publishing in non-peer-reviewed outlets for a year.[61]

The world of evolutionary psychology joined the fray. A group of sixty-nine researchers wrote that his findings were based on poor quality data, inappropriate statistical methods and a consistent failure 'to consider alternative explanations for his results'.[62] The world of race psychology rallied: twenty-three evolutionary psychologists and others put their names to a defence of his credentials.[63] These included his race science friends such as Lynn, Rushton, Vanhanen, Harpending and, most 'colourful' of the lot, Christopher Brand, an overtly racist IQ psychologist who, after he wrote a defence of paedophilia, was fired from Edinburgh University in 1997 for conduct that 'brought the university into disrepute', although he later won a case for unfair dismissal.[64] In response, Brand lambasted his 'Jew-leftie-commie' critics.[65]

RANDY THORNHILL ON DISEASE, RACE AND IQ

Another who embraced Lynn's IQ data was Randy Thornhill, a zoologist from the University of New Mexico, who doubled as a poster boy of evolutionary psychology. Thornhill came to public notice on familiar territory for evolutionary psychologists: the differences between men and women. In 2000, he co-authored a book,

A Natural History of Rape, in which he said that men had evolved to rape, and that rape was more a sexual act than an act of violence or power.[66] This delighted the evolutionary psychology community but his premises, methodology and conclusions were panned in all other academic quarters. I won't discuss Thornhill's rape premise, beyond saying that the assumption that every reasonably common variety of male or human behaviour must be the result of a specifically evolved mental module is typical evolutionary psychology fare.[67]

Thornhill's next big claim received none of the opprobrium of his rape theory. He wrote a paper (with the evolutionary psychologist Corey Fincher and his teaching assistant, Christopher Eppig) in which they argued that infectious diseases affected long-term intelligence: more bugs, fewer brains; fewer bugs, more brains. Or, as they put it: '[T]he Flynn effect may be caused in part by the decrease in the intensity of infectious diseases as nations develop.'[68] No one would dispute that certain infectious diseases can affect an individual person's brain development but the authors fail to suggest which diseases they are referring to, the length of infection, age of infection and treatment. The claim that the IQs of whole populations would be reduced is another matter. Still, while it may be a quirky view, it does not seem to be one in the same league as the stuff pumped put out by Lynn, Kanazawa and their ilk. But a closer look shows they all drink from the same bowl.

The first clue comes from the reference list. Lynn takes pride of place, with fifteen citations from eight books and articles (two from *Mankind Quarterly*). Others are also cited, including Kanazawa, Rushton and Jensen. Eppig, Fincher and Thornhill swallow Lynn's international IQ data virtually whole, apparently oblivious to most of the critique. After explaining how parasitic infection can affect brain development, they marry Lynn's data, data from another study[69] and World Health Organization figures on disease, drawing the conclusion that the 'negative relationship between infectious disease and IQ was statistically significant at the national level and within five of … six world regions'. They say the data show that infectious disease 'was a significant predictor of average national IQ …' Only in South America was the relationship insignificant. Eppig added, in another paper: '[T]he evidence suggests that infectious disease is a primary cause of the global variation in human intelligence.'[70]

But Eppig, Fincher and Thornhill go significantly further, suggesting long-lasting *evolutionary* implications: 'People living in areas of consistently high prevalence of infectious disease ... may possess adaptations that favour high obligatory investment in immune function at the expense of ... intelligence.'[71] If periods of health are rare, the strategy of allocating more energy to brain development during periods of health 'would be lost, evolutionarily'. If your body has to fight childhood infections, this will use energy that might otherwise go into IQ building and, somehow, this would be genetically inherited by future generations; an argument that sounds more Lamarckian than Darwinian. They say their findings suggest 'the heritable variation in intelligence may come from two sources: brain structure and immune system quality' and add that infectious disease may be 'the unknown factor linking skin colour and IQ'.[72] What looked like an environmental case turns into a genetic one: that people in sub-Saharan Africa, and other parasite-heavy areas, have evolved to be less intelligent than those from parasite-light areas. After this, they become bizarre, claiming that 'body symmetry' and IQ are also positively correlated 'because they are both affected negatively by exposure to high infectious disease'.[73] In other words, people with strapping, symmetrical builds are brighter than those with wonky builds. Citing Jensen, they add asthma and allergies to their intelligence-diminishing list (more inhalers, lower IQ).

As so often, the problem begins with correlation and causation. Even if we accept Lynn's data, the fact that low IQ and bad health bump along together does not imply a causal relationship. The prime causal factors of high or low IQ relate to education and wealth, even though there may be high correlation between low IQ, poor health and exposure to infectious disease. But Eppig and his colleagues have not controlled for education or wealth. More educated people do better on IQ tests, and they are usually people whose families are wealthier. Wealthier families usually have better nutrition, cleaner environments and more access to immunisation and medical care, so they are less likely to be debilitated by infectious disease. Incidentally,

Eppig says they controlled for education in an American study, finding that IQs were highest in Massachusetts, New Hampshire and Vermont and lowest in California, Louisiana and Mississippi. The reason? 'Again, infectious disease was an excellent predictor of average state IQ'.[74] Were the foetid malarial swamps of California producing a dimmer breed of human than the disease-free eastern seaboard?

The authors' most serious error concerns the supposed evolutionary implications. If those in disease-ridden areas evolved to have lower intelligence, this would imply genetic differences in their brains, yet no such differences have been found. If the authors were right, you would expect that parasite-heavy countries would be permanently locked in a low IQ cycle, because of the hereditary effects, even if the parasites disappeared. The evidence is against this: IQs rose faster in Kenya than elsewhere, yet Kenya suffers the full range of infectious diseases and has a higher than average child mortality rate, as does Sudan, where IQ scores are also rising rapidly. The claim that disease patterns prompted evolution of lower intelligence turns out to be a neo-Lamarckian thumb-suck.

RICHARD LYNN'S IMPACT

Just as there was an initially respectful press response to Kanazawa's research, the same applied to Eppig, Fincher and Thornhill's. Uncritical reports appeared in publications from *The Economist* to the *Guardian*, features in popular science magazines such as *New Scientist* and *Scientific American* and a phalanx of support on the World Wide Web.

What of Lynn himself? Among his least critical readers are mainstream hereditarian IQ advocates such as Richard Herrnstein, Charles Murray, Randy Thornhill and Corey Fincher, all of whom relied on his data. Their writing reaches a much wider readership and has been magnified by backing from the likes of Andrew Sullivan, Sam Harris, Steven Pinker and Jordan Peterson. Unlike Jensen and Rushton, who were given a torrid time by students, Lynn had it easy at the

University of Ulster, where protests were half-hearted, despite his acknowledgement of his racism. The British media also treated him gently. In the 1990s, he was billed as a world expert on intelligence on the BBC's flagship science programme, *Horizon*, and when *The Bell Curve* was published he wrote a piece in *The Times* headlined, 'Is man breeding himself back to the age of the apes?'[75] A decade later, *The Times* devoted a prominent report to his claims and he was allowed to pontificate on the BBC Radio 4 *Today* programme and on BBC Radio 5 Live. In 2010, the *Daily Mail* described Lynn as 'one of Britain's top dons' in the headline of a feature written by him.[76] A year earlier, Channel 4 television ran a series of programmes under the title 'Race: Science's Last Taboo'. Two of their 'experts' were Lynn and Rushton, prompting the geneticist Steve Jones to write in the *Telegraph* that while science 'knows no taboos', any link between skin colour and brain power was long ago disproved. He accused Channel 4 of dredging up 'a number of living fossils – elderly exponents of racial difference' who 'roll out a series of hoary, dubious predictable claims about the abilities of different racial groups'.[77] As we shall see in the final chapters, the frequently insidious impact of these fossils is a long way from being extinguished.

15

IS RACE SCIENCE MAKING
A COMEBACK?

Stefan Molyneux has a soothingly authoritative presence for those who yearn for soothing authority. Those without this urge might reach for a different adjective: smug. This balding, bearded fifty-four-year-old Irish-born Canadian is a failed actor with a measured tenor voice. He discovered his calling, and a source of income, when he founded a cult-like online 'community' he calls 'Freedomain Radio'. Freedomain has alt-right Web forums, psycho-babble, leave your family advice and Molyneux's YouTube channel, which he uses to promote his once-libertarian but now pro-Trump, anti-feminist and virulently pro-race science views. This is Molyneux on European and African brains: 'My ancestors were driven out of Africa and struggled to survive winter and hunger. Over thousands of years they became smarter and wiser through suffering. They made the modern world. Now the Africans say we are "privileged" and thieves. No – suffering made us. No more guilt.'[1]

Molyneux finds like-minded authors and academics (if you're a race science author you can expect a call), and butters them up with ingratiating questions: we're-in-this-together: you, me and my 100,000 young white male viewers. He displays just enough knowledge of their topics to give his followers the impression of a learned truth-teller going into battle against a devious, deluded, politically correct liberal-left conspiracy. Charles Murray, Richard Lynn, Linda Gottfredson and Jordan Peterson have all played along nicely, most more than once.

One of his favourites is Nicholas Wade, a seventy-six-year-old

English journalist who, I imagine, still makes *New York Times* editors blush; he went from being the newspaper's science correspondent to penning the most discredited book on race science (*A Troublesome Inheritance: Genes, Race and Human History*) of the last twenty-five years. When it was published, in 2014, 139 of the world's leading geneticists and evolutionary theorists signed a letter to the *New York Times* refuting its case. This led to an all-too-brief spell of relative silence, until the alt-right surge in the wake of Trump's election campaign gave Wade another innings, which he used to lambast his 'left-wing' academic critics: 'That attack on my book was purely political. It had no scientific basis whatever and it showed the more ridiculous side of this herd belief,' he said, later adding: 'There aren't any mistakes in my book.' He later acknowledged much of it was 'speculative'.[2]

WADE CONDENSED

In one sense, *A Troublesome Inheritance* is old hat; Wade repeats the usual race science shibboleths. But he adds some new, particularly incendiary, material: the Industrial Revolution began in England because natural selection blessed the English with genes for working hard and respecting the law; Africans are genetically inclined towards tribalism, the Chinese towards authoritarianism, and the super-smart Jews evolved for capitalism.

Wade, who insists he's not a racist, does not explore alternative explanations of why the Industrial Revolution began in Britain, why tribalism is more prevalent in Africa and the Arab world or why Jews developed a strong intellectual tradition. He argues instead that environmental changes prompted rapid genetic changes: 'The rise of the West is an event not just in history but also in human evolution,' he writes,[3] adding that if the differences between tribal society and modern states were purely cultural, 'it should be easy to modernize a tribal society by importing Western institutions. American experience in Haiti, Iraq and Afghanistan generally suggests otherwise.'[4]

Like Lynn, Wade believes people continued to evolve through natural selection after leaving Africa, and this included evolution for racial variations in personality and intellect. '[I]t is reasonable to assume that if traits like skin colour have evolved in a population, the same may be true of its social behaviour,' he says.[5] This, he claims, is the key to understanding modern humanity because 'the broad general theme of human history is that each race has developed the institutions appropriate to secure survival in its particular environment.'[6] The English 'evolved' from being the violent peasants of 1200 to the law-abiding citizens of 1800 who launched the Industrial Revolution. How? Wade says it was because the rich had more surviving children than the poor. 'Since the size of the English population remained fairly constant, many children of the rich must have dropped in social status, diffusing the genes and values that had made their parents wealthy into the wider population.'[7]

No one disputes that human populations evolved for skin colour, lactose tolerance, altitude tolerance, defences against malaria and the rest, but no scientist has provided evidence of population-specific evolution for wealth-making, authoritarianism, tribal loyalty or, indeed, intelligence. Wade simply assumes that if the one happened, then so must the other, even though he's forced to admit that the 'genetic basis of human social behaviour is still largely opaque'.[8] He conflates physical adaptations with national stereotypes that have no demonstrable genetic connection, telling us that Africans live in poverty because they have the genes for trusting tribalism too much. 'Tribal behaviour is more deeply ingrained than mere cultural prescriptions. Its longevity and stability point strongly to a genetic basis,'[9] he says, drawing no distinction between, for example, genetic adaptions for dry earwax and supposed adaptions for trust.[10]

Now and then, Wade throws in research-based nuggets, which on closer inspection fail to provide the desired ballast. For instance, he cites an academic paper identifying genes that might have been affected by natural selection over the last few thousand years.[11] However, it barely supports his thesis. The authors note that with these genes 'it is likely that we tend to underestimate the degree of sharing across populations' and conclude that 'there was not an overall enrichment for neurological genes,'[12] which counters Wade's view on, for example, advanced Ashkenazi intelligence.

Wade's slant on genetics is out of kilter with contemporary biology. In defining three main races (East Asian, African, Caucasian), he overstates the variance between them, understates the overlap, plays down genetic drift and attributes it all to natural selection. As Agustin Fuentes, a biological anthropologist from the University of Notre Dame put it, when lambasting Wade in a public debate, humans share 99.9 per cent of their genes, limiting variation to 0.1 per cent, and 'most variation in human genetics is due to gene flow and genetic drift.' Fuentes added that while Wade chose to highlight genetic differences between people from Nigeria, Beijing, Tokyo and Western Europe, we'd find similar differences between Liberians, Somalians and South Africans, because nearly all human genetic variation is found in Africa.[13] Likewise, the 139 population geneticists and other scientists, later joined by more, who wrote to the *New York Times* to refute the scientific basis of *A Troublesome Inheritance* said:

> Wade juxtaposes an incomplete and inaccurate explanation of our research on human genetic differences with speculation that recent natural selection has led to worldwide differences in IQ test results, political institutions and economic development. We reject Wade's implication that our findings substantiate his guesswork. They do not. We are in full agreement that there is no support from the field of population genetics for Wade's conjectures.[14]

Wade claimed their objections were politically motivated and there was no scientific case against his claims but the geneticists have hit back. One of the signatories, Jerry Coyne, from the University of Chicago, described Wade's book as a 'speculative house of cards' and 'simply bad science'.[15] Another signatory, David Reich, devoted five pages of his 2018 book, *Who We Are and How We Got Here*, to Wade's approach, accusing Wade of making racist claims with no scientific evidence, claims that are 'essentially guaranteed to be wrong'. Reich sums it up by referring to Wade's idea about genes that influence behaviour: 'In a written version of a nod and a wink, Wade is suggesting that popular racist ideas about the differences that exist among populations have something to them.'[16] Kevin Mitchell, a geneticist from Trinity College, Dublin, added to this argument in his 2018 book, *Innate*: 'It is a complete non-sequitur to claim that any cultural differences between populations must be caused by genetic differences,' he wrote. 'There is in fact no evidence at all that observed or supposed differences in behavioural patterns between populations reflect anything but cultural history.'[17]

There's little purpose in going through the rest of Wade's argument again. Enough to say that his approach amounts to naming a cultural generalisation and assuming it's valid throughout that population, race, ethnic group or country (he blurs these), then deciding, without evidence, that the explanation lies in natural selection. Along the way he parrots the full range of fallacies discussed in previous chapters, including a misreading of twin studies. All this, and more, left Stefan Molyneux immensely impressed.

PSYCHOLOGY AND THE PULL OF SCIENTIFIC RACISM

Wade is a journalist and Molyneux is a YouTube star but most of the advocates of race science over the past fifty years have been psychologists. Some, such as Linda Gottfredson and Jordan Peterson, are academic psychologists who do not attach the adjective 'evolutionary' to their work, but several do call themselves evolutionary psychologists. This applies not only to extremists such as Richard Lynn and Satoshi Kanazawa but also to those more firmly ensconced in the evolutionary psychology mainstream, such as Randy Thornhill, Corey Fincher and Steven Pinker. Those in the mainstream frequently draw on the data pumped out by fringe players such as Lynn.

As we saw in Chapter 9, the twentieth-century link between psychology and race science emerged from IQ testing but it soon enveloped most branches of psychology. I've already mentioned Jung's views on the animal-like primitiveness of Africans; he made similar observations of the 'American Negro', writing about their 'childlikeness' which, he felt, had an impact on white American naivety, although he concluded this was not too dangerous because blacks were in the minority.[18] Jung's racism was forged in nineteenth-century Europe, where few alternatives were available, but echoes of notions of animal-like primitiveness could be heard long after the exposure of the Holocaust had muted them in scientific circles. The leading American psychologist Henry Edward Garrett (past president of the American Psychological Association and head of psychology at Columbia University) opposed desegregation of schools, which he blamed on Jews, and argued that blacks were innately stupider than whites. In 1963, he explained that 'the Negro' had less 'abstract intelligence ... He functions at a lower level.' However, he noted that '[t]hose black Africans are fine muscular animals when they are not diseased.'[19]

The student and academic reaction to Arthur Jensen's claims meant mainstream public expression of race science was muted until the 1990s but private racist perspectives lingered among a substantial proportion of American psychologists. A questionnaire sent to 1,020 American psychologists in the mid-1980s (completed by 661), showed most were open to racist concepts of intelligence. A mere 15 per cent felt that the differences between white and black IQs were entirely due to environmental factors.[20] Twenty-five years ago, when researching his book *The Race Gallery*, the science writer Marek Kohn interviewed the biologist Steve Jones, who told him that race would not return to science. Kohn replied that it was still thriving in psychology. 'Yes', said Jones, 'but not in science'. Kohn reflected that he had a good deal of sympathy for this view when he began working on his book. 'Now that it is finished, I have a good deal more.'[21]

Why do journalists routinely describe psychologists who pontificate on race, gender and intelligence as scientists when 'real scientists', such as Jones, draw a stark distinction? I blame Isaac Newton. Not a very nice fellow and certainly one who barked up some strange trees, diverted by theological debate, his quest for the Philosopher's Stone, by occult studies and his urge to hang counterfeiters. But none of this is counted today, because Newton was so smart at the stuff he got right: physics, astronomy and the mathematics underpinning it, calculus, gravity, motion, optics, colour, sound. He failed to turn lead into gold but set the gold standard for scientific methodology. Whether seen as the first scientist or the last of the magicians, as Keynes put it, Newton locked down the idea of scientific proof: make observations, form an hypothesis to explain them, make predictions, look for confirmation of the predictions. And if empirical observation contradicts the hypothesis, form a fresh one. As one of his better biographers, James Gleick, put it: 'A worry nags at his descendants: that Newton may have been too successful; that the power of his methods gave them too much authority'[22] – so much authority that in the centuries since Newton, his scientific methodology has pushed other areas of learning to the fringes. It's hardly surprising that anyone delving into the world of research would seek the certainty of science even when what they study belongs outside the scientific realm.

Here, I'm referring to the 'social sciences'. I'm not suggesting these non-scientific areas of study are worthless. My academic training was in economic history, law and politics, with smatterings of psychology, sociology, economics and media theory along the way – all worthwhile but none of them sciences. The ways people communicate, individually or collectively, how they survive, the stories they tell of themselves, of each other and of their pasts, their emotions, dreams and thoughts; these are not reducible to scientific laws. There is no reason why the motion of particles of carbon, of atoms, quarks or strands of DNA, should be replicated in how people behave. Yet the only humanities that seem to have embraced this truth are history and philosophy – perhaps because they alone are secure in their status.

Psychology has the most severe case of 'hard science'-envy and has therefore been the most eager of the humanities to shelter under its umbrella. Its pioneers were medical doctors who wanted their theories to be accepted as scientifically valid. Freud was particularly eager and yet the id, ego, Oedipus complex and penis envy have no connection with science. These descriptions are not subject to scientific methodology, they are not measurable, they cannot usefully be correlated with neurobiology. Theories of the workings of the human mind may or may not be valid but are not the same thing as neurobiology, because minds don't exist without environments. Yet this urge to couple psychology with science has persisted in the assumption that observations about the working of particular minds must be universally applicable. From the psychoanalyst Freud's accounts of Viennese female patients to pot-luck surveys of first-year university students, we find the conclusions drawn from tiny samples being offered as proof of an assertion, and picked up by newspapers around the world. And we find psychologists striving to promote their theories according to an – often limited – understanding of scientific principle, by framing the ephemeral in terms of simplified laws and statistics.

This desire was expressed particularly cravenly by Robert M. Yerkes, who we met in Chapter 9. He yearned for psychology to be accepted, and well funded, as a 'hard science' rather than being relegated to the humanities. This Harvard psychologist saw IQ testing as the route to hard science *nirvana*. 'We must ... strive increasingly for the improvement of our methods of mental measurement,'[23] he wrote in 1917. He later declared that IQ testing 'helped to win the war'. It also helped win his own war for recognition. '[I]t has ... established itself among the other sciences and demonstrated its right to serious consideration in human engineering,' he declared.[24] Charles Spearman, the English psychologist who introduced factor analysis to IQ theory and invented *g*, expressed the same urge in 1923: 'We must venture to hope that the so long missing genuinely scientific foundation for psychology has at last been supplied, so that it can henceforward take its due place along with the other solidly founded sciences.'[25] Henceforward had arrived by 1937, prompting his delight that 'this Cinderella among sciences has made a bold bid for the level of triumphant physics itself.'[26]

This is not to say that psychology cannot validly use scientific methodologies, or that there can't be overlaps with other sciences. Cognitive psychologists would get nowhere without neuroscience, just as economists use statistical tools and mathematical models.

The problem arises when scientific methods give a faux-scientific gloss to unscientific assumptions, which is where IQ theory comes in. By applying factor analysis to find *g*, IQ proselytisers felt they'd achieved their bid for respectability. But it was lipstick on the pig. Applying valid mathematical methods to an invalid precept does not magically validate the original fallacy. Yet, despite the devastating critiques levelled at its assumptions, American psychology clings to IQ. This is Jordan Peterson on its status: 'If any social science claims whatsoever are correct then the IQ claims are correct because the IQ claims are more psychometrically rigorous than any other phenomenon that has been discovered by social scientists by a factor of about three.'[27] A whole industry would go up in smoke if IQ were dropped. In particular, the entire edifice of scientific racism would collapse.

Evolutionary psychologists are the latest to drink from the race science stream, so it's worth devoting them a few extra lines. Evolutionary psychology explains behaviour in terms of biology. Its founder-theorists, Leda Cosmides and John Tooby, describe it as 'the long-forestalled scientific attempt to assemble, out of the disjointed, fragmentary and mutually contradictory human disciplines, a single, logically integrated research framework for the psychological, social and behavioural sciences'.[28] Like sociobiologists, Cosmides and Tooby subsume the humanities under the evolutionary umbrella, downgrading culture, which is seen as a by-product of natural selection. The specifics of behaviour are as a result of gene-prompted adaptations that evolved in the Pleistocene. But evolutionary psychology has an additional element: the idea of the instinct-packed modular mind. Because there are no lengths of DNA for universal behaviours, evolutionary psychologists use the notion of 'mechanisms' encoded in the genes, leading to a Swiss army knife model of a mind equipped with a collection of discrete behaviour modules, each evolved to allow us to face Stone Age challenges.[29] Men and women evolved different modular minds because of sexual selection, leading to adaptations that prompt men to compete for women and spread their genes as far and wide as possible, and for women to pickily choose high-status or high-earning men.

Evolutionary psychologists talk of hundreds of thousands of independently evolved modules, requiring hundreds of thousands of genetic mutations, although the human genome is thought to have only around twenty thousand genes, many of which have nothing to do with cognition or emotions. Instead, the brain's development is constantly mediated by the world it encounters, not via evolved 'triggers' that activate modules that don't exist. Neurologically, human brains are plastic: highly flexible, full of non-adaptive capacity, forever sensitive to environmental input, and our 'domain general' minds are moulded into specialisation by experience.[30]

How could evolutionary psychology be so wrong? The Stanford evolutionary biologist Paul Ehrlich suggests an answer: evolutionary psychology's 'knowledge of genetics and evolution tends to trail far behind the knowledge of psychology'.[31] Instead of embarking on quixotic quests to find genes implicated in specific behaviour, evolutionary psychology works backwards, starting with the assumption that the behaviour is innate. They devise questionnaires to test whether most respondents (often their own students) think this way. When a majority confirms the hypothesis, they claim this proves the behaviour is 'hardwired', though it is invariably easy to think of independent variables, unrelated to genetics, that could have prompted their results.[32] They then embark on 'Just-So Story' trawls to find the prehistoric behaviours behind these mental modules. In this way, today's habits are put under the timeless seal of Human Nature. The biological anthropologist Ian Tattersall notes that evolutionary psychology's monolithic ancestral environment 'is little more than a figment of their nostalgia for an idealised past that never existed'[33] and adds that its reductionist notion of the relation between genes and ancestral behaviour 'betrays a misunderstanding of the fundamental realities of the evolutionary process'.[34]

We know little of our ancestors' behaviour in the Pleistocene, let alone whether this had any connection to the random genetic mutations that supposedly prompted these traits. We presume they lived in small, hunter-gatherer groups but know next to nothing about their kinship relations, social structures and relations and belief systems. The British neurobiologist Steven Rose despairs at the variety of human attitudes and habits evolutionary psychologists such as Steven Pinker attribute to natural selection in the African savannah. 'The grander such assertions, the flimsier and more anecdotal becomes the evidence on which they are based. Has Pinker ever seen savannah, one wonders?' he asks. You get the picture. Evolutionary psychology's fundamentalists might attract attention in the media, where they are portrayed as scientists, but their status in the scientific world is less secure. This is summed up well by David Buller, professor of philosophy of science at North Illinois University. Starting as a fan, he spent several years investigating evolutionary psychology's claims, but ended his 550-page book with the conclusion: 'evolutionary psychology is wrong in almost every detail.'[35]

You might assume that evolutionary psychology's belief in discrete mental modules flies in the face of the general intelligence idea of IQ testing. But evolutionary psychologists don't question the notion of intelligence as a single thing, captured through factor analysis as g. You might also assume that evolutionary psychology would be antithetical to racism, because of its belief that the human mind was moulded in the Pleistocene, hundreds of thousands of years ago. And, for a while, this was so. Sociobiologists muddied their boots but early evolutionary psychology seemed immune to its attractions. As Cosmides and Tooby put it: 'Our modern skulls house a stone age mind.'[36]

However, there were forces pulling in the opposite direction. Evolutionary psychology big-hitters such as Pinker and Thornhill, like the bad boys such as Lynn and Kanazawa, were attracted by the scent. And they have gone further than entertaining the idea that some human populations continued to evolve (or de-evolve) in terms of IQ. It is as though the mental machinery for preferring pink, wearing lipstick, relishing shopping trips, being chatty and empathetic and marrying richer, older suitors (women), or for being mean to step-children, good at parallel parking, reading maps upside down, preferring blue and wanting to get their leg over everything that moves (men) all evolved hundreds of thousands of years ago and remain unchanged, but intelligence is something else; a different category of mental machinery that has continued to evolve not just over the last 200,000 years but over the last thousand. Their core genetic determinism binds them to the belief that nature trumps nurture; that genes explain far more about how we think and act than upbringing, class or culture. They have drawn on genetic determinism to argue that men and women have evolved substantially different mental machinery, so when they saw the same evolutionary arguments used to serve claims about different mental machinery for racial population groups, they were tempted.

All psychologists, all over the world, learn about IQ testing, but it is ubiquitous in North America, penetrating all levels of education and many professions. Evolutionary psychologists particularly relish the 'scientific' label, liking to claim that their calling belongs among the biological sciences. It is hardly surprising they embrace the certainty of IQ theory, methodology and practice, which they imbibed with the mother's milk of their training. Confronted with arguments in favour of a genetic view of IQ, their inclination is to be sympathetic, even when the arguments take on a racial slant.

WHAT MOTIVATES RACE SCIENCE?

Race science likes to present itself as pure science but the politics behind its claims are seldom far from the surface. Interviewing Nicholas Wade, Stefan Molyneux dangled a lure, saying that different social outcomes are the result of different innate IQs among the races: 'high IQ Ashkenazi Jews and low IQ black people'. Wade took the bait, saying that the 'role played by prejudice' in shaping black people's social outcomes 'is small and diminishing', before slipping in an overtly political point of his own, condemning 'wasted foreign aid' for African countries.[37] This is a common theme of many race science advocates: that race confines the potential of groups as well as individual people, and therefore trying to improve their lot is a waste. Arthur Jensen made the same point about the Head Start early intervention scheme; Charles Murray and Andrew Sullivan said the same about affirmative action; Murray added it was based on the false idea that 'everybody is equal above the neck'. Sullivan warned that denying race differences in cognition could prompt us to 'overshoot and over-promise' on social policy. There is a common theme: smaller state, less taxation, less social intervention, less foreign aid.

In 1981, Stephen Jay Gould's book on genetic determinism and racism, *The Mismeasure of Man*, which focused on the founding fathers of race science, was published.[38] But as Gould noted when he revised his book fifteen years later, 'the same bad arguments recur every few years with a predictable and depressing regularity.'[39] Why? One answer is that each outbreak relates to recurring political imperatives. Gould noted that *The Bell Curve* coincided with the Republican Party's focus on slashing social services for people in need and giving tax relief to the rich: 'Can we doubt the consonance of this new mean-spiritedness with an argument that social spending can't work because, *contra* Darwin, the misery of the poor does result from the laws of nature and from the innate ineptitude of the disadvantaged?'[40]

The academic publishing industry, like the rest of the media, is sensitive to the *zeitgeist*. The spate of 'Mars and Venus'-type books, both self-help and faux-scientific, which emerged in the 1990s, dovetailed with a wider backlash against feminism. The periodic revival of race science can be seen in a similar light. If, like Murray, Sullivan, Wade and the rest, you oppose big government, welfare spending and the 'nanny state', then it must be reassuring to hear scientific-sounding claims that suggest your ideas are in lock-step with nature. If the poor are poor because they're stupid, what's the point of trying to uplift them? It's only a small step to saying that black people are poor because they're inherently stupid and that Asian Americans, or Ashkenazi Jews, or white people generally, do well because they're inherently smart.

Gould pointed to the political leanings of those behind race science, specifically referring to Murray's employment in right-wing think tanks.[41] But he missed another step: Murray's data were drawn from the work of an even more virulent group, some of whom were activists, with links to the radical far-right. Lynn and Rushton led the white supremacist Pioneer Fund, which provided money to the racist psychologists mentioned in these pages. Several wrote for the

Pioneer-funded *Mankind Quarterly* and for another race science journal, *Personal and Individual Differences* (founded by Eysenck, with Lynn as an editorial board member), and these writers were backed by Washington Summit Publishers, led by Louis Andrews, who wrote from a similar perspective on race and intelligence. More recently, they have also published in the online 'journals' of the OpenPsych website, co-founded by the white nationalist and paedophile defender Emil Kirkegaard and the alt-right enthusiast David Piffer, which has editorial links to *Mankind Quarterly*. They provide positive reviews, endorsements and public support for each other's views. Murray embraced the data of Lynn, Jensen and Rushton. Sullivan, in turn, embraced Murray and defended Jensen. Pinker aired the views of Murray, Cochran, Hardy and Harpending, who cite Lynn, Jensen, Gottfredson and Eysenck. Thornhill used Lynn's research for his paper on disease and African intelligence. Many of them came together – far right and near right – in Gottfredson's document 'Mainstream Science on Intelligence'.

Where Gould missed the mark was in his suggestion of a direct relation between political climate and fresh waves of racist science. The hardcore far-right IQ psychologists pump out perpetual streams of data, which are published by their house journals and publishers. This is a given, a constant. They are immune to the criticism that occasionally comes their way when independent academics lower themselves to review one of their books or papers. They reinforce each other, using the Web as an echo chamber in which they can be sure of aggressive backing from racist trolls the world over.

Now and then an hereditarian from outside this tight circle will decide it's time to pitch in. Their paper or book will take off, be picked up by the media and reverberate on the Web. Why the media might be interested cannot be read from immediate political imperatives. Editors like stories with bite, with controversial angles. Stories on race provide just this, particularly if combined with data that sound scientific. If the data have a genetic tint, so much the better; the editors know a significant proportion of readers believes that genes explain everything. Claims of innate differences between race groups ring editorial bells. At worst, they make good fillers for inside news pages. At best, they generate splash headlines and become running stories, with feature and comment spin-offs. The process goes like this: a race science academic or journalist writes a paper or book that is fed to their university's press office or their publisher's publicity team. A press release containing the most newsworthy bits is e-mailed to news and science editors. The press release is worked up into a news report. This prompts fervent debate on the newspaper's website and Twitter feed. The alt-right trolls get to work and a feature or a comment piece might follow. With enough momentum and enough clicks, likes and retweets, the rest of the news media will follow: radio, television, YouTube. By the time the more critical academics put together their ripostes, the story is dead.

The process is influenced by the political climate, as illustrated by the proliferation of race science on social media in the wake of Trump's election campaign and since. The rise of the alt-right in the USA, and the success of far-right and authoritarian-nationalist political parties and leaders in Hungary, Poland, Sweden, France, Italy, Austria, Germany, Brazil, Turkey, the Philippines and elsewhere has been linked to the combination of the economic fallout from the 2008 banking crash, the decline of manufacturing and mining jobs in the West, the recalibration of the world economy as information technology changes the world, and to the wars in Syria and elsewhere that have prompted the refugee crisis. Those who feel they have no stake in this rapidly changing, uncertain, Web-comment-driven, post-industrial world, and fear their positions are threatened by foreigners, darker-skinned people, urban 'elites' or women are more likely to join online hate forums and follow assertive, self-assured, vituperative media figures such as Jordan Peterson or Stefan Molyneux. It helps that there are media owners and editors who share an hereditarian agenda, although the process doesn't have to rely on that.

Therefore, we can expect more of the same in years to come. The building blocks are always in place: a group of dedicated far-right IQ missionaries pumping out race-tinged stories, a wider group of psychologists and other writers deeply committed to an hereditarian intellectual agenda and willing to draw on data from the racist core, an even-wider network of politicians, editors, academics, authors and publishers with anti-welfare agendas, leaning in a libertarian direction, and media decision-makers who know that any story combining race, genetics and intelligence is likely to fly. This, combined with a social media *milieu* in which people with racist inclinations find each other so easily, guarantees its survival.

It must be said that not all the backers of race science have right-wing agendas. Sam Harris and Steven Pinker are notable exceptions. However, what they all have in common – the true believers and the facilitators – is a fervent belief that the liberal-left media and academia are suppressing science, due to a politically correct queasiness about the truth. Wade told Molyneux he was pilloried because the academic community was in a 'very leftist mode'. When debating with Ezra Klein, Harris said:

> What I am noticing here, and what I've called a moral panic, is that there are people who think that if we don't make certain ideas, certain facts, taboo to discuss, if we don't impose a massive reputational cost in discussing these things, then terrible things will happen at the level of social policy; that the only way to protect our politics is to be – again, this is a loaded term but this is what is happening from my view scientifically – is to be intellectually dishonest.[42]

Their martyr complex – the sense that they are the intrepid truth-tellers, following the scientific breadcrumbs wherever they might lead, without fear or favour – is a strong motivating force that draws people like Harris to those with a harder race science agenda. It reverberates on the Web, playing into conspiracy theories about truth-suppression and drowning the calmer voices of science who explain that there is no evidence that IQ differences between populations are genetically rooted. Reporting an absence of evidence is never as thrilling as transforming speculation into fact and, carrying the sword of justice and the shield of truth, riding a white horse against devious, fact-denying conspirators.

RACE SCIENCE, REDDIT AND THE ALT-RIGHT

One element makes the current wave of race science different: the Web-based 'alt-right' (although the 'alt' part of it is hardly apposite when the President of the United States echoes all its themes and his former campaign chief, Steve Bannon, is its most public face). The alt-right particularly flourishes in the United States and Canada but its tentacles reach to several parts of the world. Since the early 2000s, it has emerged as a kind of movement, in the loosest sense of the word, with several overlapping zones of passion, anger and interest. Its home is on the World Wide Web, although it is not averse to street action, non-Web political campaigning and terrorism.

Race is certainly the major part of the alt-right's identity and sense of mission; its core component. Its expression ranges from the visceral white supremacist racism of Jared Taylor's *American Renaissance* magazine to the faux-scientific racism that is the focus of this book. There's a fair amount of overlap, since race science is swallowed whole by white supremacists, giving ballast to their flakiest views. Gender also features prominently; a hatred of feminism and the #MeToo campaign, a conviction that white males experience discrimination, an angry belief in traditional roles and a conviction, often drawing on the perspectives of evolutionary psychologists and others from the 'Mars and Venus' solar system, that differences between the brains and minds of men and women are huge and unbridgeable. Some marry this denigration of feminism with a hatred of Islam, which prompts them to write and tweet in defence of oppressed Islamic women; the only group of women they seem to support.

A third theme involves a conspiratorial view of the world. Conspiracy theories are by no means the exclusive preserve of the right, but its members are particularly attracted to the idea that powerful, shadowy forces are shaping the world. Bannon is not alone in finding the hand of the Jewish philanthropist investor George Soros behind every disruptive event. One expression of the conspiratorial view comes from the 'white genocide' notion, the fear of a conspiracy (often said to involve Soros) to make white Americans the minority.

This 'replacement theory' has versions in other countries, including South Africa where it focuses on white farmers, whose cause has been adopted by several on the American right, including Trump. It also inspired the Australian terrorist behind the 2019 Mosque massacre in Christchurch, New Zealand. Other proponents also include the FoxTV talk show host Sean Hannity, and his conspiracy theories about Barack Obama's birthplace and the murder of the Democratic National Committee employee Seth Rich, is a particular favourite, and many dip into Alex Jones's ultra-right conspiracy site Infowars.

Most members of the American alt-right combine are fervently anti-federal state, despite their nationalism, or at least, opposed to its welfare and foreign aid components. They fear and despise the United Nations and other world bodies, which are often seen in terms of international conspiracy. In American politics they tend to be pro-gun, anti-migrant, climate change sceptics, who either welcome the idea of human-driven global warming as a good thing or, more commonly, view it as a fact-suppressing left-wing conspiracy. A further binding theme is one that used to settle on the hatred of 'political correctness' and is now more likely to be framed by free speech: the idea that legitimate views on race, sex, politics and identity are suppressed by a liberal-left elite which dominates the universities and the mainstream media and 'no platforms' those who disagree with them. University-based 'snowflakes', with their safe spaces, and also their lecturers, who are invariably dubbed 'cultural Marxists' (even though most of those accused have no truck with Marxism), are a particular source of agitation.

These issues draw the alt-right together but there is no single, coherent agenda or programme, or formal leadership structure. And there are contradictory elements: anti-tax, small-state libertarians rub shoulders with migrant-bashing, flag-waving nativists, evangelical Christians with evangelical atheists, anti-Semites with Netanyahu-backing Zionists and pro-science empiricists with anti-science and anti-vaccination conspiracists. There is no focal point, although they tend to congregate pseudonymously on Reddit, finding communities of like-minded people in sub-reddit forums,[43] and on the imageboard websites 4chan and 8chan and the right-wing social media site, Gab. They dip into news-related sites such as *Breitbart News*, get their daily news and views from FoxTV and follow the same YouTube stars, such as Molyneux and Peterson. It is notable that the right has far more big-hitting YouTubers than the left, partly because they were early adopters who saw the potential of the platform.

Some are drawn to the alt-right discourse through particular interests, often explored through sub-reddit communities such as the 'Manosphere' – an informal Web network of misogynistic white men who are obsessed with male identity, and advocate anything from men's rights to 'pick-up artistry'. One branch is the 'Incel' sub-culture – a collective of bitter white men who define themselves as 'involuntary celibates', unblessed with the genetic make-up for pulling power. Much of their output relates to their belief that men are owed sex and to a biologically determinist view of sex and gender drawn from evolutionary psychology which suggests that women are only interested in men (called 'Chads'), who have power and certain physical assets. Some advocate that women should be treated as sexual objects, with few rights. Their favourite Web-based star is Jordan Peterson, who says the solution to the problems faced by the Incels is enforced monogamy of everyone else.[44]

The 'carnivorists' are perhaps the most bizarre new identity group of the alt-right. Their conviction is that a diet of red meat will beef up masculinity, and irritate vegetarian liberals and environmentalists. Again, Peterson is its reigning guru, after boasting, in one of his many appearances on the Joe Rogan podcast, that he was on a carnivorous diet. This has prompted Peterson followers, among many others, into following an all-meat diet.[45]

The '#GamerGate' community emerged in 2014, when a game developer, Zoe Quinn, released *Depression Quest*, which drew from her own experience of depression. It received positive reviews, prompting a vicious backlash from male gamers, who hated its departure from the usual fare of skill-based violence and saw a conspiracy behind the positive reviews. Quinn was forced to flee her home, after a Twitter campaign of harassment, which included rape and murder threats. Her family was threatened, as were her supporters in the media, and other female game developers. The harassment was coordinated through anonymous message boards on Reddit, 4chan and 8chan.[46]

People in these online communities follow others with links to different parts of the alt-right and navigate their way to its racist component. They find common enemies – leftists, anti-racists, feminists or women seen as rising above their station (as with #GamerGate) – and swarm on them, bombarding their Twitter feeds with *ad hominem* attacks, insults and site-clogging 'shitposts', hacking into their computers, going for their families; all in the cause of silencing them and discouraging others. The right complains bitterly about no-platform actions against its luminaries but it could be argued that the campaigns against people such as Charles Murray and Nicholas Wade have boosted their profiles and given them new media openings. In contrast, the trolling harassment meted out to those campaigning against the alt-right or for feminism offers no positive outcomes and makes people think twice before going public. Trolling is not confined to the right, but left-wing trolls tend to target centrists, liberals, the soft-left and feminists who are viewed as transphobic, rather than the big hitters of the alt-right.

Right-wing trolls know their enemy, as I have discovered. When I wrote a comment feature on Richard Lynn for the *Guardian* in 2003,[47] his pal, J. Philippe Rushton, leapt to his defence but only Christopher Brand and a handful of others took the bait.[48] In 2010, I wrote a comment piece in the same newspaper about Randy Thornhill's research.[49] By the next morning, there were a couple of hundred responses, many from men writing under pseudonyms, trolling their way through the night in support of race science. Between 2015 and 2018 I wrote several magazine and newspaper features and gave television interviews on the revival of race science.[50] The answering deluge of pseudonymous comments and personal abuse came not only via the publications' websites but also on social media, particularly Twitter and YouTube.[51] It was clear that a community of hereditarians was finding each other on the Web, magnifying the impact of race science and raising the cost of opposing it.

16

RACE, GENES AND INTELLIGENCE: WHERE NEXT?

I spent most of my first three decades living in a country in which a minority believed the majority was unfit to rule, because its members lacked the intelligence, character and culture for the job – but mainly the intelligence. It took tens of thousands of deaths, a thirty-year armed struggle, an international sanctions campaign and the emergence of Nelson Mandela as the right man for the moment before that minority was prepared to start changing its mind. I had distant relatives who met their deaths in the gas ovens of the Reich because they were Jewish. It took a world war to dislodge the ideas that led to the Holocaust. I have family members who are gentile and Jewish, white and black, of Irish, Danish, Welsh and English descent. I would hate any of us to be treated differently on account of our ethnic or national origins. Racism is an issue I feel strongly about. It upsets me, deeply and personally, and I believe that its faux-scientific expression is its most insidious and dangerous version.

The argument that ethnic groups or races are blessed by nature with different intellectual capacities perches on several wobbly pillars, none of which need the strength of Samson to dislodge. I have presented the arguments and counter-arguments but it is worth reiterating the core case, and its antidote, before looking ahead.

RACE 'SCIENCE' REVISITED

Race science starts with a view of evolution that has been rotting on the intellectual vine, a view that overplays natural selection at the expense of genetic drift, and assumes evolutionary change is a perpetual process of fine-tuning. It also holds that our behaviour, including our ability to perform in IQ tests, is programmed by our genes and leaves little room for evolutionary spandrels. What I tried to show in the opening chapters is that we cannot assume an intimacy between genetic inheritance and behaviour, and that a great deal of the way we act is not primarily genetically prompted. Rather, the brain's development is perpetually sculpted by its interrelationship with the world it encounters. Our thoughts, emotions and actions, including the way we use intelligence, are influenced by this inter-action. Our environment plays a huge role in the way we turn out, both in framing our cultural options and in moulding our brains.

Contemporary scientific racism clings to a particular slant on the evolution of brain power: the assumption that intelligence continued to evolve in the same way as skin colour or ethnic diseases and that, among those who migrated to Eurasia, particular environmental conditions were a spur to further cognitive evolution. This is contradicted by the fossil record, which suggests that evolution has been characterised by punctuated equilibrium: long periods of evolutionary stasis followed by sudden bursts of change.

The idea that our brains might not have changed much over the past 100,000 years or more is unsurprising. It took humans two million years to go from standing upright to creating stone tools, and another million before tools with cutting edges emerged. What makes this version of the human story more likely is that ever since humans started migrating around the world, they have shared genes – even supposedly isolated populations shared more genes than previously recognised. And the more genes mix, the less chance there is of genetic innovations taking root.

This is reinforced by new archaeological evidence about our shared heritage. Discoveries in parts of Africa have pushed back the origin of humanity 120,000 years and there is increasing evidence that symbolic behaviour dates back at least 100,000 years. Paint workshops, geometric carvings, beads and body paints provide evidence of self-adornment. Tools, implements and weapons of a kind previously thought to have been invented in Europe are insignia of self-consciousness and creativity. Evidence of trade, burial rituals and trap-based hunting are the insignia of people capable of invention, of imagining worlds beyond their own, of planning, and of using language to communicate thoughts, fears and desires. The people who migrated in the direction of Australia 70,000 years ago, and in the direction of Europe a bit later, were clearly people just like us.

Race science draws direct analogies between naturally selected physical changes (such as skin colour and ethnic diseases) and the evolution of intelligence. But this is problematic for several reasons. First, intelligence is not a single thing and second, many genes – possibly thousands acting on different parts of the brain – are implicated. While geneticists have had no trouble finding the relevant alleles for the risk of developing prostate cancer, they've had trouble isolating genes for intelligence – even the most fervent of genetic determinists have failed to find single genes for intelligence. The closest geneticists have come is in identifying clusters of scores or hundreds of genes that appear to have a small impact on the ability to do IQ tests.

This doesn't imply there are no differences between populations on all aspects of cognition. Given that some ancestral populations were separated from others for tens of thousands of years, it would be unlikely that there would be no variation regarding all aspects of cognition. In Chapter 9, I looked at three examples: two gene variants that under dire social conditions can make men more depressed and potentially more violent, a longevity-related gene variant that appeared to boost the cognition of older people and an allele that made some men more inclined to violence when drunk. In each case the proportions of different populations carrying these alleles varied, but in each only a small minority carried them and the presence of the alleles would not have made a significant difference to the behaviour of the whole population.

The next argument from race science comes from the view of the relationship between genes, intelligence and IQ. Its adherents insist there's such a thing as general intelligence, which can be accurately measured by IQ tests, the results of which are primarily governed by genes, and that all these imply that IQ differences between population groups reflect differences in innate ability. Even if we accept the shaky assumption that general intelligence exists and the even shakier assumption that IQ tests can measure it, the third assumption – that general intelligence is governed mainly by genes – doesn't hold up. Twin and adoption studies suggest that IQ can be hugely influenced by an individual person's environment. Separated identical twins brought up in similar environments have similar IQs but those brought up in very different class or educational environments can have widely differing IQs. A child adopted from a struggling working-class background into a stimulating middle-class family will have their IQ enriched. In other words, when we shift the environment substantially, IQs will tend to vary substantially, contradicting the assumption that IQ is mainly a reflection of biologically based ability.

Comparisons become absurd if we switch from the individual to the population level. As Jim Flynn has shown, the reason IQs rise from generation to generation relates not to genes but to environmental factors; especially exposure to abstraction. Those with less exposure to the scientific way of viewing the world will have lower IQ scores; those brought up to view the world through scientific spectacles will do better. This is why IQs rise more quickly among some populations than others: different starting points, different levels of exposure to abstraction. The gap between black and white American IQs is narrowing. Asian-American IQs are rising more quickly than white American IQs and Dutch IQs more quickly than British. IQs in some Scandinavian countries have hit a buffer, while those in Kenya are galloping ahead.

You cannot compare group IQ scores on the assumption that you are comparing genetic intelligence. This point has been proven beyond serious dispute and yet, as we have seen, there remains a significant group of psychologists and journalists, and a miniscule bunch of scientists, who have continued to preach racist science, regardless of the weight of evidence against them. In the case of the scientists – such as James Watson – we can only assume their views emerge from a deeply ingrained racism, so deeply ingrained that, like religious belief, it is immune to logic. Some of the psychologists, such as Richard Lynn, fit in the same box. But for others, I hope the source of their misconception relates more to their attachment to genetic determinism and their faith in the immutability of IQ than to a deep-seated racism.

I have tried to expose the poisonous brew that comes from blending a skewed notion of race with a flawed notion of intelligence under the umbrella of IQ. This is what twentieth- and twenty-first-century racial 'science' is all about. And there is nothing at all scientific about it.

THE FUTURE OF RACE SCIENCE

It would be lovely to believe that this will go away and that good arguments could put an end to race science. But that's not the way things work. As we've seen, race science comes in waves. The current wave is particularly strong and persistent for reasons, as I touched on the previous chapter, that relate to the rise of ethnic nationalism, which in turn is partly prompted by the existential insecurity, particularly of young white men, in response to a rapidly changing social and economic *milieu*. The current wave has also spread deeper and wider via the Web, crossing national boundaries.

In most respects the current wave replicates the arguments of the old. What is new is the way its proponents latch on to, and distort, genetic research. In 2018, David Reich wrote a thoughtful piece on genetics and race for the *New York Times*. It was based on his recently published book, in which he suggested it is probable that 'genetic influences on behaviour and cognition will differ across populations.'[1] Several race science advocates, including Charles Murray, Sam Harris, Andrew Sullivan and Nicholas Wade leapt in, assuming he shared their views. But if they'd read less selectively, they'd have seen that he was critiquing their position. Reich cautioned strongly against leaping to unsubstantiated conclusions about the nature of these differences, attacking three race science 'pluggers' – James Watson, Henry Harpending and Nicholas Wade – who had made that error. Regarding Wade, Reich wrote in the article that 'there is no genetic evidence to back up any of the racist stereotypes he promotes,' adding that all three provide 'rhetorical cover for hateful ideas and old racist canards'. In his book he went further, specifically attacking the Ashkenazi Jewish intelligence arguments mounted by Harpending, Wade and others and showing that the evidence is against them.[2]

Reich's point is that while we should discuss facts openly, we should also consider that 'whatever we currently believe about the genetic nature of differences among populations is most likely wrong'. He warns against unwarranted speculation and, in his book, offers an example of where differences might be found. Genome analysis shows significantly more genetic variation in West Africa than in Europe, and the implication is there could be more variation in both athletic talent and cognition in sub-Saharan Africa. As a result, Reich says, 'there is expected to be a higher proportion of sub-Saharan Africans with extreme genetically predicted abilities.' He notes that this could include cognitive traits. In other words, it is possible that Africa might have a higher proportion of geniuses, which is a point unlikely to be embraced by the current promoters of race science.

Another who is sometimes misinterpreted is the IQ theorist Jim Flynn, who avoids stating categorically that IQ differences between populations have no genetic component. Some have latched on to this to suggest he is giving support to a race science view.[3] In fact, Flynn is cautious because, like Reich, he won't make definitive statements if he lacks the scientific evidence to back them up. Flynn says the narrowing gap between African-American and white American IQs is 'probably' environmental and he provocatively raises the possibility that this environmental disadvantage, which he details meticulously in his writing, might be even wider than the IQ gap; for example a 12 per cent environmental gap and a 10 per cent IQ gap, which would suggest an African-American innate advantage of 2 per cent.[4] Again, not a point we hear race science followers repeating.

A clear majority of geneticists and other scientists have a slightly different emphasis, noting that the evolutionary factors acting against population differences in cognition are far stronger than those favouring them. This would mean there were no significant population-based differences in average intelligence and therefore that no population group is inherently smarter than the rest. This perspective is shut out by those promoting race science. I have never heard them acknowledging this argument, let alone debating it, even though it is the perspective of most people working in this area. Race science advocates simply pretend it doesn't exist because it is so directly contrary to their prejudices.

We can expect more of the same from race science regarding genetics and IQ: cherry-picking statements such as Reich's to draw conclusions that may be the opposite of those intended; cherry-picking research such as that on Finnish drunken violence genes; over-interpreting research and using it to make unwarranted suppositions that fit racist agendas. Anything from the world of genetics relating to different gene frequencies among different populations will be used in the service of the race science cause, which will find new Web-based outlets and new alt-right audiences and offer them huge leaps of logic to tie it all to supposedly hardwired IQ. This will remain the weapon of choice.

LAST WORDS ON IQ

Intelligence is real but it remains hard to pin down because it is at once an abstract notion and one that becomes concrete when we break it into smaller units. Intelligence in general is opaque but specific forms of intelligence – 'brainy', 'streetwise', 'perceptive', 'intuitive', 'artistic', 'wise' – are easy to grasp.

For much of this book I've attacked the central conceits bound up in the notion of IQ. To summarise:

- IQ reflects a form of abstract verbal and non-verbal reasoning
- The form of reasoning measured by IQ tests has a genetic component but is strongly influenced by environment
- Variation in the scores of different population groups is probably entirely due to environment

- There's no evidence of innate racial differences in intelligence of any kind
- 'Intelligence Quotient' and 'intelligence' are not the same thing (and therefore the abbreviation IQ is inappropriate)
- There are other forms of intelligence with little connection to the IQ form of reasoning
- Intelligence is ultimately not reducible to a single number.

Over the twentieth century, IQ testing did more harm than good. It was used, with devastating effect, to rank individuals and population groups on the false assumption that it reflects biologically hardwired intelligence. These assumptions have long since embedded themselves in the public consciousness; IQ is still widely equated with inherited intelligence, particularly in places where IQ testing is commonplace. When we talk of people being 'high-IQ types' or 'low-IQ types', those words are loaded with sweeping notions of intelligence and also with the sense that those chosen for these epithets were born with it: naturally smart; naturally thick.

The question remains of whether there's anything worthwhile to be salvaged from this form of testing. Clearly, it's not hocus pocus. IQ testing is still very widely used, particularly in North America, precisely because it is useful in gauging aptitude for academic performance and certain careers. If you do well in an IQ test, you are likely to do well in maths, science, computer programming and several other areas. There's not much chance of IQ fading away any time soon. It could, however, be framed differently.

The abbreviation 'IQ' is a misnomer; it is not an 'intelligence quotient'. It's a test of abstract reasoning, which relies on an individual's previous exposure to abstract reasoning. That is all. We dumped terms (such as spastic, retarded, crippled, moron, lunatic, half-caste) when they no longer reflected our perception of the world, and replaced them with more appropriate words. We could do likewise with IQ.

Testing for aptitude is not in itself a bad idea. It can help discover people's potentials and shortcomings in certain areas and might suggest where remedial help is called for. But if we're going to test for aptitude, we might consider dispensing with *g* altogether. It confuses the picture, suggesting it is more than a general mathematical factor between batteries of reasoning tests. Instead, we might move towards testing for multiple intelligences, perhaps along the lines proposed by Howard Gardner (see Chapter 9). This could provide a better sense of children's potential: they might do poorly in abstract reasoning but score highly in emotional, linguistic or artistic intelligence. Or perhaps we could go one step further, and remove the word 'intelligence' altogether, which would have the additional impact of helping to detoxify tests when average scores are married to population groups.

A FINAL WORD ON RACE AND RACISM

It would be a happy delusion to assume a book like this could be more than a slow puncture in the next bubble of racist science. And one thing we can be sure of is that more bubbles will blow our way. We can also be certain that the battle over ideas won't be a fair fight: the claims made by race science invariably get more of a hearing than their antidotes. Yet it is a battle always worth fighting. As soon as we allow the baseless idea that some races are genetically endowed to be smarter than others, the corollary – some are stupider – applies. And if some have evolved greater or lesser intelligence, what else comes with the package? What other qualities do they have or lack?

Dumping the idea that races are real, in a biological sense, is a good start. Although we use the term loosely in various contexts, it is worth remembering that race is little more than a convenient linguistic shorthand. Yes, geneticists can tell from which geographical area or population group people come, using DNA analysis that can detect tell-tale markers. Yes, particular diseases and physical capacities or liabilities are more prevalent in some population groups than others. But population groups are never 'racially pure' and what we consider to be races in fact comprise many population groups. 'Race' isn't a solid scientific category, not least because each of the commonly defined races contains almost the full human genetic range, which means that the differences within a 'race' are more profound than those between 'races'. An alternative idea – deeply rooted in science – embraces a different concept of race, of a human race with ever-looser populations that flow one into the next, sharing their genes ever more quickly as international travel becomes easier and borders more porous. In those eight billion bodies and minds, regardless of nation, region, tribe or ethnicity, can be found the entire range of human capacity, albeit a capacity mediated by wealth and poverty, by family custom and religious belief, by education, by heat and cold, trees, mountains, deserts and buildings, by viruses and microbes, by nutrition and hunger and by everything that falls within the cultural realm. Despite these huge cultural and environmental differences, we can be sure of one thing: in any population we choose to name we'll find, in roughly equal proportions, the full spread of human nature: cruelty and kindness, violence and gentleness, madness and sanity, inventiveness and unoriginality, idiocy and genius.

This idea of the human race is worth fighting for. But we should never assume that we have won the fight and can go home and put our feet up. When I wrote and broadcast about race science five years ago, I was regularly told I was wasting my time because no one believes that stuff any more. I don't hear that now. With the rise of the alt-right, fascists taking to the streets all over Europe, populist, nativist right-wingers winning power in several parts of the world; far-right terrorism on the increase; it is clear that racism, and the ideas that feed it, are more resilient than we hoped. The twentieth century showed us where bad ideas about race can lead. If we don't want the twenty-first to echo those themes, bad ideas need to be countered whenever and wherever they appear. This book has been my effort. I hope you will consider taking up the fight.

ACKNOWLEDGEMENTS

Many people have helped me with this book through giving their time in interviews, discussions on science and points of information – too many to list. I would particularly like to thank my tenacious agent Andrew Lownie, who showed faith in this project from the beginning, and my enthusiastic editor at Oneworld, Alex Christofi, whose ideas for tweaks and changes, additions and subtractions, always made sense.

NOTES

FOREWORD

1 See Saul Dubow, *Scientific Racism in Modern South Africa*, Cambridge University Press, 1995.
2 Nicholas Wade, *A Troublesome Inheritance: Genes, Race and Human History*, Penguin, New York, 2014.
3 Gavin Evans interviewed by Jennifer Sanasie, 'Black Brain, White Brain', author interview, *News24*, 14 May 2015 https://www.youtube.com/watch?v=DLl1dwvgpso

1. WHAT IS SCIENTIFIC RACISM?

1 Richard Lynn, quoted in Marek Kohn, *The Race Gallery*, Jonathan Cape, London, 1995, p. 148.
2 James Watson, quoted in the *Sunday Times*, 14 October 2007.
3 James Watson, 'To question genetic intelligence is not racism', *Independent*, 19 October 2007.
4 James Watson, quoted in David Reich, *Who We Are and How We Got Here: Ancient DNA and the New Science of the Human Past*, Oxford University Press, 2018, p. 263.
5 James Watson, quoted in Amy Harmon, 'James Watson Won't Stop Talking About Race', *New York Times*, 1 January 2019.
6 Richard Lynn, *Race Differences in Intelligence: An Evolutionary Analysis*, Washington Summit Books, Augusta, GA, 2016, p. 143.
7 Richard Dawkins, *The Ancestor's Tale: A Pilgrimage to the Dawn of Evolution*, Mariner Books, 2005, p. 413.
8 'It's 2001, do you know where your genes have been?'. Interview with Ian Tattersall by Courtney Shedd, *Artzar*, 2001.

9   Craig Venter, 'Scientists react to comments by James Watson on race and intelligence', *Science Media Centre*, 18 October 2007.

10 J. Craig Venter, *A Life Decoded: My Genome, My Life*, Penguin, 2008.

11 Cahal Milmo, 'Fury at DNA pioneer's theory: Africans are less intelligent than Westerners', *Independent*, 17 October 2007.

2. ARE WE SMARTER THAN OUR ANCESTORS?

1   Jean-Jacques Hublin, quoted in Carl Zimmer, 'Oldest Fossils of Homo Sapiens Found in Morocco, Altering History of Our Species', *New York Times*, 7 June 2017.

2   Daniel Richter, Rainer Grün, Renaud Joannes-Boyau, Teresa E. Steele, Fethi Amani, Mathieu Rué, Paul Fernandes, Jean-Paul Raynal, Denis Geraads, Abdelouahed Ben-Ncer, Jean-Jacques Hublin and Shannon P. McPherron, 'The age of the hominin fossils from Jebel Irhoud, Morocco, and the origins of the Middle Stone Age', *Nature* 546, 8 June 2017, p. 293.

3   Rick Potts, quoted in Ed Yong, 'A Cultural Leap at the Dawn of Humanity', *The Atlantic*, 15 March 2018.

4   Ibid.; quote from Alison Brooks.

5   Chris Stringer, quoted in Ian Sample, 'Skull of *Homo erectus* throws story of human evolution into disarray', *Guardian*, 17 October 2013.

6   Mark Maslin, quoted in 'Human evolution driven by climate change', *UCL News*, 17 October 2013.

7   Susanne Schultz, ibid.

8   See Milford Wolpoff and Rachel Caspari, *Race and Human Evolution: A Fatal Attraction*, Basic Books, New York, 1998.

9   Milford Wolpoff in an interview, www.grahamhancock.com/interviews/ MilfordHWolpoff.php March 2012.

10 Chris Stringer in an interview with Jim Al-Khalili for *The Life Scientific*, BBC Radio 4, 14 February 2012.

11 Alan Templeton, 'The "Eve" Hypothesis: A Genetic Critique and Reanalysis, Contemporary Issues Forum: A Current Controversy in Human Evolution', *American Anthropologist* 95 (1), March 1993.

12 Eleanor M. L. Scerri, et al., 'Did Our Species Evolve in Subdivided Populations across Africa, and Why Does It Matter?', *Trends in Ecology and Evolution* 33 (8), August 2018, pp. 582–94.

13 David Reich, *Who We Are and How We Got Here: Ancient DNA and the New Science of the Human Past*, Oxford University Press, 2018, p. 17.

14 Scerri, et al., op. cit.

15 Ibid.

16 M. F. Hammer, et al., 'Genetic Evidence for Archaic Admixture in Africa',

Proceedings of the National Academy of Sciences of the USA 108 (2011), pp. 15123–8.

17 J. Lachance, et al., 'Evolutionary History and Adaptation from High-Coverage Whole-Genome Sequences of Diverse African Hunter-Gatherers', *Cell* 250, 2012, pp. 457–69.

18 David Reich, op. cit., p. 209.

19 Ibid., pp. 40–1.

20 S. Sankaaraaraman, et al., 'The Genomic Landscape of Neanderthal Ancestry in Present-Day Humans', *Nature* 507, 2014, pp. 354–7.

21 Emilia Huerta-Sánchez, et al., 'Altitude adaptation in Tibetans caused by introgression of Denisovan-like DNA', *Nature* 512 (7513), 2014, pp. 194–7.

22 Christopher Henshilwood, quoted in John Noble Wilford, 'In African cave, signs of an ancient paint factory', *New York Times*, 11 October 2011.

23 Christopher Henshilwood, et al., 'Engraved ochres from the Middle Stone Age levels at Blombos Cave, South Africa', *Journal of Human Evolution* 57 (2009), pp. 27–47.

24 Ian Tattersall, 'Africa: continent of origins'. Lecture delivered at the Metropolitan Museum of Art for the Symposium 'Genesis: Exploration of Origins', New York, 2004.

25 Christopher Henshilwood, quoted in Ian Sample, 'Earliest known drawing found on rock in South African cave', *Guardian*, 12 September 2018.

26 Francesco d'Errico, et al., 'An Engraved Bone Fragment From c. 70,000-Year-Old Middle Stone Age Levels at Blombos Cave, South Africa: Implications for the Origin of Symbolism and Language', *Antiquity* 75 (2001), pp. 309–18.

27 Christopher Henshilwood, et al., 'Middle Stone Age shell beads from South Africa', *Science*, 304 (2004), p. 404.

28 Steve Connor, 'Stone-Age humans began using lethal technology 71,000 years ago to fight Neanderthals', *Independent*, 7 November 2012.

29 For a detailed account of this African archaeological record, see Sally McBrearty and Alison Brooks, 'The revolution that wasn't: a new interpretation of the origin of modern human behavior', *Journal of Human Evolution* 39 (5), 2000.

30 Ibid., p. 49.

31 Ibid.

32 Yuval Noah Harari, *Sapiens: A Brief History of Humankind*, Vintage, 2014, p. 13.

33 Ibid., p. 22.

34 Ibid., p. 24.

35 Ibid., p. 23.

36 Ibid.

37 Jared Diamond, *Guns, Germs, and Steel: A Short History of Everybody for the Last 13,000 Years*, Vintage, 1998, p. 39.

38 Ibid., p. 40.

39 See J. Doebley, 'Mapping the genes that made maize', *Trends in Genetics* 8, 1992, pp. 302–7; Reich, op. cit., pp. 7–8.

40 Reich, op. cit., p. 16.

41 Ibid., p. 17.

42 G. Cochran and H. Harpending, *The 10,000 Year Explosion: How Civilization Accelerated Human Evolution*, Basic Books, New York, 2009, p. 31.

43 H. Harpending, 'The Biology of Families and the Future of Civilization', (minute 38), Preserving Western Civilization 2009 Conference.

44 Robinson Meyer, 'Ancient Humans Lived in China 2.1 Million Years Ago', *The Atlantic*, 11 July 2018.

45 C. S. Henshilwood, F. d'Errico, R. Yates, Z. Jacobs, C. Tribolo, G. Duller, N. Mercier, J. C. Sealy, H. Valladas, I. Watts and A. Wintle, 'Emergence of modern human behavior: Middle Stone Age engravings from South Africa', *Science*, 295 (2002), pp. 1278–80.

46 McBrearty and Brooks, op. cit., p. 453.

47 Ibid., pp. 453–563.

48 Ian Tattersall, op. cit.

49 Daniel Everett, *Don't Sleep, There Are Snakes: Life and Language in the Amazonian Jungle*, Profile Books, 2008.

50 Barry Hewlett, *Intimate Fathers: The Nature and Content of Aka Pygmy Paternal Infant Care*, University of Michigan Press, Ann Arbor, 1991.

51 Ibid., p. 31.

3. WHY DID HUMANS MIGRATE?

1 Robin McKie, 'Neanderthals kept our early ancestors out of Europe', *Guardian*, 17 October 2015.

2 Yuval Noah Harari, *Sapiens: A Brief History of Human Kind*, Vintage, 2014, p. 23.

3 Carleton Coon, *The Origin of Races*, Jonathan Cape, London, 1962, quoted in Marek Kohn, *The Race Gallery*, Jonathan Cape, London, 1995, p. 64.

4 Ibid., p. 66.

5 John R. Baker, *Race*, Oxford University Press, Oxford, 1974, cited in Kohn, ibid., pp. 60–1.

6 John P. Jackson, 'In ways unacademical: the reception of Carleton S. Coon's "The Origin of Races"', *Journal of the History of Biology* 34 (2), 2001, pp. 247–85.

7 Chris Standish and Alistair Pike, 'It's Official: Neanderthals Created Art', *Sapiens*, 22 May 2018 (originally published in *The Conversation*).

8 Melissa Hogenboom, 'Neanderthals could speak like modern humans, study suggests', *BBC News Science and Environment*, 20 December 2013.

9   Ian Tattersall, *The Monkey in the Mirror*, Oxford University Press, Oxford, 2002, pp. 126–8.

10 Klaus Schmidt, 'Zuerst kam der Tempel, dann die Stadt', Vorlaugiger Bericht zu den Grabungen am Göbekli Tepe und am Gurcutepe 1995–1999', *Istanbuler Mitteilungen* 50, 2000, pp. 5–41.

11 E. de Wit, et al., 'Genome-wide analysis of the structure of the South African Coloured Population in the Western Cape', *Human Genetics* 128 (2), 2010, pp. 145–53.

12 David Reich, *Who We Are and How We Got Here: Ancient DNA and the New Science of the Human Past*, Oxford University Press, 2018, p. xv.

13 Ibid., pp. 109–16.

14 Ibid., pp. 109–12 and 267.

15 Ibid., p. 267.

16 Israel Finkelstein and Neil Asher Silberman, *The Bible Unearthed: Archaeology's New Vision of Ancient Israel and the Origin of Its Sacred Texts*, Touchstone, 2002.

17 Mark Haber, et al., 'Continuity and Admixture in the Last Five Millennia of Levantine History from Ancient Canaanite and Present-Day Lebanese Genome Sequences', *American Journal of Human Genetics* 101 (2), 3 August 2017, pp. 274–82.

18 Marta Costa, et al., 'A substantial prehistoric ancestry amongst Ashkenazi maternal lineages', *Nature Communications* 4 (2543), 8 October 2012 http://www.nature.com/ncomms/2013/131008/ncomms3543/full/ncomms3543.html

19 Ibid.

20 Ibid.

21 See Martin Gershowitz, 'New Study Finds Most Ashkenazi Jews Genetically Linked to Europe', *Jewish Voice*, 16 December 2013.

22 David Reich, op. cit., p. 237.

23 Marija Gimbutas, 'The Prehistory of Eastern Europe, Part 1: Mesolithic, Neolithic and Copper Age Cultures in Russia and the Baltic Area', American School of Prehistoric Research, Harvard University, Bulletin No. 20. See also Reich, ibid, pp. 236–40.

24 T. Zerjal, et al., 'The Genetic Legacy of the Mongols', *American Journal of Human Genetics* 72 (2003), pp. 717–21.

25 K. Bryce, et al., 'The Genetic Ancestry of African Americans, Latinos and European Americans Across the United States', *American Journal of Human Genetics* 96 (2015), pp. 37–53.

26 David Reich, op. cit., pp. 231–4.

27 Neus Font-Porterias, et al., 'The genetic landscape of Mediterranean North African populations through complete mtDNA sequence', *Annals of Human Biology* 45 (1), 2018, pp. 98–104 https://doi.org/10.1080/03014460.2017.1413133

28 Andrea Manica, quoted in Sarah Knapton, 'Africans carry huge amount of European DNA', *Telegraph*, 8 October 2015.

29 Pontus Skoglund, quoted in Ewen Callaway, 'Error Found in Study of First Ancient African Genome', *Nature*, 29 January 2016.

30 See Ed Yong, 'The Ancient Origins of Both Light and Dark Skin', *The Atlantic*, 12 October 2017.

4. IS AFRICA REALLY 'BACKWARD'?

1 Captain G. A. Gardner, *Mapungubwe. Volume II. Report on Excavations at Mapungubwe and Bombandyanalo in Northern Transvaal from 1935 to 1940*, University of Pretoria, 1963, p. 26.

2 Quoted in Julie Frederikse, *None But Ourselves*, Anvil Press, London, 1982, pp. 10–11.

3 J. Theodore Bent, quoted in G. Casely-Hayford, *The Lost Kingdoms of Africa*, Bantam Press, London, 2012.

4 Shadreck Chirikure and Innocent Pikirayi, 'Inside and outside the drystone walls: revisiting the material culture of Great Zimbabwe', *Antiquity* 82 (2008), pp. 976–93.

5 Henrika Kuklick, 'Contested monuments: the politics of archaeology in southern Africa', in George W. Stocking (ed.) *Colonial Situations: Essays on the Contextualization of Ethnographic Knowledge*, University of Wisconsin Press, 1991, pp. 135–70.

6 Quoted in L. Casson, *The Periplus Maris Erythraei: Text with Introduction, Translation, and Commentary*, Princeton University Press, 1989, p. 61.

7 G. Casely-Hayford, *The Lost Kingdoms of Africa*, Bantam Press, London, 2012, pp. 137–40.

8 See Gavin Evans, *Black Brain, White Brain: Is Intelligence Skin Deep?* Jonathan Ball, 2014, pp. 31–3.

9 Hume wrote 'negroes' and all non-whites were 'naturally inferior' to whites and had never produced a civilised nation. Cited in R. H. Popkin, 'The philosophical basis of modern racism', in C. Walton and J. P. Anton (eds) *Philosophy and the Civilizing Arts*, University of Ohio Press, Athens, 1974, p. 143.

10 Toynbee wrote that only the black race 'has not made a creative contribution to any one of our twenty-one civilisations ...'. Arnold Toynbee, *A Study of History* (1934 edition) cited in I. A. Newby, *Challenge to the Court: Social Scientists and the Defense of Segregation, 1954–1966*, Louisiana State University Press, Baton Rouge, 1969, p. 217.

11 This concerns his changing of dates and other details in his bid to use 'solar phallus' man to prove the universality of his archetypes. See Richard Noll,

The Jung Cult: Origins of a Charismatic Movement, Princeton University Press, 1995, pp. 181–7.

12   See Gavin Evans, *Mapreaders & Multitaskers: Men, Women, Nature, Nurture*, Thistle, 2017, pp. 82–4.

13   See F. Dalal, 'The racism of Jung', *Race and Class* 29 (3), 1988. See also Michael Ortiz Hill, 'C. G. Jung – In the Heart of Darkness', www.gatheringin.com/moh_jung.html

14   Carl Jung, *Memories, Dreams, Reflections*, A. Jaffe (ed.), Collins, New York, 1973, p. 272.

15   Carl Jung, *Collected Works 10: Civilisation in Transition*, Princeton University Press, 1970, p. 121.

16   Ibid., p. 507.

17   Jordan Peterson, 'Jordan Peterson on Studies on Race and IQ', *YouTube*, 16 April 2018, https://www.youtube.com/watch?v=qT_YS

18   John R. Baker, *Race*, Oxford University Press, Oxford, 1974.

19   Richard Lynn, quoted in Marek Kohn, *The Race Gallery*, Jonathan Cape, London, 1995, p. 152.

20   Richard Lynn, *Race Differences in Intelligence: An Evolutionary Analysis*, Washington Summit Books, Augusta, GA, 2006, p. 146. See also www.velesova-sloboda.org/archiv/pdf/lynn-race-differences-in-intelligence.pdf

21   Quoted in the *Independent*, 10 August 2013.

22   See for example Yeye Akilmali Funua Olade, *When Black People Ruled the World*, The Clegg Series, melanet.com and http://yeyeolade.wordpress.com/2007/04/11/when-blacks-ruled-the-world/

23   See for example, Kohn, op. cit., pp. 152–65.

24   Leonard Jeffries, quoted in Massimo Calabresi, 'Dispatches skin deep 101', *Time Magazine* 143 (7), 14 February 1994.

25   Leonard Jeffries, interviewed by T. L. Stanclu and Nisha Mohammed, *Rutherford Magazine*, May 1995.

26   Genesis 9:25.

27   Joshua 9:23.

28   A. H. Keane, *Man: Past and Present*, Cambridge University Press, Cambridge, 1899, p. 83, cited in Saul Dubow, *Scientific Racism in Modern South Africa*, Cambridge University Press, Cambridge, 1995, pp. 84–5.

29   H. H. Johnston, *The Opening Up of Africa*, Henry Holt and Company, New York, 1911, pp. 134–5, quoted in Dubow, ibid., p. 82.

30   Charles Seligman, *The Races of Africa*, Thornton Butterworth, London, 1930, p. 96.

31   G. S. Preller, *Day-dawn in South Africa*, Wallachs P&P Co., Pretoria, 1938, pp. 43–4, quoted in Dubow, op. cit., p. 74.

32   Preller, ibid., p. 149, quoted in Dubow, ibid., p. 269.

33   Both quoted in *Reader's Digest Illustrated History of South Africa*, D. Oakes

(ed.), Reader's Digest Association of South Africa, Cape Town, 1988, p. 207.

34 See G. Casely-Hayford, op. cit., pp. 15–45.

35 S. Munro-Hay, *Aksum: An African Civilization of Late Antiquity*, Edinburgh University Press, 1991, 1991, p. 57.

36 He died in 1337.

37 Olivia Fleming, 'Meet the 14th century African king who was richest man in the world of all time (adjusted for inflation!)', *Daily Mail*, 15 October 2012.

38 G. Casely-Hayford, op. cit., pp. 231–2.

39 John R. Baker, in G. Casely-Hayford, ibid.

40 Ibid., p. 117.

41 Quoted in E. R. Sanders, 'The Hamitic hypothesis: its origin and functions in time perspective', *Journal of African History* 10 (4), 1969, p. 528.

5. WHERE DID 'SCIENTIFIC' RACISM COME FROM?

1 George Eliot, *Middlemarch*, Penguin, London, 1871, p. 226.

2 Vasily Grossman, *Life and Fate*, Vintage, London, 2006, p. 468.

3 Charles Darwin, *The Voyage of the Beagle*, Cosimo, New York, 2008, p. 503.

4 Charles Darwin, *The Descent of Man, and Selection in Relation to Sex*, 1st edition, John Murray, London, 1874, p. 201.

5 Charles Darwin, *The Descent of Man, and Selection in Relation to Sex*, 2nd edition, John Murray, London, 1874, pp. 563–4.

6 Charles Darwin, *The Voyage of the Beagle*, op. cit., p. 148.

7 Erasmus Darwin, *Zoonomia; or the Laws of Organic Life* (4th American edition), Edward Earle, p. 397.

8 Wallace wrote: '[T]he position of woman ... will be far higher and more important than any which has been claimed for or by her in the past. While she will be conceded full political and social rights on an equality with man, she will be placed in a position of responsibility and power which will render her his superior, since the future moral progress of the race will so largely depend upon her free choice in marriage.' (Alfred Russel Wallace, *Social Environment and Moral Progress*, Cassell, London, 1913, pp. 147–8.)

9 Alfred Russel Wallace, *The Malay Archipelago*, Dover Publications, New York, 1962, pp. 456–7.

10 Thomas Jefferson, *Notes on the State of Virginia*, 1794, query 14.

11 Abraham Lincoln in *The Lincoln-Douglas Debates*, 4th Debate, Part 1, 18 September 1858.

12 Steve Jones, 'Cousins under the skin', Reith Lecture: 'The language of the genes', BBC Radio 4, 4 December 1991.

13 S. G. Morton, *Crania Americana*, John Pennington, Philadelphia, 1839, p. 90.

14 Stephen Jay Gould, *The Mismeasure of Man*, Penguin, 1996.

15 Ibid., p. 86.

16 J. Lewis, et al., 'The Mismeasure of Science', *PLoS Biology* 9 (6), 2011.

17 'Mismeasure for mismeasure', editorial in *Nature* 474, 23 June 2011, p. 419.

18 Nicholas Wade, 'Scientists Measure the Accuracy of Racism Claim', *New York Times*, 11 June 2011.

19 Sam Harris in 'The Sam Harris debate: Ezra and Sam Harris debate race, IQ, identity politics, and much more', *Vox*, 9 April 2018 (from phone interview with Klein, 2 April 2018) https://www.vox.com/2018/4/9/17210248/sam-harris-ezra-klein-charles-murray-transcript-podcast

20 Roy Unz, 'Race, IQ, and Wealth', *Unz Review*, 18 July 2012.

21 *Nature*, op. cit.

22 S. J. Gould, op. cit., p. 86.

23 Ibid., p. 101.

24 Paul Broca, 'Sur les cranes de la caverne de l'Homme-Mort (Lozère)', *Revue d'Anthropologie* 2, 1873, p. 38, quoted in Gould, op. cit., p. 119.

25 Paul Broca, 'Sur les projections de la tête et sur un nouveau procédé de céphalométrie et d'anthropométrie', *Bulletins de la Société d'Anthropologie de Paris*, 1862, p. 16, quoted in Gould, ibid., p. 133.

26 Paul Broca, 'Anthropologie', in *Dictionnaire Encyclopédique des Sciences Médicales*, A. Dechambre (ed.), Masson, Paris, 1866, p. 280, quoted in Gould, ibid., pp. 115–16.

27 S. J. Gould, op. cit., pp. 119–33.

28 Eighty years after Bean a view emerged that women had a larger corpus callosum than men, meaning there was greater connectivity between the hemispheres of the brain: men attack problems using just one side of their brain whereas women use both, can multitask and are more 'holistic'. The claim, first made in 1982, was based on a tiny sample of just five female brains. A 2009 review of all the scientific literature in this area (more than fifty studies), by the neurobiologist Mikkel Wallentin, concluded that 'the alleged sex-related corpus callosum size-difference is a myth.'

29 Robert Bennett Bean, 'Some racial peculiarities of the Negro brain', *American Journal of Anatomy* 5, 1906, p. 380.

30 Editorial, *American Medicine*, April 1907, quoted in Gould, op. cit., p. 112.

31 Richard Lynn, *Race Differences in Intelligence: An Evolutionary Analysis*, Washington Summit Publishers, Augusta, GA, 2006.

32 Thomas Huxley, quoted by Gould, op. cit., p. 105.

33 Ernst Haeckel, *The Pedigree of Man and Other Essays*, trans. Edward Aveling, Freethought, London, 1883, p. 85.

34 Herbert Spencer, quoted by Gould, op. cit., p. 146.

35 Francis Galton, *Hereditary Genius: An Inquiry into Its Laws and Consequences*, Macmillan, London, 1892, p. 330.

36   Alfred Russel Wallace, 'Human Selection', *Fortnightly Review* 48, September 1890.

37   Winston Churchill, quoted in Gustavus A. Ohlinger, 'WSC: A Midnight Interview, 1902', *Michigan Quarterly Review*, February 1966.

38   Winston Churchill, *War Memorandum of May 12 1919*.

39   Winston Churchill, quoted in Ishaan Tharoor, 'The dark side of Winston Churchill's legacy no one should forget', *Washington Post*, 3 February 2015.

40   Winston Churchill, quoted in Johann Hari, 'Not his finest hour: the dark side of Winston Churchill', *Independent*, 28 October 2010.

41   Winston Churchill, quoted in Ishaan Tharoor, op. cit.

42   Both quotes in Joseph Lelyveld, *Great Soul: Mahatma Gandhi and his Struggle with India*, Alfred A. Knopf, New York, 2011, p. 57.

43   Ibid., p. 54.

44   Ibid., p. 74.

45   Ronald Fisher, quoted in M. Kohn, op. cit., p. 46.

46   Kohn, ibid.

47   Konrad Lorenz, *On Aggression*, Methuen, London, 1966, p. 5.

48   With Niko Timbergen and Karl von Frisch.

49   Robert Ardrey, *The Territorial Imperative: a Personal Inquiry into the Animal Origins of Property and Nations*, Collins, London, 1967, p. 183.

50   Desmond Morris, *The Naked Ape*, Vintage, London, 2005, p. 163.

51   Ibid., p. 5.

52   Ibid., p. 99.

53   Ibid., p. 119.

54   Ibid., pp. 5–6.

55   Ibid., p. 28.

56   Ibid., p. 119.

57   E. O. Wilson, *Sociobiology: The New Synthesis*, Harvard University Press, Cambridge, MA, 1975.

58   *On Human Nature*, Harvard University Press, 1979; *Genes, Mind and Culture: The Co-evolutionary Process*, Harvard University Press, 1981; and *Promethean Fire: Reflections on the Origin of Mind*, Harvard University Press, 1983.

59   E. O. Wilson, quoted in P. T. Staff, 'E. O. Wilson is on Top of the World', *Psychology Today*, 1 September 1998.

60   E. O. Wilson, *Consilience: The Unity of Knowledge*, Alfred Knopf, New York, 1998, p. 225.

61   Ibid., pp. 127–8.

62   Elizabeth Allen, Stephen Jay Gould, Richard Lewontin, et al., 'Against "Sociobiology"', [letter], *New York Review of Books* 22 (182), 13 November 1975, p. 182.

63   E. O. Wilson, quoted in Ed Douglas, 'Darwin's natural heir', *Guardian*, 17 February 2001.

64 E. O. Wilson, quoted in P. Knudson, *A Mirror to Nature: Reflections on Science, Scientists and Society*, Stoddart Publishing, Toronto, 1991, p. 190.

65 E. O. Wilson, back cover quote for Nicholas Wade's *A Troublesome Inheritance*, Penguin, New York, 2014.

66 Dawkins called Islam the 'greatest force for evil in the world today' and tweeted: 'All the world's Muslims have fewer Nobel Prizes than Trinity College, Cambridge. They did great things in the Middle Ages, though.' He criticised the *New Statesman* for publishing stories by the journalist Mehdi Hassan because of his Islamic beliefs. These comments drew furious responses on Twitter, with Dawkins accused of intolerance, bigotry, xenophobia and racism; charges he refuted, tweeting: 'I don't think Muslims should segregate sexes at University College London events. Oh NO, how very ISLAMOPHOBIC of me. How RACIST of me.' (All by @RichardDawkins, 2013.) He was back at it in 2018, tweeting a picture of himself outside Winchester Cathedral, with the words: 'Listening to the lovely bells of Winchester, one of our great mediaeval cathedrals. So much nicer than the aggressive sounding "Allahu Akhbar". Or is it just my cultural upbringing?' (tweet by @RichardDawkins, 2018). Renewed accusations of Islamophobia and racism followed.

67 Richard Dawkins, interviewed for the BBC's *Belief* programme in April 2004, republished 22 October 2009, www.bbc.co.uk/religion/religions/atheism/people/dawkins.shtml

68 Stephen S. Hall, 'Darwin's Rottweiler', *Discover* magazine, 9 September 2005.

69 Ibid.

70 Richard Dawkins interviewed on *Midweek*, BBC Radio 4, 11 December 2013.

71 Ibid.

72 Richard Dawkins, on the BBC's *Belief* programme, op. cit.

73 Richard Dawkins, *The Selfish Gene*, Oxford University Press, London, 1976.

74 E. O. Wilson, interviewed on *Newsnight*, BBC, 6 November 2014.

75 In a review of Richard Lynn's pro-eugenics book *Dysgenics*, Hamilton praises Lynn for showing that 'almost all of the worries of the early eugenicists were well-founded' (W. D. Hamilton, A review of 'Dysgenics: Genetic Deterioration in Modern Populations', *Annals of Human Genetics* 64 (4), 2000, pp. 363–74.

76 Richard Dawkins, *The Selfish Gene*, op. cit., pp. 2–3.

77 Ibid., p. 252.

78 Mary Midgley, 'Why Memes?' in *Alas, Poor Darwin*, H. Rose and S. Rose (eds), Vintage, London, p. 68.

79 Richard Dawkins, 'It's all in the genes', *Sunday Times*, 12 March 2006.

80 E. O. Wilson, *Newsnight*, BBC, op. cit.

81 Paul Ehrlich, quoted in 'Do genes dictate behaviour?', *USA Today* (Society for the Advancement of Education), April 2001.

82 Motoo Kimura, *The Neutral Theory of Molecular Evolution*, Cambridge University Press, Cambridge, 1983, p. i.

83 See Michael Shermer, 'This view of science', *Social Studies of Science* 32 (4), August 2002, p. 496.

84 Stephen Jay Gould, 'More things in heaven and earth'. In, H. Rose and S. Rose, op. cit., p. 90.

85 Ibid.

86 A spandrel is an area of masonry between arches supporting a dome, which arose as a by-product rather than being planned by the architect. Stephen Jay Gould and Richard Lewontin, 'The spandrels of San Marco and the Panglossian paradigm: a critique of the adaptationist programme', *Proceedings of the Royal Society of London*, Series B, 205 (1161), 1979.

87 S. J. Gould in H. Rose and S. Rose (eds), op. cit., p. 7.

88 S. J. Gould, 'More Things in Heaven and Earth'. In, H. Rose and S. Rose, ibid., p. 104.

89 Geoff Bunn, *A History of the Brain*, episode 10, BBC Radio 4, 18 November 2011.

90 Gustav Preller, *Day-dawn in South Africa*, Wallachs P&P Co., Pretoria, 1938, pp. 149–51, quoted in Saul Dubow, *Scientific Racism in Modern South Africa*, Cambridge University Press, Cambridge, 1995, p. 269.

91 Richard Dawkins, *The Selfish Gene*, Oxford University Press, London, 1989, p. 100.

92 Richard Dawkins, 'Race and creation', *Prospect*, 23 October 2004.

93 Ibid.

94 D. J. Kelly, et al., 'Cross-race preferences for same-race faces extend beyond the African versus Caucasian contrast in three-month-old infants', *Infancy* 11 (1), 2007, pp. 87–95.

95 P. C. Quinn, et al., 'Representation of the gender of human faces by infants: a preference for female', *Perception* 31 (9), 2002, pp. 1109–21.

96 L. Castelli, et al., 'Learning social attitudes: children's sensitivity to nonverbal behaviours of adult models during interracial interactions', *Personality and Social Psychology Bulletin* 34 (11), 2008, pp. 1512.

97 L. Castelli, et al., 'The transmission of racial attitudes within the family', *Developmental Psychology* 45 (2), 2009, pp. 586–91.

6. ARE RACE GROUPS REAL?

1 Her story was made into Anthony Fabian's award-winning feature film *Skin* (2008), starring Sophie Okonedo as Sandra, and Sam Neill and Alice Krige as her parents.

2 Steve Jones, *The Language of the Genes*, Flamingo, London, 2000, introduction to Chapter 7.

3 Richard Dawkins, *The Ancestor's Tale: A Pilgrimage to the Dawn of Evolution*, Mariner Books, Boston, 2005, p. 413.

4 Mark Stoneking, quoted in Steve Connor, 'The great taboo of genetics and race – a most unnatural selection of the facts', *Independent*, 13 August 2014.

5 Paul R. Ehrlich, *Human Natures: Genes, Cultures, and the Human Prospect*, Penguin, 2002, p. 49.

6 Ibid., pp. 49–50.

7 Sandra Wilde, quoted in 'Natural selection has altered the appearance of Europeans over the past 5,000 years', *UCL News*, 11 March 2014.

8 Ed Yong, op. cit.

9 Ibid.

10 See I. Mathieson, et al., 'Genome-Wide Patterns of Selection in 230 Ancient Eurasians', *Nature* 528, 2015, pp. 499–503.

11 Barbary Corsairs are estimated to have seized and enslaved more than one million Europeans between 1500 and 1800. The men were used for labour and the women both for labour and as concubines throughout the Ottoman Empire. (Peter Holland, *Barbary Slave*, Brigand, London, p. 11.)

12 See David Reich, op. cit., pp. 140–51.

13 See Gavin Evans, *The Story of Colour: An Exploration of the Hidden Messages of the Spectrum*, Michael O'Mara Books, 2017, pp. 178–80.

14 Chimamanda Ngozi Adichie, *Americanah*, HarperCollins, 2017, p. 214.

15 Ibid., pp. 296–7.

16 Sarah Tishkoff, quoted in Ed Yong, op. cit.

17 See Marek Kohn, *The Race Gallery*, Jonathan Cape, London, 1995, p. 82.

18 R. C. Lewontin, 'The apportionment of Human Diversity', *Evolutionary Biology* 6, 1972, pp. 381–98.

19 See David Reich, *Who We Are and How We Got Here: Ancient DNA and the New Science of the Human Past*, Oxford University Press, 2018, pp. 249–50.

20 N. A. Rosenberg, et al., 'Genetic Structure of Human Populations', *Science* 298, 2002, pp. 2381–5.

21 Melissa Ilardo, et al., 'Physiological and Genetic Adaptations to Diving in Sea Nomads', *Cell* 173 (3), 19 April 2018, pp. 569–80.

22 For a fuller explanation of genetic drift see Robert G. Bednarik, 'Genetic Drift in Recent Human Evolution?', in Kevin V. Urgano (ed.), *Advances in Genetics Research* 6, Nova Science Publishers, 2011, pp. 109–60.

23 D. Wagener, et al., 'Ethnic variation of genetic disease: roles of drift for recessive lethal genes', *American Journal of Human Genetics* 30 (3), May 1978, pp. 262–70.

24 L. Peltonen, 'Founder Effect', in *Encyclopaedia of Genetics*, 2001, pp. 724–6.

25 M. Myles-Worsley, et al., 'Genetic epidemiological study of schizophrenia in Palau, Micronesia: prevalence and familiarity', *American Journal of Medical Genetics* 88 (1), 5 February 1999, pp. 4–10.

26 David Reich, op. cit., pp. 248–50.

27 David Reich, 'How Genetics is Changing Our Understanding of "Race"', *New York Times Sunday Review*, 23 March 2018.

28 A. Langer-Gould, et al., 'Incidence of multiple sclerosis in multiple racial and ethnic groups', *Neurology* 19, 7 May 2013.

29 D. A. Lane and G. Y. H. Lip, 'Ethnic differences in hypertension and blood pressure control in the UK', *QJM* 94, 1 July 2001, pp. 391–6.

7. CAN WHITE MEN JUMP AND BLACK MEN SWIM?

1 The athletes, in order, were: Dave Bedford, Brendan Foster, Dave Moorcroft, Steve Jones, Steve Ovett, Sebastian Coe, Steve Cram, Peter Elliott, Liz McColgan, Paula Radcliffe and Kelly Holmes.

2 Jerry Odlin, interviewed by Gavin Evans, 19 August 2018.

3 Tim Noakes, *Challenging Beliefs*, Zebra Press, 2012, p. 365.

4 John H. Manners, 'Kenya's running tribe', essay quoted in Lori Shontz, 'Fast forward: the rise of Kenya's women runners', *International Reporting Project*, 5 June 2009.

5 Noakes, op. cit., p. 365.

6 Ibid., p. 356.

7 Ibid., p. 333.

8 Ibid., p. 315.

9 Ibid., p. 366.

10 Dr Yannis Pitsiladis, quoted in Matthew Syed, op. cit., p. 253.

11 Bengt Saltin, unpublished paper, cited by Marek Kohn, *The Race Gallery*, Jonathan Cape, London, 1995, pp. 80–1. See also, 'Genetics in sport: what research tells us about African runners: are they really genetically more gifted?' *Peak Performance*, www.pponline.co.uk/encyc/0056b.htm

12 Robert Scott, quoted in Tim Harris, 'Black men CAN swim', *Prospect*, 21 July 2010.

13 Bengt Saltin, quoted in Kate Kelland, Reuters, 'Does nature or nurture make a top sprinter?', *Chicago Tribune*, 31 July 2012.

14 Michael Johnson and Usain Bolt, quoted in Kate Kelland, Reuters, 'Does nature or nurture make a top sprinter?', *Chicago Tribune*, 31 July 2012.

15 Robert Scott, quoted in Tim Harris, op. cit.

16 Daniel MacArthur, 'Nature or nurture to make a sprinter?', *IOL Sport*, 31 July 2012.

17 Ken van Someren, quoted in Kate Kelland, op. cit.

18 David Reich, *Who We Are and How We Got Here*, Oxford University Press, 2018, p. 264.

19 Ibid., p. 265.

20 K. Anders Ericsson, et al., 'The Role of Deliberate Practice in the Acquisition of Expert Performance', *Psychological Review* 100 (3), 1993, p. 400.

21 Malcolm Gladwell, *Outliers: The Story of Success*, Penguin, 2009.

22 K. Anders Ericsson, 'The Danger of Delegating Education to Journalists: Why APS Observers Needs Peer Review When Summarizing New Scientific Developments', self-published paper, 28 October 2012, p. 3.

23 Malcolm Gladwell, quoted in Ben Carter, 'Can 10,000 hours of practice make you an expert?', *BBC News*, 1 March 2014.

24 Mark Hannen, quoted in Gavin Evans, 'White men can't', *Medicine Cabinet, Guardian*, 19 September 1997.

25 Ibid. Alex Anzelmo quote.

26 For example, African-American high school graduation rates rose from 67% in 2010/11 to 71% two years later. The percentage of black students among all students attending college rose from 11.7% to 13.7% between 2000 and 2016. This represented 58.2% of black high school graduates and 43% of all black people of high school graduating age – up from 5% in 1963. The rate of increase is faster than among white students. (Bureau of Labor Statistics, '69.7 percent of 2016 high school graduates enrolled in college in October 2016', United States Department of Labor, 22 May 2017.)

27 Nearly 5% of black American men are in jail, compared with 0.7% of white men, according to figures from the US Bureau of Justice.

28 H. Richard Weiner, quoted in Adam Hadhazy, 'What makes Michael Phelps so good?', *Scientific American*, 18 August 2008.

29 The first was a black Dutch woman, Enith Brigitha, who won two bronze medals in the 1976 Olympics, in the 200m freestyle and 100m freestyle (behind two East Germans who, it later emerged, had taken performance-enhancing steroids). First to win Olympic gold was Anthony Nesty of Suriname (born in Trinidad and Tobago), who won the 100m butterfly at the 1988 Seoul Olympics, and took bronze in Barcelona in 1992. In 2000, Anthony Ervin became the first American of African heritage to make a US Olympic team, returning from Sydney with one gold (50m freestyle) and one silver. He later won two World Championship golds. Also, in 2000, Edvaldo Valerio Silva, a black Brazilian, was part of the bronze-medal-winning 4 x 100 m relay team. The first black woman to win an Olympic medal for America was Maritza Correia, a Puerto Rican of Guyanan heritage, who won relay silver in Athens in 2004. She also broke two US freestyle records. The most impressive was Cullen Jones, who briefly held the US record for 50m freestyle and the world record for being part of the US 4 x 100m freestyle relay. He won two Olympic golds and two silvers in 2004 and 2008 and was one of three African Americans to represent the USA in swimming in London in 2012. The others were Ervin and the seventeen-year-old Lia Neal, who won a bronze medal and silver in Rio four years later. In 2016, Simone Manuel

became the first African American to win Olympic swimming gold; her 100m time set a new Olympic record. She won a second gold in the 4 x 100m medley and two silvers in the 50m freestyle and 4 x 100m freestyle relay.

30 US Center for Disease Control data, cited in Finlo Rohrer, 'Why don't black Americans swim?', *BBC News US & Canada*, 3 September 2010.

31 See Jeff Wiltse, *Contested Waters: A Social History of Swimming Pools in America*, University of North Carolina Press, Raleigh, 2010.

32 Finlo Rohrer, op. cit.

8. ARE THERE RACE-BASED INTELLIGENCE GENES?

1 Andrew Sullivan, 'Science is rescuing us from our moral mazes', *News Review*, *Sunday Times*, 25 November 2007.

2 Bruce Lahn, 'Could interbreeding between humans and Neanderthals have led to an enhanced human brain?', *Howard Hughes Medical Institute News*, 6 November 2006.

3 Michael Balter, 'Evolution: are human brains still evolving? Brain genes show signs of selection', *Science*, 9 September 2005.

4 Chris Brand, 'Race realism takes a step forward', *American Renaissance* 16 (12), December 2005, www.amren.com/ar/2005/12/

5 http://www.chimpout.com/forum/showthread.php?114076-DR-BRUCE-LAHN-U-of-Chicago-Geneticist-proves-NIGGERS-not-fully-HUMAN!

6 John Derbyshire, 'The specter of difference: what science is uncovering, we will have to come to grips with', *National Review*, 7 November 2005.

7 Bruce Lahn, quoted in Antonio Regalado, 'Scientist's study of brain genes sparks a backlash', *Wall Street Journal*, 16 June 2006.

8 Bruce Lahn and Lanny Ebenstein, 'Let's celebrate human genetic diversity', *Nature* 461, 8 October 2009, p. 726.

9 Bruce Lahn, quoted in Catherine Gianaro, 'Lahn's analysis of genes indicates human brain continues to evolve', *University of Chicago Chronicle* 25 (1), 22 September 2005.

10 Bruce Lahn, quoted in Sarah Richardson, op. cit., p. 428.

11 Bruce Lahn, quoted in *Howard Hughes Medical Institute News*, op. cit.

12 Bruce Lahn, quoted in Antonio Regalado, op. cit.

13 Sarah Richardson, op. cit., p. 432.

14 Ibid., p. 429.

15 Ibid., p. 428.

16 Bruce Lahn, quoted in Antonio Regalado, op. cit.

17 Steven Pinker, 'Genetic tests said I would be intelligent, swayed by novelty and bald. Two out of three ain't bad', *Sunday Times*, 1 February 2009.

18 Daniel Kosman, quoted in Val Dusek, 'Sociobeology sanitized: evolutionary psychology and gene selectionism', *Science as Culture* 8 (2), 1999, pp. 129–69.

19 Craig Venter, quoted in 'GE fantasy shattered by human genome project', 13 February 2001, www.btinternet.com/~nlpwessex/Documents/GEfantasy. htm

20 Craig Venter, quoted in 'Nature or nurture', *BBC News*, 11 February 2001.

21 Paul Ehrlich, quoted in 'Do genes dictate behaviour?', *USA Today* (Society for the Advancement of Education), April 2001.

22 Ibid.

23 The phenotype is the traits – the outward or observed manifestations – expressed by an individual organism as a result of its genotype (its genetic make-up).

24 C. A. Cooney, et al., 'Maternal methyl supplements in mice affect epigenetic variation and DNA methylation of offspring', *Journal of Nutrition* 132 (8 Suppl) 2393S–2400S, 2002.

25 Cited in 'Grandma's curse: some of the effects of smoking may be passed from grandmother to grandchild', *The Economist*, 3 November 2012.

26 Emma Young, 'Rewriting Darwin: the new non-genetic inheritance', *New Scientist*, 8 July 2008.

27 Brian G. Dias and K. J. Ressler, 'Parental olfactory experience influences behaviour and neural structure in subsequent generations', *Nature Neuroscience* 17, pp. 89–96.

28 K. J. Ressler, quoted in Steve Connor, 'Fear can be inherited via father's sperm, says study', *Independent*, 2 December 2013.

29 See for example Ti-Fei F. Yuan, et al., 'Transgenerational Inheritance of Paternal Neurobehavioral Phenotypes: Stress, Addiction, Ageing and Metabolism', *Molecular Neurobiology* 53 (9), 16 November 2015, pp. 6367–76.

30 Study referred to in Robert Winston, 'The science delusion', Greatest Minds Lecture, University of Dundee, 2009.

31 B. T. Heijmans, et al., 'Persistent epigenetic differences associated with prenatal exposure to famine in humans', *Proceedings of the National Academy of Sciences of the United States of America* 105 (44), 2008, pp. 17046–9.

32 David Dobbs, 'The social life of genes', *Pacific Standard*, 3 September 2013.

33 Ibid., Steve Cole quoted.

34 'Mainstream Science on Intelligence', public statement, *Wall Street Journal*, 13 December 1994.

35 Robert Plomin, quoted in Jonathan Leake, 'Check ... science closes in on intelligence gene test', *Sunday Times*, 19 September 2010.

36 Robert Plomin and Oliver S. P. Davis, 'The future of genetics in psychology and psychiatry: microarrays, genome-wide association and non-coding RNA', *Journal of Child Psychology and Psychiatry* 50 (1–2), 2009, pp. 63–71.

37 Ibid., Ian Deary, quoted.

38 W. D. Hill, et al., 'A combined analysis of genetically correlated traits identifies 187 loci and a role for neurogenesis and myelination in intelligence', *Molecular Psychiatry* 24, 11 January 2018, pp. 169–81.

39 David Hill, quoted in 'Global study identifies hundreds of genes linked to intelligence', University of Edinburgh press release, 13 March 2018.

40 Op. cit, Ian Deary quoted.

41 Jeanne E. Savage, et al., 'Genome-wide association meta-analysis in 269867 individuals identifies new genetic and functional links to intelligence', *Nature Genetics* 50, 25 June 2018, p. 912019.

42 Dena Dubal, et al., 'Life Extension Factor Klotho Enhances Cognition', *Cell Reports* 7, 22 May 2014, p. 12.

43 Jonathan Leake, 'IQ-boosting gene offers hope to dementia victims', *Sunday Times*, 11 May 2014.

44 Ibid.

45 'The 3 per cent solution', *The Economist*, 10 May 2014.

46 Ibid.

47 B. Morar, et al., 'The longevity gene Klotho is differentially associated with cognition in subtypes of schizophrenia', *PubMed* 193, March 2018, pp. 348–53.

48 Javad Tavakkoly-Bazzaz, et al., 'Absence of kl-bs variant of klotho gene in Iranian cardiac patients (comparison to the world populations)', *Disease Markers* 31, 2011, p. 211.

49 See Richard Lynn and Tatu Vanhanen, *IQ and the wealth of nations*, Praeger, Westport, CT, 2002.

50 Javad Tavakkoly-Bazzaz, et al., op. cit., p. 213.

51 David Reich, *Who We Are and How We Got Here: Ancient DNA and the New Science of the Human Past*, Oxford University Press, 2018, pp. 260–1.

52 Ibid., p. 264.

53 Ibid.

54 Ibid., p. 256.

55 Ibid., pp. 259–60.

56 Ibid., p. 267.

57 Ibid., p. 265.

58 Kevin Mitchell, 'Why genetic IQ differences between "races" are unlikely', *Guardian*, 2 May 2018.

59 Kevin Mitchell, *Innate – How the Wiring of Our Brains Shapes Who We Are*, Princeton University Press, 2018, p. 263.

60 Ibid.

61 Kevin Mitchell, *Guardian*, op. cit.

62 Ibid.

63 Kevin Mitchell, *Innate*, op. cit., p. 263.

64 Ibid.

65 Ibid., p. 261.

66 Ibid.

67 James Flynn, quoted in 'Women overtake men in IQ stakes for the first time since tests began', *Daily Record*, 6 July 2012.

68 James Flynn, quoted by Ezra Klein, 'The Sam Harris debate: Ezra and Sam Harris debate race, IQ, identity politics, and much more', *Vox*, 9 April 2018 (from phone interview with Klein, 2 April 2018). https://www.vox.com/2018/4/9/17210248/sam-harris-ezra-klein-charles-murray-transcript-podcast

69 Richard Lynn, 'Race and Psychopathic Personality: Racial differences in "average personality"'. *American Renaissance* 13 (7), 20 July 2002.

70 J. Christoff Erasmus, et al., 'Allele frequencies of AVPR1A and MAOA in the Afrikaner population', *South African Journal of Science* 111(7), July/August 2015.

71 D. M. Fergusson, et al., 'Moderating role of the MAO-A genotype in antisocial behaviour', *Br J Psychiatry* 200 (2), February 2012, pp. 116–23.

72 E. Shumay, et al., 'Evidence that the methylation state of the monoamine oxidase A (MAOA) gene predicts brain activity of MAOA enzyme in healthy men', *Epigenetics* 7 (10), October 2012, pp. 1151–60.

73 R. A. Philibert, et al., 'MAOA methylation is associated with nicotine and alcohol dependence in women', *American Journal of Medical Genetics*, Part B, 147B (5), July 2008, pp. 565–70.

74 Cited in David Dodds, 'The social life of genes', *Pacific Standard*, 3 September 2013.

75 D. Choe, et al., 'Interactions between monoamine oxidase A and punitive discipline in African American and Caucasian men's antisocial behaviour', *Clinical Psychological Science* 2 (5), 1 September 2014, pp. 591–601.

76 Michael E. Roettger, et al., 'The association between MAOA 2R genotype and delinquency over time among men: the interactive role of parental closeness and parental incarceration', *Criminal Justice Behaviour* 43(8), 2016, p. 1076.

77 Ibid. Kevin Beaver quoted.

78 Irving M. Reti, et al., 'MAOA regulates antisocial personality in Caucasians with no history of physical abuse', *Comparative Psychiatry* 52 (2), March–April 2011, pp. 188–94.

79 Wade, op. cit., p. 57.

80 R. Tikkanen, et al., 'Impulsive alcohol-related risk-behavior and emotional dysregulation with a serotonin 2B receptor stop codon', *Translational Psychiatry* 5 (681), 17 November 2015.

81 Tikkanen, quoted in Paivi Lehtinen, 'Gene mutation linked to reckless drunken behaviour', University of Helskinki Press Release, 11 November 2015.

82 H. Chen, et al., 'The MAOA gene predicts happiness in women', *Progress in Neuro–Psychopharmacology & Biological Psychiatry*, 10 January 2013, pp. 122–5.

83 Y. W. Yu, et al., 'Association study of a monoamine oxidase gene promoter polymorphism with major depressive disorder and antidepressant response', *Neuropsychopharmacology* 30 (9), September 2005, pp. 1719–23.

84 Reich, op. cit., pp. 263–4.

9. WHAT IS IQ AND IS IT HARDWIRED?

1   Alfred Binet, quoted in Stephen Jay Gould, *The Mismeasure of Man*, Penguin, 1997, p. 181.

2   Ibid.

3   Ibid., p. 190.

4   Ibid., p. 197.

5   Ibid., p. 212.

6   Ibid., p. 220.

7   Ibid., p. 221.

8   Arthur Jensen, *Bias in Mental Testing*, Free Press, New York, 1979, p. 113.

9   Ibid.

10 Quoted in Gould, op. cit., p. 227.

11 See Diane B. Paul, *Controlling Human Heredity: 1865 to the Present*, Humanity Books, Toronto, 1995, pp. 65–7, 109; and Gould, op. cit., pp. 229–30.

12 Gould, op. cit., pp. 229–37.

13 C. C. Brigham, *A Study of American Intelligence*, Princeton University Press, Princeton, NJ, 1923, p. xx.

14 Ibid., p. 92.

15 Ibid., pp. 110–11.

16 Ibid., p. 178.

17 Ibid., p. 202.

18 Ibid., p. 204.

19 Gould, op. cit., p. 222.

20 Interview with E. G. Boring, quoted in D. J. Kevles, 'Testing the army's intelligence: psychologists and the military in World War 1', *Journal of American History* 55, 1968, pp. 565–81.

21 C. C. Brigham, 'Intelligence tests of immigrant groups', *Psychological Review* 37, 1930, p. 164.

22 I. J. Deary, et al., 'A conversation between Charles Spearman, Godfrey Thomson, and Edward L. Thorndike: The International Examinations Inquiry Meetings 1931–1938: Correction to Deary, Lawn, and Bartholomew (2008)', *History of Psychology* 11 (3), 2008, p. 157.

23 L. L. Thurstone, 1940, quoted in Gould, op. cit., pp. 327–8.

24 See Chapter 8. These 'Primary Mental Abilities' were V (verbal comprehension), W (word fluency), N (computational number skills), S

(special visualisation), M (associative memory), P (perceptual speed) and R (reasoning).

25 L. L. Thurstone, quoted in Gould, op. cit., p. 335.

26 Deary, et al., op. cit., p. 157.

27 Quoted in Gould, op. cit., p. 301.

28 Charles Murray, letter to the *New Yorker*, 26 December 1994.

29 Spearman, quoted in Gould, op. cit., p. 298.

30 Cyril Burt, 'Factor analysis and its neurological basis', *British Journal of Statistical Psychology* 14, 1961, p. 57.

31 Cyril Burt, 'Ability and income', *British Journal of Educational Psychology* 13, 1943, p. 141.

32 Cyril Burt, *The Backward Child*, D. Appleton, New York, 1937, p. 110.

33 Ibid., p. 186.

34 Anthony Sampson, *Anatomy of Britain Today*, Hodder and Stoughton, London, 1965, p. 195.

35 James Flynn, *What is Intelligence?*, Cambridge University Press, Cambridge, 2009, p. 18.

36 Quoted in Robert Kaplan and Dennis Saccuzzo, *Psychological Testing: Principles, Applications, and Issues* (7th edition), Wadsworth, Belmont, CA, 2009, p. 256.

37 See Alan S. Kaufman, *IQ Testing 101*, Springer Publishing Company, New York, 2009.

38 Raymond B. Cattell, *Beyondism: Religion from Science*, Praeger, New York, 1987.

39 Howard Gardner in an interview with Dipin Damodharan of OneIndia News, 'The circuitry of multiple intelligence', *Education Insider*, 29 December 2012.

40 Adam Hampshire, Roger Highfield, Beth Parkin and Adrian Owen, 'Fractionating human intelligence', *Neuron* 76 (6), 20 December 2012, pp. 1225–37.

41 Roger Highfield, quoted in Steve Connor, 'IQ tests are "fundamentally flawed" and using them alone to measure intelligence is a "fallacy", study finds', *Independent*, 21 December 2012.

10. WHAT CAN TWINS TELL US ABOUT NATURE AND NURTURE?

1 I cite several examples of such claims in my book *Black Brain, White Brain: Is Intelligence Skin Deep?*, Jonathan Ball, 2014.

2 Study by Nancy Segal cited in Erika Hayasaki, 'Identical Twins Hint at How Environments Change Gene Expression', *The Atlantic*, 15 May 2018.

3   Jim Springer, quoted in '"Jim Twins" return to university for more tests', *Associated Press*, 5 November 1987.

4   Erika Hayasaki, op. cit.

5   Cited in '"Jim Twins" return to university for more tests', *Associated Press*, 5 November 1987.

6   Nancy Segal, quoted in Erika Hayasaki, op. cit.

7   Nancy Segal, quoted in Susan Dominus, 'The Mixed-Up Brothers of Bogota', *New York Times Magazine*, 9 July 2015.

8   Francis Galton, 'The history of twins as a criterion of the relative powers of nature and nurture', *Fraser's Magazine* 12, 1875, pp. 566–76.

9   Matt Ridley, *Nature via Nurture*, Fourth Estate, London, 2003, p. 75.

10 See N. Langstrom, et al., 'Genetic and environmental effects on same-sex sexual behaviour: a population study of twins in Sweden', *Archives of Sexual Behavior* 39 (1), February 2010, pp. 75–80.

11 Ian Sample, 'Genetics accounts for more than half of variation in exam results', *Guardian*, 11 December 2013.

12 Nicholas Shakeshaft, et al., 'Strong Genetic Influence on a UK Nationwide Test of Educational Achievement at the End of Compulsory Education at Age 16', *PLoS One* 8 (12) e80341, 11 December 2013, p. 5.

13 Ibid., p. 8.

14 This is a high number (today there are 33 twin births per 1,000; more than when I was a child), and the ratio is unusual (one quarter of multiple births are identical twins).

15 B. Devlin, et al., 'The heritability of IQ', *Nature* 388 (6641), 1997, pp. 468–71.

16 Leon Kamin, *The Science and Politics of IQ*, Penguin, London, 1977.

17 Letter from Cyril Burt to William Shockley, 10 April 1971, quoted in Ronald Fletcher, *Science, Ideology and the Media: The Cyril Burt Scandal*, Transaction Publishers, 1991, p. 381.

18 Ibid., p. 378.

19 Ronald Fletcher, op. cit.

20 H. J. Eysenck, quoted by Gould, op. cit., p. 265.

21 Robert Joynson, *The Burt Affair*, Routledge, 1989.

22 Ronald Fletcher, op. cit.

23 Leslie Hearnshaw, *Cyril Burt: Psychologist*, Hodder and Stoughton, London, 1979.

24 'Sir Cyril Burt', *Encyclopaedia Britannica* entry, 2013, http://www.britannica.com/related-places/85886/related-places-to-Sir-Cyril-Burt

25 See for example Rob Newman, 'Total Eclipse of Descartes: The Inheritance', BBC Radio 4, 14 September 2018 https://www.bbc.co.uk/radio/play/b0bh5hp2

26 M. Savastano, et al., 'Psychological characteristics of patients with Meniere's disease compared with patients with vertigo, tinnitus, or hearing loss', *Ear Nose & Throat Journal* 86 (3), 2007, pp. 148–56.

27 M. R. Rosenweig, et al., 'Effects of environmental complexity and training on brain chemistry and anatomy: A replication and extension', *Journal of Comparative and Physiological Psychology* 55 (4), 1962, pp. 429–37.

28 Gould, op. cit., p. 266.

29 Wickliffe Preston Draper, quoted in Ridley, op. cit., p. 81.

30 The Pioneer Fund website, www.thepioneerfund.org

31 Linda Gottfredson, et al., 'Mainstream Science on Intelligence', public statement published in the *Wall Street Journal*, 13 December 1994.

32 Ridley, op. cit., p. 81.

33 Susan Farber, *Identical Twins Reared Apart*, Basic Books, New York, 1981.

34 Cited by Professor Jack Kaplan, 'How to inherit IQ: an exchange', *New York Review of Books*, 30 November 2006.

35 Leon Kamin, quoted in 'A study of twins bred apart produces some curious discoveries', *The Hour*, 14 October 1981.

36 Thomas J. Bouchard Jr, et al., 'Sources of human psychological differences: the Minnesota Study of Twins Reared Apart', *Science* 250, 1990, p. 225.

37 Robert Plomin, *Blueprint: How DNA Makes Us Who We Are*, Allen Lane, 2018, pp. 54–5.

38 Ibid., p. 57.

39 James Flynn, *What Is Intelligence?* Cambridge University Press, Cambridge, 2009, p. 39.

40 Ibid., p. 90.

41 Eric Turkheimer, quoted in David L. Kirp, 'After the Bell Curve', *New York Times Magazine*, 23 July 2006.

42 Steven Rose, 'Commentary: heritability estimates – long past their sell-by date', *International Journal of Epidemiology* 35 (3), June 2006, p. 527.

43 Paul Ehrlich, *Human Natures: Genes, Cultures and the Human Prospect*, Penguin, 2002, p. 11.

44 Robert Plomin, 2018, op. cit., p. 55.

45 Alexander Young, et al., 'Relatedness disequilibrium regression estimates heritability without environmental bias', *Nature Genetics* 50, 13 August 2018, pp. 1304–10.

46 Bouchard, et al., 1990, op. cit., p. 224.

47 Ibid.

48 Ibid., p. 227.

49 Richard Nisbett, quoted in John-Paul Flintoff and Jonathan Leake, 'How to make your child more intelligent', *Sunday Times*, 17 May 2009.

50 Thomas Bouchard, quoted in Arthur Allen, 'Nature and Nurture', *Washington Post Magazine*, 11 July 1998.

51 Robert Plomin, 2018, op. cit., p. 55.

52 Leon Kamin, quoted in N. Mackintosh, *IQ and Human Intelligence*, Oxford University Press, Oxford, 1998, pp. 78–9.

53 Ridley, op. cit., p. 90.
54 Ibid.
55 Research by Christiane Capron and Michel Duyme, *New York Review of Books*, 2006, op. cit.
56 Ridley, op. cit., pp. 90–1.
57 *New York Review of Books*, op. cit.
58 See Eric Turkheimer, 'Individual and group differences in adoption studies', *Psychological Bulletin* 110 (3), 1991, p. 398.
59 Linda Gottfredson, et al., op. cit.
60 Robert Plomin 2018, op. cit., pp. viii and ix.
61 Richard Nisbett 2009, op. cit.
62 Stephen Jay Gould, *The Mismeasure of Man*, Penguin, 1997, p. 185.
63 Ibid.
64 Ibid., p. 186.
65 Borrowed from John Loehlin, 'On Schönemann, on Guttman, on Jensen, via Lewontin', *Multivariate Behavioral Research* 27 (2), 1992, p. 261.

11. DO RISING IQS SUGGEST GENETIC OR ENVIRONMENTAL CHANGE?

1 Gavin Evans, interview with Colonel Peter Ashton-Wickett, 15 May 2000.
2 Gavin Evans, interview with Brian Deal, 6 October 2013.
3 James Flynn, quoted in Ulric Neisser, 'Rising scores on intelligence tests', *American Scientist* 85, 1997, pp. 440–7.
4 James Flynn, quoted in Marek Kohn, *The Race Gallery*, Jonathan Cape, London, 1995, p. 105.
5 James Flynn, *What is Intelligence?*, Cambridge University Press, Cambridge, 2009, p. 36.
6 T. C. Daley, et al., 'IQ on the rise: the Flynn effect in rural Kenyan children', *Psychological Science* 24, 2003, p. 218.
7 Ron Unz, 'Race, IQ, and Wealth', *The American Conservative*, 18 July 2012.
8 James Flynn, op. cit., pp. 10–11.
9 Ibid., pp. 24–5.
10 Ibid., p. 41.
11 R. Colom, et al., 'Generational changes on the Draw-a-Man test: a comparison of Brazilian urban and rural children tested in 1930, 2002 and 2004', cited in Flynn, 2009, op. cit., pp. 170–1.
12 Steven B. Johnson, *Everything Bad Is Good for You: How Today's Popular Culture Is Actually Making Us Smarter*, Riverhead Books, New York, 2005.
13 Study cited in Dan Hurley, 'Can you make yourself smarter?', *New York Times Magazine*, 18 April 2012.

14 Susanne M. Jaeggi, et al., 'Improving fluid intelligence with training on working memory', *Proceedings of the National Academy of Sciences*, 18 March 2008, p. 1. www.pnas.org/content/early/2008/04/25/0801268105. abstract

15 Interview with Martin Buschkuehl, in Alvaro Fernandez, 'Can intelligence be trained? Martin Buschkuehl shows how', *SharpBrains*, 13 May 2008.

16 Ibid., Susanne Jaeggi quoted.

17 E. A. Maguire, et al., 'Navigation-related structural change in the hippocampi of taxi drivers', *Proceedings of the National Academy of Sciences* 97, 2000, pp. 4398–403.

18 Glenn Schellenberg, 'Music lessons enhance IQ', *Psychological Science* 15 (8), 2004, pp. 511–14.

19 E. Glenn Schellenberg, 'Music lessons, emotional intelligence and IQ', *Music Perception: an Interdisciplinary Journal* 29 (2), 2011, pp. 185–94.

20 Frances H. Rauscher and Gordon L. Shaw, 'Listening to Mozart enhances spatial-temporal reasoning: towards a neurophysiological basis', *Neuroscience Letters* 185, 1995, pp. 44–7.

21 John-Paul Flintoff and Jonathan Leake, 'How to make your child more intelligent', *Sunday Times*, 17 May 2009.

22 See L. Melton, 'Use it, don't lose it', *New Scientist*, 17 December 2005, pp. 32–5.

23 M. Qian, et al., 'The effects of iodine on intelligence in children: a meta-analysis of studies conducted in China', *Asia Pacific Journal of Clinical Nutrition* 14 (1), 2005, pp. 32–42.

24 Speech by Elise Labonte-LeMoyne, delivered to a conference hosted by the Society for Neuroscience in San Diego, quoted in 'Pregnant mothers who exercise boost babies' brains, claim researchers', *Guardian*, 10 November 2013.

25 See Richard E. Nisbett, 'Heredity, environment, and race differences in IQ: a commentary on Rushton and Jensen (2005)', *Psychology, Public Policy and Law* 11 (2), 2005, p. 302.

26 Millennium Cohort Study, cited in Warwick Mansell, 'Children can fall behind as early as nine months', *Guardian*, 17 February 2010.

27 Roger Dobson, 'Smoking in late pregnancy is linked to lower IQ in offspring', *British Medical Journal* 330 (7490), 5 March 2005, p. 499.

28 L. Goldschmidt, et al., 'Prenatal marijuana exposure and intelligence test performance at age 6', *Journal of American Academic Child and Adolescent Psychiatry* 47 (3), March 2008, pp. 254–63.

29 Tia Ghose, 'Light drinking while pregnant could lower baby's IQ', *livescience*, 14 November 2012.

30 Karen Richardson, 'Smoking, low income and health inequalities: thematic discussion document', report for Health Development Agency, May 2001.

31 C. Blair, 'How similar are fluid cognition and general intelligence? A developmental neuroscience perspective on fluid cognition as an aspect of human cognitive ability', *Journal of Behavioral and Brain Science* 29, 2006, pp. 109–60.

32 See Sue Gerhardt, *Why Love Matters: How Affection Shapes a Baby's Brain*, Routledge, London, 2010. See also, interview with Dr Ronald Federici (a psychologist who founded several US relief efforts for Romanian orphans), 'Dr Ronald Federici: Romanian orphans Q&A', *Developments in Therapy*, 25 June 1999.

33 Both studies – the Millennium Cohort study and the Sutton Trust Charity study – cited in Warwick Mansell, op. cit.

34 Matthew Syed, 'Rethink the brain and stem the summer slide', *The Times*, 21 July 2011.

35 See for example, James Flynn, op. cit., pp. 85–6.

36 Cited in Cherry Norton, 'Smacking hits a child's IQ', *Sunday Times*, 2 August 1998.

37 Ilan Katz, et al, 'The relationship between parenting and poverty', Joseph Rowntree Foundation, 2007, p. 13.

38 Xin Zhang, et al., 'The impact of exposure to air pollution on cognitive performance', *PNAS*, 27 August 2018, https://doi.org/10.1073/pnas.1809474115

39 Dr Erzsébet Bukodi, quoted in Alison Kershaw, 'Background hurts poorer bright pupils', *Independent*, 17 October 2013.

40 Cited in Sun Meilin, 'UK study: Bright pupils from poor backgrounds lag two years behind rich', *Helium*, 3 July 2012.

41 James Flynn, op. cit., pp. 104–5.

42 Eric Turkheimer, Kathryn Paige Harden and Richard E. Nisbett, 'Charles Murray is once again peddling junk science about race and IQ: Podcaster and author Sam Harris is the latest to fall for it', *Vox*, 18 May 2017.

43 F. A. Campbell and C. T. Ramey, 'Effects of early intervention on intellectual and academic achievement: a follow-up study of children from low-income families', *Child Development* 65, 1994, pp. 684–98.

44 S. L. Ramey and C. T. Ramey, 'Early experience and early intervention for children "at risk" for developmental delay and mental retardation', *Mental Retardation and Developmental Disabilities Research Reviews* 5, 1999, pp. 1–10. See also, Richard E. Nisbett, 'Heredity, environment and race differences in IQ: a commentary on Rushton and Jensen (2005)', *Psychology, Public Policy and Law*, 11 (2), 2005, p. 303.

45 'A* to C in English and Maths GCSE attainment for children aged 14 to 16 (key stage 4)', *GOV.UK Ethnicity facts and figures*, Department for Education, 10 October 2017.

46 Richard Garner, 'Working-class white boys do worst in class', *Independent*, 3 September 2013.

47 *GOV.UK Ethnicity facts and figures*, op. cit.

48 'Students aged 16 to 18 achieving 3 A grades or better at A level', *GOV.UK Ethnicity facts and figures*, Department for Education, 5 March 2018.

49 In 2016, 12,180 girls achieved three A or A* grades in their A-levels compared with 11,968 boys. 'A level and other 16–18 results (provisional): 2016/17', Department for Education, 12 October 2017.

50 Graeme Paton, 'Poor white boys "lagging behind classmates at age five"', *Telegraph*, 21 November 2012.

51 'Mathematics attainments for children aged 6 and 7 (key stage 1)', *GOV.UK Ethnicity facts and figures*, Department for Education, 1 August 2018.

52 'Reading attainments for children aged 6 to 7 (key stage 1)', *GOV.UK Ethnicity facts and figures*, Department for Education, 1 August 2018.

53 Linda Gottfredson, et al., 'Mainstream Science on Intelligence', op. cit.

54 James Flynn, op. cit., p. 123.

55 Ibid.

56 Richard E. Nisbett 2005, op. cit., p. 302.

57 Richard E. Nisbett in Eric Turkheimer, Kathryn Paige Harden and Richard E. Nisbett, 'There's still no good reason to believe black-white IQ differences are due to genes: our response to criticisms', *Vox*, 17 June 2017.

58 Ibid.

59 See Mathew Kredell, 'Sociologist reports only minor progress in racial equality over last 40 years', *USCPrice*, 27 February 2015.

60 James Flynn, op. cit., p. 64.

61 Ibid.

62 Charles Murray, quoted in Malcolm Gladwell, 'None of the Above: What IQ doesn't tell you about race', *New Yorker*, 17 December 2007.

63 Eleni Karageorge, 'The unexplainable, growing black-white wage gap', *Monthly Labor Review*, November 2017.

64 US Census Bureau distribution of income, 2017.

65 US Federal Reserve data reported in Tracy Jan, 'White families have nearly 10 times the net worth of black families. And the gap is growing', *Washington Post*, 28 September 2017.

66 Jessica L. Semega, et al., 'Income and Poverty in the United States: 2016', Current Population Reports, United States Census Bureau, September 2017, p. 12.

67 3.4% for whites and 6.1% for blacks. 'Labor Force Statistics from the Current Population Survey', Bureau of Labor Statistics, US Department of Labor, 6 July 2018.

68 Marianne Bertrand and Sendhil Mullainathan, 'Are Emily and Greg more employable than Lakisha and Jamal?: a field experiment on labor market discrimination', *American Economic Review* 94, September 2004, pp. 991–1013.

69 John Gramlich, 'The gap between the number of blacks and whites in prison is shrinking', *Factank*, Pew Research Center, 12 January 2018.

70 See German Lopez, 'After legalization, black people are still arrested at higher rates for marijuana than white people', *Vox*, 29 January 2018.

71 University of Chicago National Opinion Research Center survey, cited in Marek Kohn, *The Race Gallery: The Return of Racial Science*, Jonathan Cape, 1995, p. 115.

72 For other examples, see Gavin Evans, *Mapreaders & Multitaskers: Men, Women, Nature, Nurture*, Thistle, 2017, pp. 154–7.

73 M. J. Sharps, J. L. Price and J. K. Williams, 'Spatial cognition and gender: instructional and stimulus influences on mental image rotation performance', *Psychology of Women Quarterly* 18 (3), 1994, pp. 413–25.

74 C. Good, J. Aronson and J. Harder, 'Problems in the pipeline: stereotype threat and women's achievement in high-level math courses', *Journal of Applied Developmental Psychology* 19 (1), 2008, p. 25.

75 Claude Steele, 'Thin ice: stereotype threat and black college students', *Atlantic Monthly*, August 1999.

76 Ibid.

77 K. Eyferth, 'Leistungen verschiedener Gruppen von Besatzungskindern in Hamburg-Wechsler Intelligenztest für Kinder (HAWIK)', *Archiv für die gesamte Psychologie* 13, 1961, pp. 222–41.

78 James Flynn, *Race, IQ and Jensen*, Routledge & Kegan Paul, 1980, pp. 87–8.

79 L. Willerman, A. F. Naylor and N. C. Myrianthopoulos, 'Intellectual development of children from interracial matings: performance in infancy and at 4 years', *Behavior Genetics* 4, 1974, pp. 84–8.

80 Richard Herrnstein and Charles Murray, *The Bell Curve*, Simon & Schuster, New York, 1996, p. 283.

81 James Flynn, op. cit., p. 45.

82 Ibid.

83 Herrnstein and Murray, op. cit., p. 44.

84 Nisbett, op. cit., p. 302.

85 James Flynn, quoted in Ezra Klein, 'Sam Harris, Charles Murray, and the allure of race science. This is not "forbidden knowledge". It is America's most ancient justification for bigotry and racial inequality', *Vox*, 27 March 2018.

86 Flynn, 1980, op. cit.

87 Reich, op. cit., p. 264.

12. *THE BELL CURVE*: WHAT'S IT ALL ABOUT AND WHY IS IT BACK?

1 Sam Harris in 'Forbidden Knowledge (with Charles Murray): Waking up with Sam Harris #73', *YouTube*, 23 April 2017, https://www.youtube.com/watch?v=Y1lEPQYQk8s

2   Ibid., Charles Murray.

3   The others being Richard Dawkins, Christopher Hitchens and Daniel Dennett; all wrote books advocating atheism at around the same time.

4   Ezra Klein in 'The Sam Harris Debate', The Ezra Klein Show, Vox, 9 April 2018 https://www.vox.com/2018/4/9/17210248/sam-harris-ezra-klein-charles-murray-transcript-podcast

5   A. R. Jensen, 'How much can we boost IQ and scholastic achievement?', Harvard Educational Review 39, 1969, pp. 1–123.

6   Ibid.

7   Margalit Fox, 'Arthur R. Jensen Dies at 89, Set Off Debate About IQ', New York Times, 1 November 2012.

8   Bill Blakemore, Peter Jennings and Beth Nissen, 'The Bell Curve and the Pioneer Fund', ABC World News Tonight, ABC News, 22 November 1994.

9   H. J. Eysenck, Race, Intelligence and Education, Temple Smith, London, 1971, p. 130.

10 Stephen Jay Gould, The Mismeasure of Man, Penguin, 1997, p. 159 and pp. 186–7.

11 Richard E. Nisbett, 'Heredity, environment and race differences in IQ: a commentary on Rushton and Jensen (2005)', Psychology, Public Policy and Law, 11 (2), 2005, p. 309.

12 William Shockley, quoted in 'Shockley's race view called "senile, Fascist"', St Petersburg Times, 8 September 1971.

13 J. P. Rushton, quoted in Devin Burghard, 'Stateside: inside the Preserving Western Civilization conference', Searchlight, April 2009.

14 See Charles Lane, 'Response to Daniel R. Vining, Jr', New York Review of Books, 42 (5), 23 March 1995.

15 Zack Cernovsky, 'On the similarities of American blacks and whites: a reply to J. P. Rushton', Journal of Black Studies 25, 1 July 1995, p. 672.

16 J. P. Rushton, 'Race differences in behaviour: a review and evolutionary analysis', Personality and Individual Differences 9 (6), 1988, pp. 1009–24.

17 J. P. Rushton, 'The evolution of racial differences: a reply to M. Lynn', Journal of Research in Personality 23, 1989, pp. 7–20.

18 See Marek Kohn, The Race Gallery, Jonathan Cape, London, 1995, pp. 142–3.

19 E. O. Wilson, quoted in P. Knudson, A Mirror to Nature: Reflections on Science, Scientists and Society, Stoddard Publishing, Toronto, 1991, p. 190.

20 Richard Herrnstein and Charles Murray, The Bell Curve: Intelligence and Class Structure in American Life, Free Press, New York, 1994.

21 J. L. Graves, 'What a tangled web he weaves: race, reproductive strategies and Rushton's life history theory', Anthropological Theory 2 (2), 2002, pp. 131–54.

22 D. P. Barash, 'Book review: "Race, Evolution and Behavior"', Animal Behaviour 49, 1995, pp. 1131–3.

23 Blakemore, et al., op. cit.

24 Ibid., p. 113.

25 Ibid., p. 92.

26 Ibid., p. 118.

27 Ibid., p. 315.

28 Ibid., p. 311.

29 Ibid., p. 314.

30 Ibid., p. 340.

31 Bob Herbert, 'In America; throwing a curve', *New York Times*, 26 October 1994.

32 Ibid., p. 413.

33 Ibid., p. 546.

34 Stephen Jay Gould, op. cit., p. 368.

35 Ibid., p. 367.

36 Ibid., p. 369.

37 Ibid., p. 378.

38 For a summary of the critiques of sources see Charles Lane, 'The Tainted Sources of "The Bell Curve"', *New York Review*, 1 December 2004.

39 Leon Kamin, 'Behind the curve', *Scientific American* 272 (2), February 1995, p. 99.

40 Joseph L. Graves, *The Emperor's New Clothes*, Rutgers University Press, NJ, 2001, p. 8.

41 See for example Claude S. Fischer, et al., *Inequality by Design: Cracking the Bell Curve Myth*, Princeton University Press, NJ, 1996.

42 See Janet Currie and Duncan Thomas, 'The intergenerational transmission of "intelligence" down the slippery slopes of "The Bell Curve"', *Industrial Relations: A Journal of Economy and Society* 38 (3), July 1999.

43 Claude S. Fischer, et al., 1996, op. cit. p. 55.

44 Sanders Korenman and Christopher Winship, 'A reanalysis of "The Bell Curve"', *NBER Working Paper Series*, Vol. w5230, August 1995.

45 James Heckman, 'Lessons from "The Bell Curve"', *Journal of Political Economy* 103 (5), 1995, pp. 1091–120.

46 R. M. Hauser and M. H. Huang, 'Verbal ability and socioeconomic success: a trend analysis', *Social Science Research* 26 (3), September 1997, pp. 331–76.

47 See for example Charles Tittle and Thomas Rotolo, 'IQ and stratification: an empirical evaluation of Herrnstein and Murray's social change argument', *Social Forces* 79 (1), September 2000, pp. 1–28.

48 Charles Murray, back-cover review for James Flynn, *What Is Intelligence?*, Cambridge University Press, Cambridge, 2009.

49 James Flynn, op. cit., p. 36.

50 Charles Murray, *Losing Ground: American Social Policy, 1950–1980*, Basic Books, New York, 1984.

51 Nathaniel Weyl, quoted in Jeet Heer, 'Charles Murray is a Marketing Genius', *New Republic*, 11 April 2018.

52 Charles Murray, quoted in Jason DeParle, 'Daring Research or Social Science Pornography?: Charles Murray', *New York Times Magazine*, 9 October 1994.

53 American Psychological Association, 'Intelligence: knowns and unknowns', *American Psychologist* 51 (2), 1996, pp. 77–101.

54 Stephen Jay Gould, op. cit., p. 31.

55 Ibid., pp. 31–4.

56 Andrew Sullivan, 'The study of intelligence, ctd.', *The Dish*, 29 November 2011.

57 Andrew Sullivan, *The Daily Dish*, 26 August 2005.

58 Andrew Sullivan, quoted in Johann Hari, 'Andrew Sullivan: Thinking. Out. Loud.', *Intelligent Life*, Spring 2009.

59 Alan Wolfe, 'From PC to PR (race, IQ and genetics in the Charles Murray/ Richard Herrnstein book *The Bell Curve*)', *New Republic*, 31 October 1994.

60 Andrew Sullivan, 2005, op. cit.

61 Andrew Sullivan, 'The study of intelligence, ctd.', *The Dish*, 29 November 2011.

62 Ibid.

63 Ibid.

64 Ibid.

65 Andrew Sullivan, 'Race and IQ. Again', *The Dish*, 14 May 2013.

66 Andrew Sullivan, 'Denying Genetics Isn't Shutting Down Racism, It's Fueling It', *Daily Intelligencer, New York*, 30 March 2018.

67 Ibid.

68 Ibid.

69 Dominic Cummings, quoted in 'Genetics outweighs teaching, Gove advisor tells his boss', *Guardian*, 11 October 2013.

70 Boris Johnson, quoted in Nicholas Watt, 'Boris Johnson invokes Thatcher spirit with greed is good speech', *Guardian*, 27 November 2013.

71 Rod Liddle, quoted in Simon Hattenstone, 'Rod Liddle Interview: "I'm not a bigot"', *Guardian*, 13 June 2014.

72 'Exposed: London's eugenics conference and its neo-Nazi links', *London Student*, 8 January 2018.

73 Kevin Rawlinson and Richard Adams, 'UCL to investigate eugenics conference secretly held on campus', *Guardian*, 8 January 2018.

74 'UCL were paid thousands by eugenics conference with neo-Nazi connections', *London Student*, 5 February 2018.

75 Toby Young, 'The Fall of the Meritocracy', *Quadrant online*, 7 September 2015.

76 Toby Young, quoted in Rawlinson and Adams, op. cit.

77 Claire Phipps, Kevin Rawlinson and Rowena Mason, 'Toby Young resigns from the Office for Students after backlash', *Guardian*, 9 January 2018.

78 'Open Letter: No Place for Racist Pseudoscience at Cambridge', https://docs. google.com/forms/d/e/1FAIpQLScqiHicJXPCWPZVXKdhyPTO9jHFgdX Vo8WdA6nPfo_k2j5Xvw/closedform

79 Toby Young, 'Will Noah Carl get a fair hearing?', *Spectator*, 14 December 2018.

13. ARE JEWS SMARTER THAN EVERYONE ELSE?

1 See for example Nellie Bowles, 'Jordan Peterson, Custodian of the Patriarchy', *New York Times*, 18 May 2018.

2 Jordan Peterson interviewed by Stefan Molyneux, 'The IQ Problem: Jordan Peterson & Stefan Molyneux', *YouTube*, 15 August 2017.

3 Jordan Peterson, 'Jordan Peterson: The Dangerous IQ Debate', *Think Club*, *YouTube*, 15 April 2018.

4 Ibid.

5 Ibid.

6 Peterson interviewed by Molyneux, op. cit.

7 Jordan Peterson, tweet by @jordanbperson, 19 March 2018.

8 Gavin Evans, *Mapreaders & Multitaskers: Men, Women, Nature, Nurture*, Thistle, 2016, pp. 96–102, 115–16, 127–31, 244–7.

9 Jordan Peterson, *Think Club*, op. cit.

10 Ibid.

11 Peterson interviewed by Molyneux, op. cit.

12 Stephen Jay Gould, *The Mismeasure of Man*, Penguin, London, 1997, pp. 195–6.

13 Carl Brigham, *A Study of American Intelligence*, Princeton University Press, Princeton, NJ, 1923, p. 190.

14 Thomas Sowell, 'Ethnicity and IQ', *American Spectator* 28 (2), February 1995.

15 A. G. Hughes, 'Jews and Gentiles: their intellectual temperamental differences', *Eugenics Review*, July 1928, p. 90.

16 Amia Lieblich, Anat Ninio and Sol Kugelmas, 'Effects of ethnic origin and parental SES on WPPSI performance of pre-school children in Israel', *Journal of Cross-Cultural Psychology* 3, 1972, pp. 159–68.

17 Steven Pinker in 'Steven Pinker destroys environmental explanations of race differences', *YouTube*, 21 August 2012, https://www.youtube.com/watch?v=5cnwlN1CC28

18 James Flynn, *What Is Intelligence?*, Cambridge University Press, 2009, pp. 117–20.

19 Ibid., p. 120.

20 Ibid., pp. 121–2.

21 Ibid., p. 122.

22 'Table B-5: Total Enrolment by US Medical School and Race/Ethnicity, 2017–2018', *AAMC*, 2018.

23 Figures for 1992 cited in Jennifer Senior, 'Are Jews Smarter?', *New York*, 16 October 2005.

24 Tony Judt, quoted in Sarah Baxter, 'Murray lauds "genius of Jews"', *Sunday Times*, 8 April 2007.

25 Richard Lynn, 'The Intelligence of American Jews', *Personality and Individual Differences* 36 (1), January 2004, pp. 204–5.

26 Richard Lynn and Satoshi Kanazawa, 'How to explain high Jewish achievement: The role of intelligence and values', *Personality and Individual Differences* 44, 2008, pp. 801–8.

27 Richard Lynn, *The Chosen People: A Study of Jewish Intelligence and Achievement*, Washington Summit Publishers, Washington, 2011.

28 Moshe Zeidner, 'Review of "The Chosen People: A Study of Jewish Intelligence and Achievement"', *Times Higher Education*, 9 February 2012.

29 G. Cochran and H. Harpending, *The 10,000 Year Explosion: How Civilization Accelerated Human Evolution*, Basic Books, New York, 2009.

30 Reich, op. cit., p. 262.

31 H. Harpending, 'The Biology of Families and the Future of Civilization', (minute 38), Preserving Western Civilization 2009 Conference.

32 Twenty three signatories; 'Sinned against, not sinning', *Times Higher Education*, 16 June 2011.

33 G. Cochran, J. Hardy, H. Harpending, 'Natural History of Ashkenazi Intelligence', http://web.mit.edu/fustflum/documents/papers/kim-beder.pdf p.2.

34 Ibid., p. 4.

35 Ibid., p. 5.

36 Ibid., p. 5.

37 Ibid., p. 6.

38 Ibid., p. 2.

39 Ibid., p. 12.

40 Ibid., p. 2.

41 Ibid., p. 8.

42 Ibid., p. 15.

43 The DNA repair diseases include the mutations at BRCA1 and BRCA2. Other Ashkenazi diseases they cite include Bloom's syndrome and Fanconi's anaemia.

44 Ibid., p. 19.

45 Ibid., p. 20.

46 Ibid., p. 21.

47 Quoted in Jennifer Senior, op. cit.

48 Reich, op. cit., p. 264.

49 Marta Costa, et al., 'A substantial prehistoric ancestry amongst Ashkenazi maternal lineages', *Nature Communications* 4, 2013, p. 2543, http://www.nature.com/ncomms/2013/131008/ncomms3543/full/ncomms3543.html

50 Ibid.

348 | SKIN DEEP

51 Ibid.

52 See Martin Gershowitz, 'New Study Finds Most Ashkenazi Jews Genetically Linked to Europe', *Jewish Voice*, 16 December 2013.

53 Shlomo Sand, *The Invention of the Jewish People*, Verso, London, 2010.

54 Bernard Weinryb, 'The Jews of Poland: a social and economic history of the Jewish community in Poland from 1100 to 1800', *Jewish Publication Society of America*, Philadelphia, 1972, p. 70.

55 R. Brian Ferguson, 'How Jews Became Smart: Anti-"Natural History of Ashkenazi Intelligence"', http://www.ncas.rutgers.edu/sites/fasn/files/How%20Jews%20Became%20Smart%20(2008).pdf p. 32.

56 Ibid., pp. 32–3.

57 See for example, P. F. Plamara, et al., 'Length Distributions of Identity by Descent Reveal Fine-Scale Demographic History', *American Journal of Human Genetics* 91, 2012, pp. 809–22; and M. Slatin, 'A Population-Genetic Test of Founder Effects and Implications for Ashkenazi Jewish Diseases', *American Journal of Human Genetics* 75, 2004, pp. 282–93.

58 Reich, op. cit., p. 261.

59 Ferguson, op. cit. p. 1.

60 Such as with the BRCA conditions (genetic mutations that raise the risk of some cancers).

61 Such as Gaucher's, Idiopathic torsion dystonia (ITD) and non-classic congenital adrenal hyperplasia (CAH).

62 Ferguson, op. cit., p. 21.

63 Andrew Sullivan, 'The study of intelligence', *The Dish*, 21 November 2011.

64 Steven Pinker, Werner Kalow, Harold Kalant and reply by Stephen Jay Gould, 'Evolutionary Psychology: An Exchange', *New York Review of Books*, 9 October 1997. In response to Stephen Jay Gould, 'Darwinian Fundamentalism', *New York Review of Books*, 12 June 1997.

65 Steven Pinker and Steven Rose, 'The Two Steves – Pinker vs Rose – A Debate', Institute of Education, University of London, 21 January 1998, published online by *Edge The Third Culture*, 25 March 1998, http://www.edge.org/3rd_culture/pinker_rose/pinker_rose_p1.html

66 Elizabeth Spelke in 'The Science of Gender and Science. Pinker vs Spelke: A Debate', Harvard University Mind/Brain/Behavior Initiative, 16 May 2005, *Edge The Third Culture*, http://www.edge.org/3rd_culture/debate05/debate05_index.html

67 Ibid.

68 Steven Pinker, 2005 lecture, YIVO Institute for Jewish Research, https://www.youtube.com/watch?v=Beqtt42iDW8

69 'Your Lying Eyes. 2005. Pinker on Jews, Genes and Intelligence', 2 December 2005, http://lyingeyes.blogspot.co.uk/2005/12/pinker-on-jews-genes-and-intelligence_02.html

70 Maggie Wittlin, *Seed* magazine, cited in Marek Kohn, 'The racist undercurrent in the tide of genetic research', *Guardian*, 17 January 2006.

71 Steven Pinker, 'The Lessons of the Ashkenazim', *New Republic Online*, 17 June 2006.

72 Ibid.

73 Ibid.

74 Ibid.

75 Steven Pinker, 'Groups of people may differ genetically in their average talents and temperaments', in 'The *Edge* Annual Question 2006', *Edge The World Question Centre*, 2006, http://www.edge.org/q2006/q06_print.html#pinker

76 Ibid.

77 Steven Pinker, *The Blank Slate: The Modern Denial of Human Nature*, Penguin, 2003, p. 144.

78 Steven Pinker, https://twitter.com/sapinker/status/853394916508135424?lang=en 15 April 2017.

79 Steven Pinker, https://twitter.com/sapinker/status/1051277271099674625 13 October 2018.

80 Linda Gottfredson, et al., 'Mainstream Science on Intelligence', public statement, *Wall Street Journal*, 13 December 1994.

81 Ibid.

82 Charles Murray, 'Jewish Genius', *Commentary*, 1 April 2007, http://www.commentarymagazine.com/article/jewish-genius/

83 Ibid.

84 Ibid.

85 Ibid.

86 Ibid.

87 Nicholas Wade, *A Troublesome Inheritance: Genes, Race and Human History*, Penguin, New York, 2014.

88 Gavin Evans, 'The unwelcome revival of "race science"', *Guardian*, 2 March 2018.

89 Wade, *A Troublesome Inheritance*, op. cit., p. 214.

90 Ibid.

91 Ibid.

92 Ibid.

93 Ferguson, op. cit., p. 30.

94 Eric Turkheimer in Eric Turkheimer, Kathryn Paige Harden and Richard E. Nisbett, 'There's still no good reason to believe black-white IQ differences are due to genes: our response to criticisms', *Vox*, 17 June 2017.

95 Steven Pinker, 2006, op. cit.

14. DO SOME POPULATIONS HAVE DIFFERENT BRAINS FROM OTHERS?

1 A. R. Luria, *Cognitive development: Its cultural and social foundations*, Harvard University Press, Cambridge, MA, 1976, pp. 108–9.

2 Ibid., p. 112.

3 Ibid., pp. 81–2.

4 James Flynn, *What Is Intelligence?*, Cambridge University Press, 2009, pp. 26–7.

5 Ibid., p. 29.

6 Daniel Everett, *Don't Sleep, There Are Snakes*, Profile Books, 2009.

7 Interview with Piet Rooi, 5 November 2002, for Gavin Evans, 'The Diet Secret of the Desert', *The Times*, 19 November 2002.

8 Ibid. Interview with Susanna Witbooi, 5 November 2002.

9 Brenna Henna, Christopher Gignoux and Matthew Jobin, 'Hunter-gatherer genomic diversity suggests a southern African origin for modern humans', *Proceedings of the National Academy of Sciences of the United States of America* (National Academy of Sciences) 108 (13), 2011, pp. 5154–62.

10 Nick Crumpton, 'Earliest evidence of modern human culture found', *BBC News*, 31 July 2012.

11 Marjorie Shostak, *Nisa: The Life and Words of a !Kung Woman*, New York, Vintage, 1983, p. 10.

12 Interview with Nigel Crawhall, 4 November 2002.

13 Interview with Roger Chennels, 1 November 2002.

14 Ibid. Piet Rooi.

15 Ibid. Susanna Witbooi.

16 Quoted from interview with Marek Kohn, in Marek Kohn, *The Race Gallery*, Jonathan Cape, London, 1995, p. 148.

17 Richard Lynn, *Race Differences in Intelligence: An Evolutionary Analysis*, Washington Summit Publishers, Augusta, GA, 2006. http://www.velesova-sloboda.org/antrop/lynn-race-differences-in-intelligence.html

18 Ibid.

19 Ibid.

20 Raven's Progressive Matrices.

21 Kamin, ibid.

22 Leon Kamin, 'Behind the Curve', *Scientific American* 272 (2), February 1995, p. 100.

23 Kenneth Owen, quoted in Leon Kamin, ibid.

24 Kenneth Owen, quoted in Charles Lane, 'The Tainted Sources of "The Bell Curve"', *New York Review*, 1 December 1994.

25 Ibid.

26 Ibid.

27   Richard Lynn, 'Race Differences in Intelligence: A Global Perspective',
     Mankind Quarterly 31 (3), Spring 1991, p. 284.
28   Richard Lynn, *Race Differences in Intelligence: An Evolutionary Analysis*,
     Washington Summit Books, Augusta, GA, 2006, http://www.velesova-
     sloboda.org/archiv/pdf/lynn-race-differences-in-intelligence.pdf
29   Ibid., pp. 47 and 63–4.
30   R. Lynn, E. Backhoff and L. Contreras, 'Ethnic and Racial Differences on
     the Standard Progressive Matrices in Mexico', *Journal of Biosocial Science*
     37, 2005, pp. 107–13.
31   R. Lynn, 'In Italy, north–south differences in IQ predict differences in income,
     education, infant mortality, stature, and literacy', *Intelligence* 38 (1), 2016,
     pp. 93–100.
32   R. Lynn, 2006, op. cit., p. 124.
33   Ibid., p. 123.
34   Ibid., pp. 7–9.
35   Ibid., p. 12.
36   Ibid.
37   Richard Lynn, *Eugenics: A Reassessment*, Praeger, Westport, CT, 2001.
38   R. Lynn, 2006, op. cit., p. 143.
39   Ibid., p. 144.
40   Ibid., pp. 136–9.
41   Dr Marta Lahr, quoted in Jonathan Leake, 'Farmers, you've shrunk mankind',
     Sunday Times, 12 June 2011.
42   Ian Tattersall, *The Monkey in the Mirror*, Oxford University Press, 2002, p. 68.
43   See for example, T. C. Daley, et al., 'IQ on the rise: the Flynn effect
     in rural Kenyan children', *Psychological Science* 14 (3), May 2003,
     pp. 215–9.
44   R. Lynn, op. cit., pp. 6–7.
45   James Flynn, op. cit., p. 193.
46   Richard Lynn and Tatu Vanhanen, *IQ and the Wealth of Nations*, Praeger,
     Westport, CT, 2002.
47   E. Hunt and W. Wittmann, 'National intelligence and national prosperity',
     Intelligence 36 (1), January–February 2008, pp. 1–9.
48   R. Lynn, 'IQ in Japan and the United States shows a growing disparity', *Nature*
     297, 20 May 1982, pp. 222–3.
49   H. Stevenson and H. Azuma, 'IQ in Japan and the United States', *Nature* 306,
     17 November 1983, pp. 291–2.
50   Richard Herrnstein and Charles Murray, *The Bell Curve: Intelligence and Class
     Structure in American Life*, Free Press, 1994, p. 716.
51   N. J. Mackintosh, 'Book review: "Race differences in intelligence: An
     evolutionary hypothesis"', *Intelligence* 35 (1), January–February 2007,
     pp. 94–6.

52 R. Lynn and S. Kanazawa, 'How to Explain High Jewish Achievement: The Role of Intelligence and Values', *Personality and Individual Differences* 44, 2008, pp. 801–8.

53 R. Lynn and S. Kanazawa, 'A Longitudinal Study of Sex Differences in Intelligence at ages 7, 11 and 16 Years', *Personality and Individual Differences* 53, 2012, pp. 90–3.

54 S. Kanazawa, 'Mind the gap ... in intelligence: Re-examining the relationship between inequality and health', *British Journal of Health Psychology* 11 (4), November 2006, p. 623.

55 Denis Campbell, 'Low IQs are Africa's curse, says lecturer', *Observer*, 5 November 2006.

56 Kevin Denny, 'On a dubious theory of cross-country differences in intelligence', *Journal of Evolutionary Psychology* 7 (4), December 2009, p. 341.

57 George T. H. Ellison, 'Health, wealth and IQ in sub-Saharan Africa: Challenges facing the "Savanna Principle" as an explanation for global inequalities in health', *British Journal of Health Psychology* 12 (2), May 2007, p. 191.

58 Quoted in Denis Campbell, op. cit.

59 In the National Longitudinal Study of Adolescent Health.

60 Quoted in Kate Loveys and Colin Fernandez, 'Black women are less attractive than others: Controversial LSE psychologist sparks backlash with his "scientific" findings', *Daily Mail*, 19 May 2011.

61 Quoted in Jack Grove, 'LSE scholar admits race analysis was "flawed"', *Times Higher Education*, 15 September 2011.

62 Sixty-nine signatories; 'Kanazawa's bad science does not represent evolutionary psychology', *epjournal.net*, 2011.

63 Twenty-three signatories; 'Sinned against, not sinning', *Times Higher Education*, 16 June 2011.

64 Lucy Ward, 'Lecturer sacked for saying child sex "harmless"', *Independent*, 9 August 1997.

65 Christopher Brand, 'Leftists Love Their Chains', *gfactor.blogspot.com*, 16 January 2011.

66 Randy Thornhill and Craig T. Palmer, *A Natural History of Rape: Biological Bases of Sexual Coercion*, MIT Press, MA, 2000.

67 Thornhill's claims on rape are discussed in more detail in Gavin Evans, *Mapreaders & Multitaskers: Men, Women, Nature, Nurture*, Thistle, 2016, pp. 187–90.

68 Christopher Eppig, Corey L. Fincher and Randy Thornhill, 'Parasite prevalence and the worldwide distribution of cognitive ability', *Proceedings of the Royal Society*, 30 June 2010. http://rspb.royalsocietypublishing.org/content/early/2010/06/29/rspb.2010.0973.full

69 J. M. Wicherts, et al., 'A systematic literature review of the average IQ of sub-Saharan Africans', *Intelligence* 38, 2010, pp. 1–20.

70 Christopher Eppig, 'Why Is Average IQ Higher in Some Places?', *Scientific American*, 6 September 2011.

71 Eppig, Fincher and Thornhill, op. cit.

72 Ibid.

73 Ibid.

74 Eppig, 2011, op. cit.

75 Richard Lynn, 'Is man breeding himself back to the age of the apes?', *The Times*, 24 October 1994.

76 Professor Richard Lynn, 'Sorry men ARE more brainy than women (and more stupid too!) It's a simple scientific fact, says one of Britain's top dons', *Mail Online*, 8 May 2010.

77 Steve Jones, 'It's time to lay this race issue to rest', *Telegraph*, 26 October 2009.

15. IS RACE SCIENCE MAKING A COMEBACK?

1 Stefan Molyneux, tweet, 3 December 2018.

2 Nicholas Wade interviewed by Stefan Molyneux, 'Troublesome Inheritance – Genetics and Race: Wade', *Freedomain Radio*, *YouTube*, 1 March 2017.

3 Nicholas Wade, *A Troublesome Inheritance: Genes, Race and Human History*, Penguin, New York, 2014, p. 238.

4 Ibid., p. 241.

5 Ibid., p. 41.

6 Ibid., p. 136.

7 Nicholas Wade, 'The genome of history', *Spectator*, 17 May 2014.

8 Wade, *A Troublesome Inheritance*, op. cit., p. 235.

9 Wade, *A Troublesome Inheritance*, op. cit., p. 177.

10 Ibid., p. 91.

11 B. F. Voight, et al., 'A Map of Recent Positive Selection in the Human Genome', *PLOS Biology* 4 (3), March 2006.

12 Ibid., p. 454.

13 Agustin Fuentes, 'Things to Know When Talking About Race and Genetics', *Psychology Today*, 13 May 2014.

14 'A Troublesome Inheritance', Letters, *New York Times*, 8 August 2014.

15 Jerry Coyne, quoted in Steve Connor, 'The great taboo of genetics and race – a most unnatural selection of the facts', *Independent*, 13 August 2014.

16 David Reich, *Who We Are and How We Got Here: Ancient DNA and the New Science of the Human Past*, Oxford University Press, 2018, p. 262.

17 Kevin Mitchell, *Innate – How the Wiring of Our Brains Shapes Who We Are*, Princeton University Press, 2018, p. 262.

18 Carl Jung, *Collected Works 10: Civilisation in Transition*, Princeton University Press, 1970, p. 45.

19 Henry Edward Garrett, quoted in Matthew Syed, *Bounce*, Fourth Estate, London, 2001, p. 260.

20 Mark Snyderman and Stanley Rothman, *The IQ Controversy: The Media and Public Policy*, Transaction, New Brunswick, NJ, 1988.

21 Marek Kohn, *The Race Gallery*, Jonathan Cape, London, 1995, p. 7.

22 James Gleick, *Isaac Newton*, Fourth Estate, London, 2003, p. 193.

23 Robert M. Yerkes, quoted in Stephen Jay Gould, *The Mismeasure of Man*, Penguin, London, 1997, p 223.

24 Robert M. Yerkes, quoted ibid., p. 224.

25 Charles Spearman, quoted ibid., p. 293.

26 Ibid.

27 Jordan Peterson, 'Jordan Peterson: The Dangerous IQ Debate', *Think Club*, YouTube, 15 April 2018.

28 Leda Cosmides and John Tooby, quoted in David Buss (ed.), *The Handbook of Evolutionary Psychology*, John Wiley & Sons, NJ, 2005, p. 5.

29 Steven Pinker, *How the Mind Works*, W. W. Norton & Company, 1999.

30 See David J. Buller *Adapting Minds*, Bradford Books, MIT Press, Cambridge, MA, 2006, pp. 133–4.

31 Paul Ehrlich, quoted in *USA Today* magazine, published by the Society for the Advancement of Education, April 2001, and argued in more depth in his book *Human Natures: Genes, Cultures and the Human Prospect*, Shearwater Books/Island Press, 2000.

32 Examples include claims that women evolved to prefer pink and to like shopping more than men, and that we all evolved to favour our maternal cousins over our paternal cousins. See Gavin Evans, *Mapreaders & Multitaskers: Men, Women, Nature, Nurture*, Thistle, 2016, pp. 9–13 and 247–9.

33 Tattersall, op. cit., p. 171.

34 Ibid., p. 171.

35 David J. Buller, op. cit., p. 481.

36 Leda Cosmides and John Tooby, *Evolutionary Psychology: A Primer*, Center for Evolutionary Psychology, University of California, Santa Barbara, 13 January 1997, http://www.cep. ucsb.edu/primer.html

37 Nicholas Wade interviewed by Stefan Molyneux, *www.freedomainradio.com*, YouTube, 7 December 2016.

38 Stephen Jay Gould, *The Mismeasure of Man*, Penguin, 1997, pp. 22–3.

39 Ibid., p. 27.

40 Gould, op. cit., p. 32.

41 Ibid., pp. 37–8.

42 Sam Harris in 'The Sam Harris Debate', The Ezra Klein Show, *Vox*, 9 April

2018. https://www.vox.com/2018/4/9/17210248/sam-harris-ezra-klein-charles-murray-transcript-podcast

43 For a comprehensive picture of how Reddit works, see Adrienne Massanari, *Participatory Culture, Community, and Play: Learning from Reddit*, Peter Lang, 2015.

44 Nellie Bowles, 'Jordan Peterson, Custodian of the Patriarchy', *New York Times*, 18 May 2018.

45 'The Bizarre Fad Diet Taking the Far Right by Storm', *Mother Jones*, 7 September 2018.

46 See Adrienne Massanari, '#Gamergate and The Fappening: How Reddit's algorithm, community, and culture support toxic technocultures', *New Media & Society*, 19 (3), 9 October 2015, pp. 329–46..

47 Gavin Evans, 'He'll be weighing brains next', *Guardian*, 14 November 2003.

48 J. Philippe Rushton, 'Yes, "He'll be measuring brains next"', letter to the *Guardian*, published in Chris Brand, *IQ & PC* 1/11,2003–1/12/2003.

49 Gavin Evans, 'Linking nationality and IQ is wrong', *Guardian*, 5 July 2010.

50 Gavin Evans, 'Race science rears its ugly head', *New Internationalist*, March 2015; Gavin Evans, 'Analysis: Black Brain, White Brain – The new wave of racist science', *Africa Check*, 12 March 2015; Gavin Evans, 'The unwelcome return of "race science"', *Guardian*, 2 March 2018; Gavin Evans, 'Die Ruckkehr einer Bullshit-Wissenschaft', *Zeit Online, Die Zeit*, 31 March 2018.

51 See for example comments on https://www.youtube.com/watch?v=DLl1dwvgps0&t=89s and https://www.youtube.com/watch?v=6VoKBuhRQho

16. RACE, GENES AND INTELLIGENCE: WHERE NEXT?

1 David Reich, 'How Genetics is Changing Our Understanding of "Race"', *New York Times*, 23 March 2018.

2 David Reich, *Who We Are and How We Got Here: Ancient DNA and the New Science of the Human Past*, Oxford University Press, 2018, p. 265.

3 See for example Sam Harris in 'The Sam Harris Debate', The Ezra Klein Show, *Vox*, 9 April 2018, https://www.vox.com/2018/4/9/17210248/sam-harris-ezra-klein-charles-murray-transcript-podcast

4 James Flynn, quoted by Ezra Klein in ibid.

INDEX